# Advances in Legumes for Human Nutrition

# Advances in Legumes for Human Nutrition

Editors

**Stuart Johnson**
**Rewati Bhattarai**

MDPI

Basel • Beijing • Wuhan • Barcelona • Belgrade • Novi Sad • Cluj • Manchester

*Editors*
Stuart Johnson
Curtin University
Perth, Australia

Rewati Bhattarai
Curtin University
Perth, Australia

*Editorial Office*
MDPI
St. Alban-Anlage 66
4052 Basel, Switzerland

This is a reprint of articles from the Special Issue published online in the open access journal *Nutrients* (ISSN 2072-6643) (available at: https://www.mdpi.com/journal/nutrients/special_issues/Advances_in_Legumes_for_Human_Nutrition).

For citation purposes, cite each article independently as indicated on the article page online and as indicated below:

| Lastname, A.A.; Lastname, B.B. Article Title. *Journal Name* **Year**, *Volume Number*, Page Range. |
|---|

**ISBN 978-3-0365-9620-4 (Hbk)**
**ISBN 978-3-0365-9621-1 (PDF)**
**doi.org/10.3390/books978-3-0365-9621-1**

# Contents

**Francesca Sparvoli, Silvia Giofré, Eleonora Cominelli, Elena Avite, Gianluca Giuberti, Diomira Luongo, et al.**
Sensory Characteristics and Nutritional Quality of Food Products Made with a Biofortified and Lectin Free Common Bean (*Phaseolus vulgaris* L.) Flour
Reprinted from: *Nutrients* **2021**, *13*, 4517, doi:10.3390/nu13124517 . . . . . . . . . . . . . . . . . . 1

**Krystyna Szymandera-Buszka, Małgorzata Gumienna, Anna Jędrusek-Golińska, Katarzyna Waszkowiak, Marzanna Hęś, Artur Szwengiel and Anna Gramza-Michałowska**
Innovative Application of Phytochemicals from Fermented Legumes and Spices/Herbs Added in Extruded Snacks
Reprinted from: *Nutrients* **2021**, *13*, 4538, doi:10.3390/nu13124538 . . . . . . . . . . . . . . . . . . 29

**Lesley Bryant, Anna Rangan and Sara Grafenauer**
Lupins and Health Outcomes: A Systematic Literature Review
Reprinted from: *Nutrients* **2022**, *14*, 327, doi:10.3390/nu14020327 . . . . . . . . . . . . . . . . . . . 47

**Marina Carbonaro and Alessandro Nucara**
Legume Proteins and Peptides as Compounds in Nutraceuticals: A Structural Basis for Dietary Health Effects
Reprinted from: *Nutrients* **2022**, *14*, 1188, doi:10.3390/nu14061188 . . . . . . . . . . . . . . . . . . 67

**Rahil Malekipoor, Stuart K. Johnson and Rewati R. Bhattarai**
Lupin Kernel Fibre: Nutritional Composition, Processing Methods, Physicochemical Properties, Consumer Acceptability and Health Effects of Its Enriched Products
Reprinted from: *Nutrients* **2022**, *14*, 2845, doi:10.3390/nu14142845 . . . . . . . . . . . . . . . . . . 79

**Mohammad M. H. Abdullah, Jaimee Hughes and Sara Grafenauer**
Legume Intake Is Associated with Potential Savings in Coronary Heart Disease-Related Health Care Costs in Australia
Reprinted from: *Nutrients* **2022**, *14*, 2912, doi:10.3390/nu14142912 . . . . . . . . . . . . . . . . . . 95

**Jaimee Hughes, Emma Pearson and Sara Grafenauer**
Legumes—A Comprehensive Exploration of Global Food-Based Dietary Guidelines and Consumption
Reprinted from: *Nutrients* **2022**, *14*, 3080, doi:10.3390/nu14153080 . . . . . . . . . . . . . . . . . . 105

**Sandhya Subedi, Uthaiwan Suttisansanee, Aikkarach Kettawan, Chaowanee Chupeerach, Chanakan Khemthong, Sirinapa Thangsiri and Nattira On-nom**
Food Fortification of Instant Pulse Porridge Powder with Improved Iron and Zinc Bioaccessibility Using Roselle Calyx
Reprinted from: *Nutrients* **2022**, *14*, 4070, doi:10.3390/nu14194070 . . . . . . . . . . . . . . . . . . 125

**Ru Huang, Yuhan Liu, Sheng Hu, Alexander Tamalunas, Raphaela Waidelich, Frank Strittmatter, et al.**
Inhibition of $\alpha_1$-Adrenergic, Non-Adrenergic and Neurogenic Human Prostate Smooth Muscle Contraction and of Stromal Cell Growth by the Isoflavones Genistein and Daidzein
Reprinted from: *Nutrients* **2022**, *14*, 4943, doi:10.3390/nu14234943 . . . . . . . . . . . . . . . . . . 139

**Jingyu Gu, Minhao Li, Malik Adil Nawaz, Regine Stockmann, Roman Buckow and Hafiz A. R. Suleria**
In Vitro Digestion and Colonic Fermentation of UHT Treated Faba Protein Emulsions: Effects of Enzymatic Hydrolysis and Thermal Processing on Proteins and Phenolics
Reprinted from: *Nutrients* **2023**, *15*, 89, doi:10.3390/nu15010089 . . . . . . . . . . . . . . . . . . 163

**Duc Toan Do, Jaspreet Singh, Stuart Johnson and Harjinder Singh**
Probing the Double-Layered Cotyledon Cell Structure of Navy Beans: Barrier Effect of the Protein Matrix on In Vitro Starch Digestion
Reprinted from: *Nutrients* **2023**, *15*, 105, doi:10.3390/nu15010105 . . . . . . . . . . . . . . . . . . . . . **185**

**Larry A. Tucker**
Legume Intake, Body Weight, and Abdominal Adiposity: 10-Year Weight Change and Cross-Sectional Results in 15,185 U.S. Adults
Reprinted from: *Nutrients* **2023**, *15*, 460, doi:10.3390/nu15020460 . . . . . . . . . . . . . . . . . . . . . **201**

nutrients

MDPI

Article

# Sensory Characteristics and Nutritional Quality of Food Products Made with a Biofortified and Lectin Free Common Bean (*Phaseolus vulgaris* L.) Flour

**Francesca Sparvoli [1,*], Silvia Giofré [1,2], Eleonora Cominelli [1], Elena Avite [3], Gianluca Giuberti [4], Diomira Luongo [5], Edoardo Gatti [6], Marta Cianciabella [6], Giulia Maria Daniele [6], Mauro Rossi [5] and Stefano Predieri [6]**

1 Institute of Agricultural Biology and Biotechnology, National Research Council, Via Bassini 15, 20133 Milano, Italy; giofresilvia@gmail.com (S.G.); cominelli@ibba.cnr.it (E.C.)
2 Dipartimento di Scienze Agrarie, Alimentari e Ambientali, Università Politecnica delle Marche, 60131 Ancona, Italy
3 Blumen Group SPA, Corso Savona 168, 14100 Asti, Italy; elena.avite@blumen.it
4 Department for Sustainable Food Process, Università Cattolica del Sacro Cuore, Via Emilia Parmense 84, 29122 Piacenza, Italy; Gianluca.Giuberti@unicatt.it
5 Institute of Food Science, National Research Council, Via Roma 64, 83100 Avellino, Italy; diomira.luongo@isa.cnr.it (D.L.); mauro.rossi@isa.cnr.it (M.R.)
6 Institute for BioEconomy, National Research Council, Via Piero Gobetti 101, 40129 Bologna, Italy; edoardo.gatti@ibe.cnr.it (E.G.); marta.cianciabella@ibe.cnr.it (M.C.); giuliamaria.daniele@ibe.cnr.it (G.M.D.); stefano.predieri@ibe.cnr.it (S.P.)
* Correspondence: sparvoli@ibba.cnr.it

**Citation:** Sparvoli, F.; Giofré, S.; Cominelli, E.; Avite, E.; Giuberti, G.; Luongo, D.; Gatti, E.; Cianciabella, M.; Daniele, G.M.; Rossi, M.; et al. Sensory Characteristics and Nutritional Quality of Food Products Made with a Biofortified and Lectin Free Common Bean (*Phaseolus vulgaris* L.) Flour. *Nutrients* **2021**, *13*, 4517. https://doi.org/10.3390/nu13124517

Academic Editors: Rewati Bhattarai and Stuart Johnson

Received: 9 November 2021
Accepted: 14 December 2021
Published: 17 December 2021



**Abstract:** Common beans (*Phaseolus vulgaris* L.) are an important source of nutrients with beneficial effects on human health. However, they contain lectins, that limit the direct use of flour in food preparations without thermal treatment, and phytic acid, that reduces mineral cation bioavailability. The objectives of this research were: to obtain biofortified snacks and a cream using an untreated common bean flour devoid of active lectins (*lec*$^-$) and with reduced content of phytic acid (*lpa*) and to evaluate the sensorial appreciation for these products. The main results of the present work were: the products with the *lpa lec*$^-$ flour did not retain residual hemagglutinating activity due to lectins; they showed higher residual α-amylase inhibitor activity (from 2.2 to 135 times), reduced in vitro predicted glycemic index (about 5 units reduction) and increased iron bioavailability compared to the products with wild type flour; products with common bean flour were less appreciated than the reference ones without this flour, but the presence of an intense umami taste can be a positive attribute. Results confirmed that the use of the *lpa lec*$^-$ flour has important advantages in the preparation of safe and nutritionally improved products, and provide useful information to identify target consumers, such as children and elderly people.

**Keywords:** α-amylase inhibitor; biofortification; lectins; nutritional enhancement; phytic acid; predicted glycemic index; sensory analysis; temporal dominance sensation

## 1. Introduction

Legume seeds (pulses) are an excellent and inexpensive source of proteins, dietary fiber, vitamins, minerals, and bioactive components, and their consumption is associated with multiple health benefits, such as lower risk of incidence of cancers, type II diabetes, or cardiovascular diseases [1–5]. However, legume consumption per capita has not increased in the last three decades, despite massive population growth [6], and in several countries estimated average legume consumption does not meet recommendations. In recent years, public awareness of the nutritional benefits of pulses, as part of sustainable food production for food security and nutrition, has increased. As a result, attention to legume research has generally increased. Furthermore, innovative food alternatives that promote legume consumption and that can provide alternatives to satisfy consumer demand for healthy foods have emerged. These novel foods incorporate legumes into bakery products, pasta, soups, cereals, tortillas, meat substitutes, beverages, and snacks (recently reviewed by [7]).

Snacks occupy a considerable slice of the ready-to-eat food market and are appreciated by consumers of all ages, eaten during work-breaks by adults and during school recreation by children and adolescents. In some cases, snacks have become meal substitutes; the tendency to consume low-value snacks instead of having a complete and balanced meal is frequent and common, mainly among teenagers. This phenomenon has been associated with an increase in the body mass index and poor academic performance of students [8]. The increased consumption of energy-dense, nutrient-poor snacks is one of the major and growing concerns associated with the alarming trend of overweight, obesity and metabolic disorders worldwide [9]. This trend is also related to undesirable changes in snack-food patterns, including an increase in portion sizes, added sugars, total fats, saturated and trans-fats [10]. For this reason, the snack food market is currently searching for healthier products, characterized by enhanced nutritional value, a lower caloric content and a more balanced composition, to satisfy the changing habits of adults and children and to prevent the development of chronic diseases. Furthermore, overweight and obesity, which have always been considered a problem particularly in high-income countries, are currently dramatically increasing in low-income countries that now face concerns associated both with malnutrition and those associated with obesity [11].

Because of their high nutritional value and their positive performance in heat-treated products, legume flours are suitable ingredients for use in baked goods, and bean flour seems to receive a positive response in the formulation of ready-to-eat snacks [12–16]. However, legume-based snacks face challenges such as eliminating or reducing antinutritional factors (e.g., lectins, enzyme inhibitors, α-galacto-oligosaccharides, phytic acid, tannins) and meeting consumer acceptance, including texture, flavor, appearance, color, and final product approval. Sensory analysis evaluates the attributes related to perception at consumption, predicts consumer acceptance drivers, and helps identify target consumers. Sensory evaluation has been applied to bakery products containing legumes, including gluten-free [17,18] and low-fat [19] biscuits, and snacks for children [20]. Furthermore, it must be kept in mind that consumer preferences have changed over the years and nowadays food products that could provide health benefits are preferred by consumers [21].

The common bean (*Phaseolus vulgaris* L.) is an essential source of proteins and many macro and micronutrients. Therefore, its consumption has beneficial effects on human health and related pathologies, such as reducing cardiovascular diseases and diabetes mellitus, preventing different types of cancer, and controlling some metabolic functions [22]. Furthermore, because of their characteristics, beans can be introduced in gluten-free, vegetarian, and vegan diets as an optimal source of proteins. Moreover, the introduction of beans into popular industrial products, especially snacks, may be helpful to attract people and increase the use of legumes within different food formulations so that all groups of consumers can benefit from the positive effects of common beans. Moreover, they are an essential and complete staple food in many developing countries, where 'hidden hunger' is a critical challenge and common bean represents one of the most common crops [23].

Despite their positive characteristics, beans also contain some antinutritional compounds, such as lectins and phytic acid [24]. It is well known that consumption of raw or inadequately cooked beans causes poisoning, characterized by extreme nausea, vomiting, diarrhea, severe acute gastroenteritis and intestinal malabsorption [25–27]. This toxicity has been ascribed to the presence of active lectins [28], one of the major classes of common bean storage proteins. For these reasons, the presence of lectins strongly limits the direct use of common bean flour without any thermal treatment to prepare baked products, as the cooking conditions do not always ensure the complete inactivation of these toxic proteins [12]. In addition, the presence in common bean seed of high amounts of the strong cation chelator phytic acid reduces the bioavailability of important minerals such as iron, zinc, potassium, calcium, and magnesium that, at physiological pH, easily precipitate as phytate salts [29].

In recent decades, most research has sought, on the one hand, to develop approaches aimed at reducing legume seed antinutrients and improving their excellent properties and, on the other, to find possible ways to sensitize consumers to their regular consumption. For this purpose, not only well-tested technological methods (such as germination, extrusion and heat treatments) exist [13,30–32], but even genetic approaches can be employed [24,33]. Moreover, some technological modifications can modify some nutritional and antioxidative properties. For instance, migration of compounds with high antiradical activity to brine may significantly diminish bioavailability of active compounds because the brine is usually discarded before consumption of canned legumes [34]. Genetic approaches, by exploiting natural and induced biodiversity, were successfully used to isolate lines devoid of active lectins (*lec*$^-$ [35,36]) and with low seed phytic acid content (*lpa* [37,38]). In common bean seeds, major lectins, the erythroagglutinating and leucoagglutinating phytohemagglutinins (PHA-E and PHA-L, respectively), and the $\alpha$-amylase inhibitor ($\alpha$-AI), also known as phaseolamin, occur at the same genetic locus, called APA [39]. $\alpha$-AI can inhibit mammalian $\alpha$-amylases, thus inhibiting starch digestion [40] and is widely used as an active ingredient of commercial starch blocker preparations for the control of body weight [41]. Moreover, $\alpha$-AI can also reduce post-prandial glucose plasma levels, insulin C-peptide and gastric inhibitory polypeptide in healthy and diabetic subjects [42,43]. Through the screening of natural biodiversity, followed by breeding, genotypes showing differences in the presence of PHAs and $\alpha$-AI have been isolated [35,36]. For example, the cv. Lady Joy, devoid of active PHAs and with an active $\alpha$-AI, was obtained through the introgression into a commercial cultivar of an inactive PHA, the so-called "pinto lectin", present in only 10% of common bean genotypes [35].

Two allelic common bean *lpa* mutants, affecting the PvMRP1 phytic acid transporter, have been isolated and characterized to date, showing a 75–90% reduction in seed phytic acid content [37,38]. From an in vivo study, iron absorption per test meal from the *lpa1* seeds was higher than from wild type beans, showing that the *lpa1* beans are biofortified [44]. However, the following study found that the *lpa1* beans could cause adverse gastrointestinal symptoms due to a hard-to-cook (HTC) phenotype concomitant with increased thermal stability of lectins in these lines [27]. As recently shown, both traits depend on the genetic backgrounds in which the *lpa1* mutation is present [45].

Unprocessed common bean flours obtained from the Lady Joy genotype, and the *lpa1* mutant have been used to prepare biscuits [12]. Results showed that if flour derives from conventional common bean, lectin activity is still detectable after baking; consequently, it has been suggested to avoid using unprocessed traditional flour of beans to produce baked products. Conversely, the unprocessed Lady Joy flour (*lec*$^-$) is safe, as no lectin activity is present in the biscuits. Moreover, it was shown that baking did not fully inactivate $\alpha$-AI, a trait that may contribute to lowering the predicted glycaemic index (pGI) of the biscuits. When the *lpa1* flour was used to prepare biscuits, a 50% reduction of phytic acid content in the final product was observed [12].

In a previous study [12] the use of different bean flours devoid of specific antinutrients to prepare biscuits was analyzed. Moreover, different proportions of bean flours in different composite flour formulations (e.g., with wheat and/or maize flours) were used. In the current study a wider range of bean-based products, made with an advanced bean genotype that combines the reduction of phytic acid (*lpa*) with the absence of an active lectin (*lec*$^-$) while maintaining the accumulation of $\alpha$-AI, were analyzed. Crackers with two different proportions of bean flour, two different types of biscuits and a cream, with a content of bean flour ranging from 9% (cream) up to 38% (Cracker 2) were produced. Our general purpose is to promote bean consumption by developing biofortified and low antinutrients beans which are well suited for directly use as bean flours. More specifically, in this study, we aimed, on the one hand, to verify how some nutritional characteristics of the bean flours (e.g., activities of the $\alpha$-AI and the lectin, iron bioavailability) were influenced by different formulations and/or cooking conditions and, on the other hand, to assess the sensory properties and related acceptance level of bean-based products, according to a trained evaluation panel.

## 2. Materials and Methods

### 2.1. Plant Materials

Bean flours were obtained from a wild type (wt) *P. vulgaris* L. genotype, belonging to the bean market class Borlotto and an *lpa lec*$^-$ genotype with the same genetic background of Borlotto. This genotype (*lpa lec*$^-$) was a $BC_1F_5$ line derived from a cross between a $BC_2F_2$ *low phytic acid* (*lpa*) line [37] and the Lady Joy genotype [12], characterized by the absence of active lectins (*lec*$^-$).

### 2.2. Products Preparation

Composite flours, containing different percentages of the bean flour, wheat flour type 2, whole wheat flour and buckwheat flour were used to prepare biscuits, crackers and a cream. Recipes were developed with the technical assistance of the backers and pastry chefs Mr. Matteo Consolo and Mr. Ferruccio Farioli at ENAIP (Ente Acli Istruzione Professionale) Lombardia, Busto Arsizio (VA, Italy). Crackers with two different amounts of bean flour, two types of biscuits, and a cream were produced. Formulations are reported in Table 1.

**Table 1.** Formulations of bean-based food products.

| Product | Ingredients | Quantity (g) | % Total | % Bean Flour in | |
|---|---|---|---|---|---|
| | | | | Total Flour | Whole Product |
| Cracker 1 | Wheat flour type 2 * | 600 | 38.10 | 40 | 26 |
| | Bean flour | 400 | 25.40 | | |
| | Water | 550 | 34.92 | | |
| | Yeast | 10 | 0.63 | | |
| | Salt | 15 | 0.95 | | |
| Cracker 2 | Wheat flour type 2 * | 400 | 24.62 | 60 | 38 |
| | Bean flour | 600 | 36.92 | | |
| | Water | 600 | 36.92 | | |
| | Salt | 15 | 0.92 | | |
| | Yeast | 10 | 0.62 | | |
| Biscuit 1 (shortbread) | Butter | 500 | 30.30 | 50 | 20 |
| | Milk | 350 | 21.21 | | |
| | Wheat flour type 2 * | 325 | 19.70 | | |
| | Bean flour | 325 | 19.70 | | |
| | Vanilla icing sugar | 150 | 9.09 | | |
| Biscuit 2 (buckwheat biscuit) | Sugar | 350 | 27.08 | 33 | 12 |
| | Eggs | 250 | 19.34 | | |
| | Butter | 200 | 15.47 | | |
| | Whole wheat flour * | 160 | 12.38 | | |
| | Buckwheat flour | 160 | 12.38 | | |
| | Bean flour | 160 | 12.38 | | |
| | Backing | 12.5 | 0.97 | | |
| Cream | Almond milk | 1000 | 28.41 | 100 | 9 |
| | Sugar | 300 | 8.52 | | |
| | Eggs yolks | 300 | 8.52 | | |
| | Bean flour * | 150 | 4.26 | | |
| | Lemon peel | 10 | 10 | | |

* in control products equivalent amounts of these flours replaced the common bean flour. In the case of the cream, rice flour replaced bean flour.

Crackers and biscuits were both prepared in three variants, using the flours from the wt or the *lpa lec*$^-$ bean genotype or replacing the common bean flour with an equivalent amount of wheat flour type 2 or whole wheat flour (Biscuit 2). Cream was prepared with the *lpa lec*$^-$ bean flour or with rice flour. Crackers and biscuits prepared with the two different common bean flours (wt and *lpa lec*$^-$) were used for the different analyses, except the sensorial ones. For these last analyses, crackers, biscuits, and cream prepared with the *lpa lec*$^-$ flour were compared to the same product prepared without the common bean flour.

For cracker preparation the dough was produced by mixing wheat flour type 2, bean flour, with 95% of water. The dough was left to rest to activate the autolysis process. After 30 min, yeast was added and the remaining 5% of water was gradually added to the dough. The dough was divided in pieces of the same weight and each loaf was reduced to a thin sheet. Each sheet was put in a baking tin, previously oiled, brushed with olive oil and sprinkled with salt to taste. Sheets were pierced and cut in regular pieces and then cooked in an oven at 190 °C for approximately 8 min.

Shortbread biscuits (Biscuit 1) were prepared by mixing sugar with butter in a planetary mixer until they were completely homogenized. Then the other ingredients were added. The dough resulted was soft and biscuits were formed using a sac a poche. Biscuits were cooked in an oven at 190 °C for 14–18 min.

Buckwheat biscuits (Biscuit 2) were prepared by mixing all the ingredients together to produce the dough. Once all the ingredients were completely homogenized, the dough was worked by "cylinder processing" and formed. Before cooking, biscuits were sprinkled with a mixture of sugar or salt and cornmeal. They were cooked in an oven at 190 °C for 14–18 min.

For the cream, the ingredients were mixed together and cooked for approximately 15 min at low temperature (65 °C) to pasteurize the egg yolks. The final cream was stored in the fridge at about 4 °C.

### 2.3. Proximate Composition Analyses

The chemical composition of food samples was assessed according to AOAC standard methods [46]. Samples were milled and analyzed for dry matter (DM), ash, crude protein (CP), and ether extract (EE). Starch content was measured enzymatically (kit K-TSTA, Megazyme, Bray, Ireland). Total sugar determination was performed using the sucrose, D-fructose and D-glucose assay procedure (kit K-SUFRG Megazyme). The total dietary fiber (TDF) was determined using the AOAC Method 991.43 (kit K-TDFR, Megazyme) based on the sequential enzymatic digestion of 1 g of sample by heat-stable α-amylase, protease and amyloglucosidase.

The evaluation of in vitro starch digestion of food products over time was performed following the multi-enzymatic protocol detailed by Giuberti et al. [47] and as described in Sparvoli et al. [12]. Commercial fresh white bread (starch content of 72.3% DM) was used as reference and a blank was also included to correct for the glucose in the amyloglucosidase solution. The percentage of digested starch at each time interval was calculated using a factor of 0.9 to convert mono to polysaccharides. For each treatment, samples were analyzed in duplicate. After the enzyme digestion, a hydrolysis index (HI) was derived from the ratio between the area under the hydrolysis curve (AUHC) of each sample and the corresponding AUHC of the reference fresh white bread as a percentage over the same period. From the obtained HI, a pGI value was calculated with the formula $pGI = 8.198 + 0.862 \times HI$ [48].

### 2.4. Hemagglutination Test

Bean flours, or defatted food samples, were extracted with 20 volumes of phosphate buffered saline buffer (PBS: 10 mM $KHPO_4$, 15 mM NaCl, pH 7.4). Hemagglutinating activity in the extracts was determined by a serial dilution method using a human type A erythrocyte suspension. Drops of blood were drawn from the donor finger and collected inside 1.5 mL Eppendorf tube: about 150 μL of blood was collected. Erythrocytes were separated from the serum by washing with about 10 volumes (1.5 mL) of PBS. This step was repeated three times. The serum was eliminated, and the erythrocytes were stored at 4 °C in 10 volumes of PBS. At the time of use, erythrocytes were diluted 1:10 with PBS.

For each sample (PBS samples both from flours and defatted products), serial dilutions in PBS, ranging from 1:2 to 1:256, were assayed. Agglutination was visually determined after 4 h incubation at room temperature.

### 2.5. *Assay of α-Amylase Inhibitor Activity*

The analysis was based on a protocol that measures the inhibitory activity of the sample against human salivary α-amylase (EC 3.2.1.1; Type IX-A) by the increase of iodine staining after the action of the salivary α-amylase with soluble starch [49]. Briefly, different volumes (from 10 to 100 μL) of each sample extract (bean flour or defatted food products in PBS, as described above) were diluted 50 or 200-fold in 20 mM borate buffer pH 9 and were pre-incubated with a fixed amount of α-amylase (0.15 U) for 30 min at room temperature in a final volume reaction of 300 μL. Then, 200 μL of a 0.15% solution of potato starch was added and, after 5 min at room temperature, the reaction was stopped by adding 1 mL of iodine reagent [50] and absorbance was measured at 620 nm. Results were expressed as units of α-amylase inhibited per mg of flour, where one unit of inhibitor activity is the amount which will bring about 50% inhibition of the α-amylase in 30 min under the above conditions according to Marshall and Lauda [51]. The percentage of residual activity was calculated comparing the expected U of α-AI/100 mg of flour to the measured values.

### 2.6. *Preparation of the Pepsin, Pancreatin and Bile Extract Digests*

A quantity of 0.2 g porcine pepsin (800–2500 units/mg protein; Sigma-Adrich, St-Louis, MO, USA) was dissolved in 5 mL 0.1 N HCl; 0.05 g pancreatin (4 × USP specifications, Sigma-Adrich) and 0.3 g bile extract porcine (glycine and taurine conjugates of hyodeoxycholic and other bile salts, Sigma-Adrich) were dissolved in 25 mL 0.1 $NaHCO_3$. Both solutions were preliminary subjected to batch treatment with Chelex-100 (Bio-Rad Laboratories, Hercules, CA, USA) to eliminate any residual polyvalent metal ion. A quantity of 1.65 g food sample was dissolved in 10 mL water in a 50-mL screw-cap culture tube. The pH of each sample was adjusted to pH 2.0 with 5.0 N HCl. A volume of 0.5 mL of the pepsin solution was added and incubated for 60 min on a shaker at 37 °C. The pH of the sample was then raised to 6 with 1 M $NaHCO_3$ and 2.5 mL of the pancreatin-bile extract mixture was added. Finally, the pH was adjusted to 7 with 2 M NaOH, and the volume was brought to 15 mL with a 120 mM NaCl/5 mM KCl water solution.

### 2.7. *In Vitro Analysis of Iron Bioavailability and Iron Content Evaluation*

The in vitro digestion model from [52] was adopted with some modifications. Caco-2 cells (American Type Culture Collection (Rockville, MD, USA)) were used in experiments at passage 29. Cells were seeded at densities of 50,000 cells/$cm^2$ in collagen-treated 6-well plates (Costar Corp., Cambridge, MA, USA) and cultured at 37 °C in Dulbecco's modified Eagle medium containing 25 mmol/L HEPES, and 10% fetal bovine serum (complete medium) in 5% $CO_2$ atmosphere. On day 13 from cell confluence, the complete medium was removed and replaced with minimum essential medium (MEM) supplemented with 10 mmol/L PIPES, 4 mg/L hydrocortisone, 5 mg/L insulin, 5 μg/L selenium, 34 μg/L triiodothyronine and 20 μg/L epidermal growth factor (enriched MEM), to ensure background levels < 80 μg Fe/L. A Transwell insert ring (Costar Corp.) was then added to each well, in which the upper chamber was formed by fitting the bottom with a dialysis membrane (15,000 Da molecular weight cut off; Sigma-Aldrich). A volume of 1.5 mL digested sample was added to the upper chamber and incubated for 2 h. Then, inserts were removed, and 1 mL of enriched MEM was added. Cells were incubated for a further 22 h at 37 °C. Then, the monolayer was washed with 140 mM NaCl, 5 mM KCl and 10 mM PIPES pH 7 (rinse solution). The rinse solution was aspirated and 2 mL of a freshly prepared rinse solution containing 5 mM sodium hydrosulfite and 1 mM athophenanthroline disulfonic acid (removal solution) was added for 10 min. The removal solution was aspirated and the monolayer was washed with rinse solution. Next, 2 mL of deionized water was added and cells were sonicated for 15 min with a benchtop sonicator. Finally, cells were recovered by scraping, collected along with the 2 mL water and stored at −20 °C. A 10-μL sample was used for ferritin measurement using the Human Ferritin ELISA (Enzyme-Linked Immunosorbent Assay) kit (SIGMA-Aldrich). Cell protein was measured using the Bio-Rad DC protein assay kit (Bio-Rad).

For iron content evaluation, 300 mg samples were digested in Teflon tubes filled with 5 mL of 65% ($v/v$) $HNO_3$ by a microwave digester system (MULTIWAVE-ECO, Anton Paar Italia Srl., Rivoli, Italy) by applying two-step power ramps (Step 1: To 500 W in 10 min, maintained for 5 min; Step 2: To 1200 W in 10 min, maintained for 15 min). After 20 min cooling, the mineralized samples were transferred into polypropylene test tubes and diluted 1:20 with MILLI-Q water (Merck). Iron concentrations were measured by inductively coupled plasma-mass spectrometry (ICP–MS; Bruker AURORA M90 ICP–MS, Bruker Daltonik GmbH, Leipzig, Germany).

### 2.8. Sensorial Evaluation

Sensory evaluation was executed on food products prepared with bean flour (Table 1) and on reference I samples, in which bean flour was replaced with equal amounts of wheat flour type 2 (Cracker R, Biscuit 1R), whole wheat flour (Biscuit 2R), or rice flour (Cream R). Evaluated products are shown in Figure 1.

Sensory descriptive analysis (DA) and temporal dominance of sensations (TDS) [53,54], were carried out by a panel composed of 10 expert judges (5 females and 5 males, age 30–50), with significant experience in sensory descriptive evaluation and in the use of the sensorial software used to collect data. Attributes to be used in this study were selected based on literature on bakery goods and cookies [55] and proposed to the judges to familiarize with the products. Tests were performed by the panel at appropriate light and temperature conditions, according to UNI EN ISO 8589:2014 [56], in individual booths, with notebooks equipped with specific software for sensory data acquisition (FIZZ Biosystèmes, France).

Samples were distributed to the judges, coded with three-digit numbers and presented randomly. Mineral water was distributed to the judges to clean their mouths between samples.

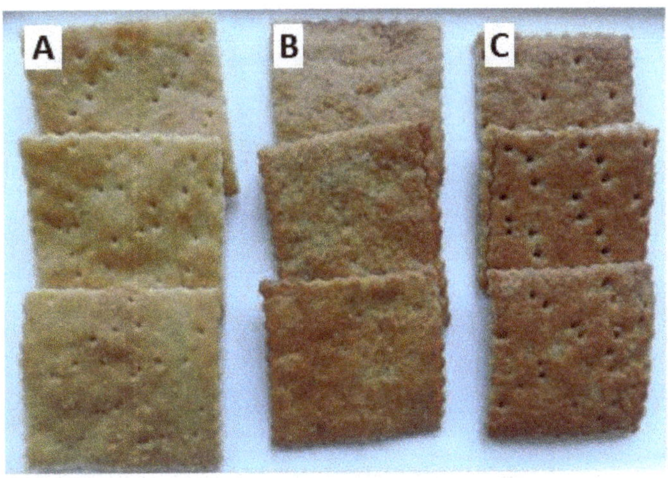


**Figure 1.** *Cont.*

**Figure 1.** Samples used for the sensorial evaluation. (**A**) Cracker R, (**B**) Cracker 1, (**C**) Cracker 2, (**D**) Biscuit 1R, (**E**) Biscuit 1, (**F**) Biscuit 2R, (**G**) Biscuit 2, (**H**) Cream R, (**I**) Bean Cream. All R samples were reference without common bean flour, while the samples Crackers 1 and 2, Biscuits 1 and 2 and Bean Cream contained 26, 38, 29, 14 and 9% of common bean flour, respectively.

### 2.8.1. Descriptive Analysis

For DA, each judge was delivered one product at a time and asked to evaluate the attributes (Table 2) using a nine-point intensity scale (1 = hardly perceptible; 9 = very intense) an overall quality score (1 = dislike extremely; 9 = like extremely) [57,58]. DA tests were carried out in duplicate.

**Table 2.** List of attributes evaluated in descriptive analysis (DA) and temporal dominance of sensations (TDS).

| | **Cracker** | **Biscuit 1** | **Biscuit 2** | **Cream** |
|---|---|---|---|---|
| Flavor of legumes | DA + TDS | DA + TDS | DA + TDS | DA + TDS |
| Flavor of wholemeal/bran | DA + TDS | DA + TDS | DA + TDS | |
| Flavor of wheat | DA + TDS | | | |
| Flavor of butter | | DA + TDS | DA + TDS | |

**Table 2.** *Cont.*

| | Cracker | Biscuit 1 | Biscuit 2 | Cream |
|---|---|---|---|---|
| Flavor of shortcrust pastry | | DA + TDS | | |
| Flavor of biscuit | | | DA + TDS | |
| Flavor of almond | | | | DA + TDS |
| Flavor of lemon | | | | DA + TDS |
| Flavor of cream | | | | DA + TDS |
| Flavor of eggs | | | | DA + TDS |
| Flavor of biscuit | | | | DA |
| Odor of legumes | DA | DA | DA | DA |
| Odor of wholemeal/bran | DA | DA | DA | |
| Odor of butter | | | DA | |
| Odor of shortcrust pastry | | DA | | |
| Odor of crackers | DA | | | |
| Odor of biscuit | | | DA | |
| Odor of eggs | | | | DA |
| Odor of cream | | | | DA |
| Odor of lemon | | | | DA |
| Snapping | DA | | | |
| Consistency | DA | DA | DA | |
| Crunchiness | DA | DA | DA | |
| Friability | DA | DA | DA | |
| Graininess | DA | DA | DA | DA |
| Viscosity | | | | DA |
| Creaminess | | | | DA |
| Adhesiveness | DA | DA | DA | |
| Humidity | DA | DA | | |
| Chewiness | DA | | | |
| Flouriness | | | | DA |
| Astringency | DA | DA | DA | DA |
| Umami | DA + TDS | DA + TDS | DA + TDS | DA + TDS |
| Bitterness | DA | DA | DA | DA |
| Salty | DA + TDS | DA | DA | DA |
| Sweetness | | DA + TDS | DA + TDS | DA + TDS |

2.8.2. Temporal Dominance of Sensations

The TDS was performed, considering only 5–7 attributes (Table 2) linked to the gustative/aromatic aspects considered relevant for each product. According to Pineau [53], during the product tasting (60 s) judges were asked to express the dominant attribute.

Results were elaborated by statistical analysis using the software SAS® (SAS 9.4, SAS Institute Inc., Cary, NC, USA). The sensorial profiles were analyzed through ANOVA and post hoc test (Tukey's HSD) and displayed in a visual plot. For the TDS, the proportion of runs for which each attribute was considered as dominant was calculated for each point in time. These proportions were traced over time using the SAS® TRANSREG procedure and named "TDS curves".

## 3. Results

### 3.1. Proximate Composition

Crackers and biscuits prepared with wt or *lpa lec*$^-$ common bean flours were analyzed for their proximate composition (Table 3). Results showed that the protein content was relatively high in all the baked products, due to the contribution of the bean flour. As expected, the highest protein content was found in Cracker 2 (21.7 g/100 g DM and 18.5 g/100 g DM for *lpa lec*$^-$ and wt, respectively), which was characterized by the highest bean flour content (38% of the total product). Biscuits contained the lowest protein content (average 11.3 g/100 g DM), according to the lower percentage of the bean flour in their formulation (20% and 12% of the total in Biscuit 1 and Biscuit 2, respectively).

**Table 3.** Proximate composition of bean-based products (g/100 g dry matter). na, non-analyzed; <LOQ, under limit of quantification.

| Sample | | Water | Crude Protein | Crude Lipid | Total Carbohydrates | | | | | Ash |
|---|---|---|---|---|---|---|---|---|---|---|
| | | | | | Starch | Saccharose | Glucose | Fructose | Total Dietary Fiber | |
| Cracker 1 | *lpa lec*⁻ | 7.3 | 19.1 | 3.3 | 44.6 | <LOQ | 0.9 | <LOQ | 16.7 | 6.3 |
| | wt | 6.3 | 17.0 | 3.1 | 46.3 | <LOQ | 0.9 | <LOQ | 17.9 | 6.7 |
| Cracker 2 | *lpa lec*⁻ | 6.5 | 21.7 | 3.4 | 36.6 | <LOQ | <LOQ | 1.0 | 21.7 | 7.5 |
| | wt | 6.9 | 18.5 | 2.2 | 38.5 | <LOQ | 1.1 | 0.5 | 22.8 | 7.8 |
| Biscuit 1 | *lpa lec*⁻ | 4.6 | 11.7 | 32.0 | 19.3 | 11.3 | 0.0 | <LOQ | 18.4 | 1.3 |
| | wt | 5.0 | 11.0 | 32.9 | 21.2 | 12.7 | 0.0 | 0.0 | 14.8 | 1.5 |
| Biscuit 2 | *lpa lec*⁻ | 4.6 | 11.6 | 15.4 | 14.8 | 32.3 | 0.1 | <LOQ | 18.2 | 1.8 |
| | wt | 4.5 | 11.0 | 15.0 | 18.6 | 28.5 | 0.0 | <LOQ | 17.7 | 1.8 |
| Bean flours | *lpa lec*⁻ | 10.6 | 27.7 | 0.6 | 36.0 | na | na | na | na | 3.5 |
| | wt | 10.6 | 23.1 | 1.1 | 44.9 | na | na | na | na | 3.8 |

The lipid content was higher in Biscuit 1 (average 32.4 g/100 g DM) and consistent in Biscuit 2 (average 15.2 g/100 g DM), since, in both cases, butter was one of the main ingredients. In contrast, crackers, which were prepared with no added fats (except for the olive oil used to sprinkle the pan), showed a lower lipid content (average 3.0 g/100 g DM).

The carbohydrate fraction appeared variable between baked products. The starch content was higher in Cracker 1 (average 45.4 g/100 g DM), but was lower in Cracker 2 (average 38.3 g/100 g DM), where the bean flour was present in a higher proportion. In contrast, starch content was higher in Biscuit 1 than in Biscuit 2 (average 20.2 g/100 g DM and 16.7 g/100 g DM, respectively), despite the fact Biscuit 1 contained a higher percentage of common bean flour. This difference might be explained by the composition of Biscuit 2, in which wheat flour was replaced by whole wheat and buckwheat flours that further contributed to lowering the starch content. On the other hand, these ingredients did not influence the content of total dietary fiber (average 16.7 and 17.9 g/100 g). The total dietary fiber also constituted a significant fraction in crackers and increased with the increasing level of bean flour in the products, being higher in Cracker 2 than in Cracker 1 (20.9 g/100 g DM against 16 g/100 g DM). In crackers, total sugars were negligible, while they were relevant in biscuits for the saccharose fraction (glucose and fructose were almost absent) that correlated to the added sugar in the recipes, that was almost three times higher in Biscuit 2 than in Biscuit 1 (average 30.4 g/100 g DM and 12.0 g/100 g DM, respectively).

Finally, the ash content was higher in crackers (average 6.5 g/100 g DM in Cracker 1 and 7.6 g/100 g DM in Cracker 2), while Biscuit 2 had a slightly higher ash fraction (average 1.8 g/100 g DM) than Biscuit 1 (average 1.4 g/100 g). These differences might have been due to the bean flour present in the products, together with the amount of less refined flours in Biscuit 2.

In general, products made with *lpa lec*⁻ bean flour had higher protein content and lower starch content compared to those made with wt bean flour, in agreement with the composition of the respective flours (27.7 g/100 g vs. 23.1 g/100 g for proteins and 36.0 g/100 g vs. 44.9 g/100 g for starch) (Table 3).

### 3.2. Evaluation of Hemagglutinating Activity

Bean seeds accumulate high amounts of toxic PHAs and the excessive consumption of raw or improperly cooked beans may cause poisoning. Therefore, since our aim was to exploit the use of a biofortified and lectin null bean flour to obtain novel fortified foods, it was important to verify the lack of any residual lectin activity in the bean-based products (crackers and biscuits in their variants). To quantify the amount of lectins in the different bean-based products, equal amounts of extracts were compared with equal amounts of extracts from flours of the wt genotype as control, and *lpa lec*⁻ genotype (Figure 2).

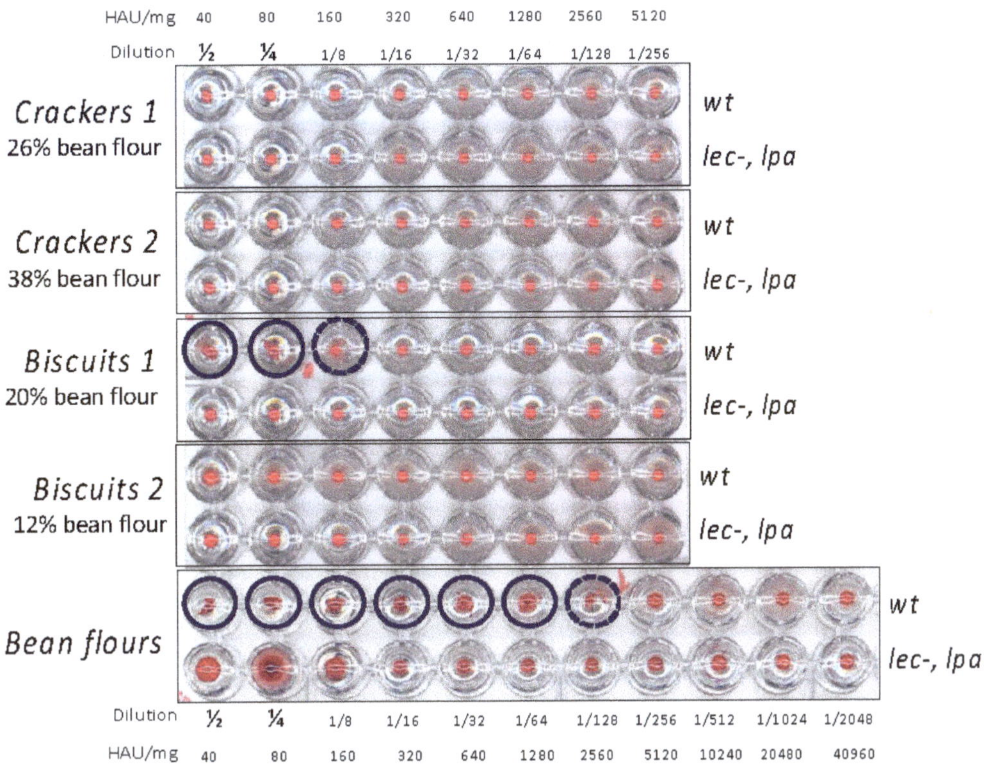


**Figure 2.** Hemagglutinating activity of bean-based products and bean flour extracts. Serial dilutions of equal amounts of bean-based products or seed flour extracts obtained from two different genotypes (wt and *lpa lec*$^{-}$) were compared. Blue circles indicate sample dilutions able to agglutinate red blood cells. The percentage of bean flour in biscuits was adjusted based on the weight of defatted samples (in defatted biscuits 1 bean flour was 29.4%).

As expected, the flour extract characterized by the absence of active PHAs (*lpa lec*$^{-}$ genotype) was not able to agglutinate blood erythrocytes at any dilution. On the contrary, wt flour extract (containing PHAs) was able to agglutinate blood erythrocytes after a serial dilution until 1/128 (dotted circle in Figure 2).

The hemagglutination test on defatted products (Figure 2) indicated the presence of residual lectin activity only in Biscuit 1 containing 20% wt bean flour (corresponding to 29.4% in defatted flour). Cracker 1 and Cracker 2, containing 26% and 38% of bean flour on the total product, respectively, were not able to agglutinate blood erythrocytes. As expected, none of the products made with *lpa lec*$^{-}$ genotype showed residual hemagglutinating activity. Comparing these agglutination results with those obtained from corresponding amounts of unprocessed bean flours, and considering that defatted Biscuit 1 contains about 1/3 of bean flour, it was possible to quantify the residual PHA activity in processed biscuit 1, prepared with wt common bean flour, being approximately 10–12% that of the corresponding bean flour.

### 3.3. Evaluation of α-Amylase Inhibitor Activity

The α-Amylase inhibitor plays an important role in lowering starch digestion and its effect is reflected in decreased postprandial plasma hyperglycemia and insulin levels, increased resistance of starch to digestion, and increased activity of the colorectal tract, as well as reduction in the glycemic index [12,59]. Processed products characterized by total, or at least partial, α-AI activity retention could be introduced in diets for people who require a low blood-glucose level (including individuals with type 1 diabetes) for the maintenance of good health status. To verify the stability and residual activity of α-AI in different processed products, α-AI activity was assessed in defatted extracts of crackers, biscuits and lyophilized cream, and compared to the α-AI activity of control raw flours.

Results of α-AI activity of tested products are reported in Table 4. An insignificant residual inhibitory activity was detected in Crackers 1 and 2 when made with wt bean flour, while a higher percentage of residual activity was reported for crackers made with *lpa lec*$^-$ genotype (16.63% and 32.34% respectively for crackers 1 and 2). Biscuit 1 was characterized by limited retention of α-AI activity when made with wt bean flour (20.44%), while biscuits made with the flour from the *lpa lec*$^-$ genotype showed higher inhibitory activity (51.19%). Biscuit 2 was characterized by higher residual α-AI activity compared to Biscuit 1, however it behaved in a similar way as regards the selected genotype for the bean flour; when *lpa lec*$^-$ bean flour was used, the residual α-AI activity was more than double that of Biscuit 2 containing wt bean flour (84.17% for *lpa lec*$^-$ and 37.33% for wt). In the case of the cream a moderate retention of α-AI activity was observed (27.28%). This product was made only with the *lpa lec*$^-$ bean flour as, if using wt flour, a high lectin activity would be expected due to the mild cooking conditions (70 °C, see Section 2).

**Table 4.** α-amylase inhibitor activity (%) in defatted bean-based product extracts. U α-AI, Units of α-amylase inhibitor.

| Sample | | % Bean Flour in the Total Sample | Expected U α-AI/100 mg Flour | Measured U α-AI/100 mg Flour | % Residual α-AI Activity |
|---|---|---|---|---|---|
| Control flours | wt | 100 | - | 1552.33 | - |
| | *lpa lec*$^-$ | | - | 1253.10 | - |
| Cracker 1 | wt | 26 | 620.93 | 4.23 | 0.68 |
| | *lpa lec*$^-$ | | 501.24 | 83.86 | 16.63 |
| Cracker 2 | wt | 38 | 931.40 | 2.20 | 0.24 |
| | *lpa lec*$^-$ | | 640.35 | 207.09 | 32.34 |
| Biscuit 1 | wt | 29 [a] | 494.73 | 101.11 | 20.44 |
| | *lpa lec*$^-$ | | 399.36 | 204.43 | 51.19 |
| Biscuit 2 | wt | 14 [a] | 232.38 | 87.77 | 37.77 |
| | *lpa lec*$^-$ | | 187.60 | 157.89 | 84.17 |
| Cream | *lpa lec*$^-$ | 24 [b] | 304.50 | 83.06 | 27.28 |

[a] percentage adjusted on the weight of defatted samples; [b] percentage adjusted on the weight of the lyophilized sample.

Interestingly, in all samples extremely high residual α-AI activity was always detected when the *lpa lec*$^-$ bean flour was used instead of the wt genotype and the highest activity was found in Biscuit 2, although it was not the one with the highest proportion of bean flour, suggesting that most probably, the food matrix and the cooking conditions are important for maintenance of α-AI stability and activity.

### 3.4. *In Vitro Predicted Glycemic Index*

The predicted in vitro glycemic index (pGI) was analyzed for crackers and biscuits using common white bread as reference. Results are reported in Table 5, where pGI was compared in the carbohydrate (starch and saccharose) fractions of each sample.

**Table 5.** In vitro predicted glycemic index (pGI) of bean-based products. The starch, saccharose and dietary fiber fractions are indicated as g/100 g. <LOQ, under limit of quantification.

| Sample | | % of Bean Flour in the Total Sample | pGI [a] | Starch | Saccharose | Dietary Fiber |
|---|---|---|---|---|---|---|
| Cracker 1 | wt | 26 | 76.6 | 46.3 | <LOQ | 17.9 |
| | *lpa lec*$^-$ | | 71.4 | 44.6 | <LOQ | 16.7 |
| | Average | | 74.0 | 45.4 | | 17.3 |
| Cracker 2 | wt | 38 | 61.6 | 38.5 | <LOQ | 22.8 |
| | *lpa lec*$^-$ | | 55.4 | 36.6 | <LOQ | 21.7 |
| | Average | | 58.5 | 37.5 | | 22.2 |
| Biscuit 1 | wt | 29 [b] | 42.9 | 21.2 | 12.7 | 14.8 |
| | *lpa lec*$^-$ | | 39.8 | 19.3 | 11.3 | 18.4 |
| | Average | | 41.3 | 20.2 | 12.0 | 16.6 |
| Biscuit 2 | wt | 14 [b] | 50.4 | 18.6 | 28.5 | 17.7 |
| | *lpa lec*$^-$ | | 45.0 | 14.8 | 32.3 | 18.2 |
| | Average | | 47.7 | 16.7 | 30.4 | 17.9 |

[a] white bread as control (pGI = 70); [b] percentage adjusted on the weight of defatted samples; LOQ = limit of quantification.

Considering the different bean-based products, Cracker 1 had the highest pGI (average 74), which correlates to its high percentage of starch (average 45.4 g/100 g), primarily deriving from wheat flour. In contrast, in Cracker 2, the higher amount of bean flour (38% in Cracker 2 vs. 26% in Cracker 1) contributed to reducing the starch fraction (37.5 vs. 45.4 g/100 g) and to increasing the dietary fiber and the crude protein contents, thus lowering the pGI value.

The biscuits had a more complex composition and a similar pGI (average 41.3 for biscuits 1 and 47.7 for biscuits 2). The starch fraction was higher in Biscuit 1 (20.2 g/100 g vs. 16.7 g/100 g), while sugars, consisting almost entirely of saccharose, were more than doubled in Biscuit 2 (30.4 g/100 g vs. 12.0 g/100 g). Saccharose was thus primarily responsible for the higher pGI of Biscuit 2. Biscuit 2 was also characterized by the presence of whole wheat and buckwheat flours that counterbalanced the effect of sugars by contributing to an increased total fiber fraction, which was quite similar to that of Biscuit 1 (17.9 g/100 g vs. 16.6 g/100 g) in which a major contribution derived from bean flour, which also contributed, together with a reduced saccharose content, to the slight reduction of pGI.

Interestingly, if data were compared considering the contribution of the bean genotype, it clearly appeared that, when products contained the *lpa lec*$^-$ bean flour, a reduction of about 5 units of the pGI, corresponding to about 7% of the reference (white bread) pGI; was detected (71.4 vs. 76.6 for Cracker 1, 55.4 vs. 61.6 for Cracker 2, 39.8 vs. 42.9 for Biscuit 1, 45.0 vs. 50.4 for Biscuit 2).

### *3.5. Assessment of Fe Bioavailability of Bean Biofortified Biscuit 1 in Caco-2 Cells*

We determined the available iron in examined food samples by analyzing ferritin formation in the intestinal cell line Caco-2. This parameter is considered a good indicator of cell Fe uptake from the food digest [52,60]. Caco-2 cell ferritin formation was shown to increase linearly in the range 50–100 μM Fe [52]. Accordingly, based on our estimated of Fe content (Table 6), 1.65 g was identified as an appropriate amount of food to digest, in order to maintain the Fe concentration within the 50–100 μM range (Fe molar mass: 55.845 g/mol). Our analysis showed very low levels of ferritin formation in untreated Caco-2 cells (Figure 3), in agreement with previous data [52]. The test was characterized by high sensitivity because it also allowed estimation of the small variations recorded among all the examined samples. On the other hand, following challenge with food sample digests, the resulting ferritin production was quite variable and required an appropriate number of biological replicates (n = 10) to detect any trend or achieve statistical significance. Interestingly, both *lpa lec*$^-$ bean and Biscuit 1 flours tended to have a higher ferritin value than their wt counterpart. In addition, ferritin levels in *lpa lec*⁻ Biscuit 1 were significantly higher compared to cells alone. (Figure 3).

**Table 6.** Iron content in wild type (wt) and *lpa lec*$^-$ bean flours and Biscuit 1 flours.

| Sample | Fe Content (mg/Kg) |
|---|---|
| wt bean flour | 27.03 |
| *lpa lec*$^-$ bean flour | 29.76 |
| wt Biscuit 1 flour | 29.38 |
| *lpa lec*$^-$ Biscuit 1 flour | 30.13 |

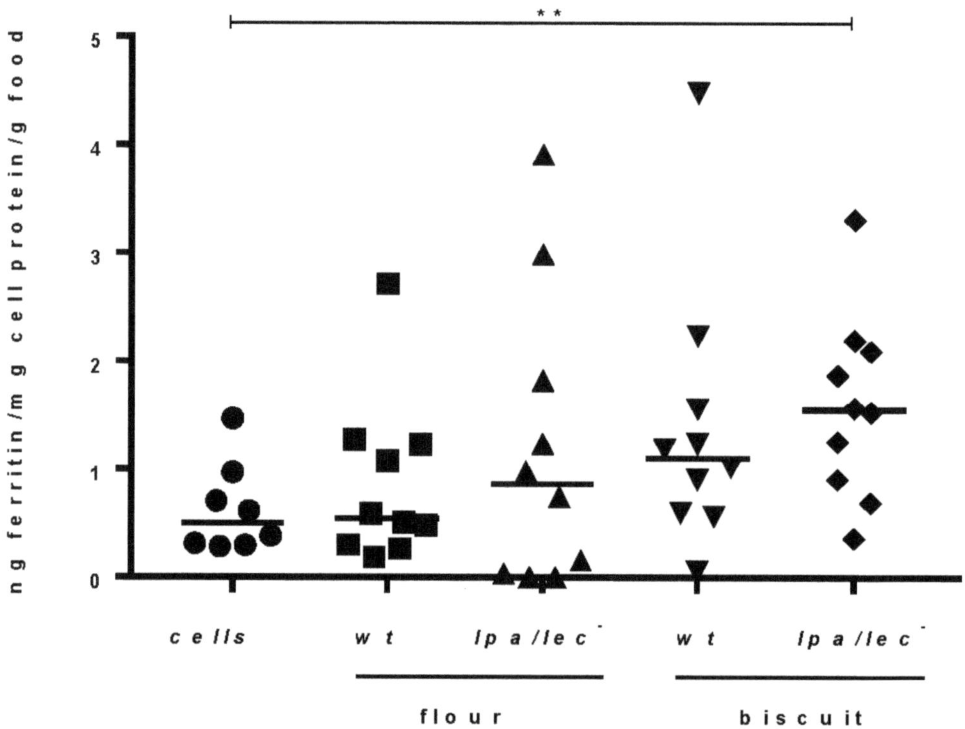


**Figure 3.** Ferritin production after incubation of Caco-2 cells with wild type (wt) flour, *lpa lec*$^-$ flour, wt Biscuit 1 or *lpa lec*$^-$ Biscuit1 samples. Samples (n = 10) were statistically evaluated by Mann-Witney test. ** $p = 0.085$.

### 3.6. Sensory Analyses of Bean-Based Products

Crackers 1 and 2 had very similar characteristics, differing only for crunchiness, higher in Cracker 1. On the other hand, both crackers significantly differed from the control product with lower snapping, crunchiness, friability, chewiness and wheat flavor, and higher intensities of legume, wholemeal/bran odor and flavor, cracker odor, mouth coating and umami. In addition, Cracker 2 was also different from the reference for lower cracker odor and firmness, and higher intensity of bitterness. Both Crackers 1 and 2 recorded an acceptance score (4.4 and 4.1, respectively) between "Neither Like nor Dislike" and "Dislike Slightly", corresponding, respectively, to 79% and 73% of the score obtained by the Cracker R (5.6) (Figure 4).

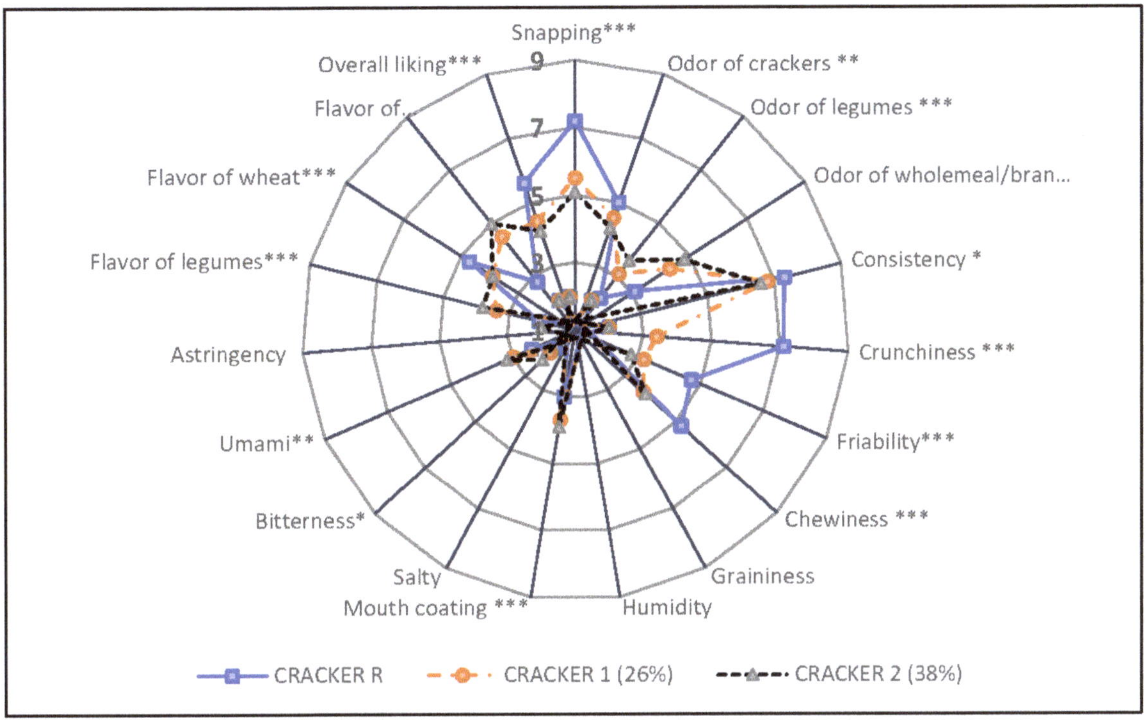


**Figure 4.** Sensorial profiles of crackers with [Cracker 1 (26%); Cracker 2 (38%)] and without [Cracker R] bean flour. Data analyzed by ANOVA and post hoc test (Tukey's HSD); *, **, *** indicate significant differences between the tested products at $p < 0.1$, $p < 0.5$ and $p < 0.01$, respectively.

Regarding the TDS analysis for flavor, in Cracker R (0% bean flour) the flavor of wheat was predominant over the whole tasting; salty showed an important role in the first phase of tasting (Figure 5A). With the addition of 26% bean flour, the flavor of legumes and whole-wheat became dominant (Figure 5B); a further addition of bean flour increased the dominance of whole-wheat flavor, together with a clear perception of umami in the central part of the tasting course (Figure 5C).

Biscuit 1 had a more intense odor and flavor of legumes and wholemeal/bran, as compared to Biscuit 1R. Moreover, Biscuit 1 was perceived as grainier, bitter and with a higher intensity of umami. The control biscuit (Biscuit 1R) was perceived as more consistent, sweeter with a stronger shortcrust pastry flavor (Figure 6). Both biscuits recorded an overall liking over "Neither Like nor Dislike", with Biscuit 1 recording an overall liking of 5.6, corresponding to 93% of the reference biscuit score.

The TDS of Biscuit 1R (Figure 7A) was mainly characterized by butter flavor and of shortcrust pastry, being the first perceived during the first and last part of tasting, and the second having higher impact in the middle part. In Biscuit 1 (Figure 7B), the flavor of wholemeal/bran and butter alternated for dominance during the tasting. Umami and flavor of legumes increased from the beginning to the central segment of tasting where they were the most perceived.

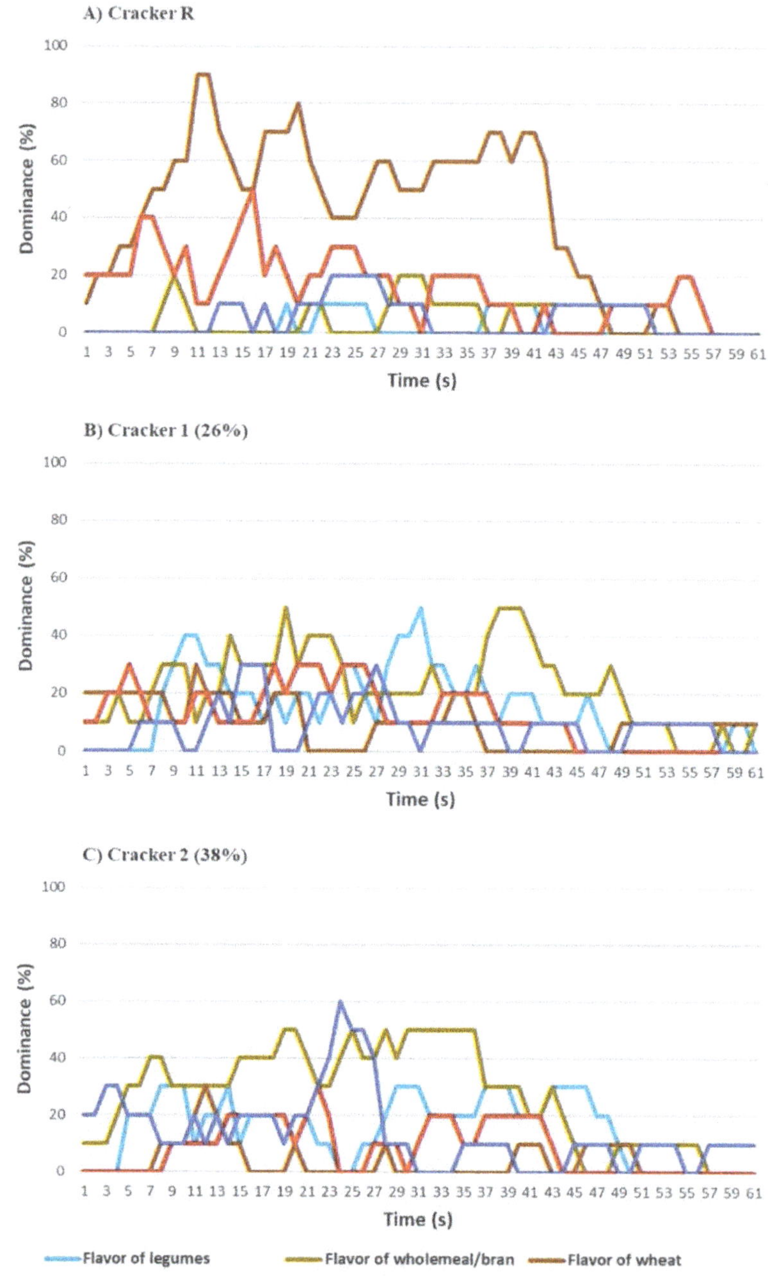


**Figure 5.** Temporal dominance of sensations (TDS) flavor of (**A**) Crackers R (reference, without common bean flour), (**B**) Crackers 1 (with 26% common bean flour) and (**C**) Crackers 2 (with 38% common bean flour).

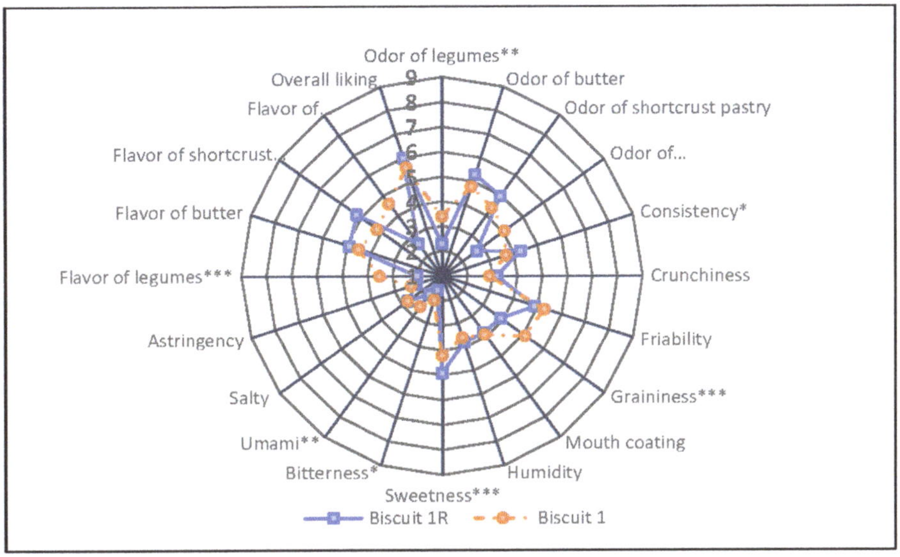


**Figure 6.** Sensorial profile of shortbread biscuits without (Biscuits 1R, reference) and with common bean flour (Biscuit 1). Data analyzed by ANOVA and post hoc test (Tukey's HSD); *, **, *** indicate significant differences between the tested products at $p < 0.1$, $p < 0.5$ and $p < 0.01$, respectively.

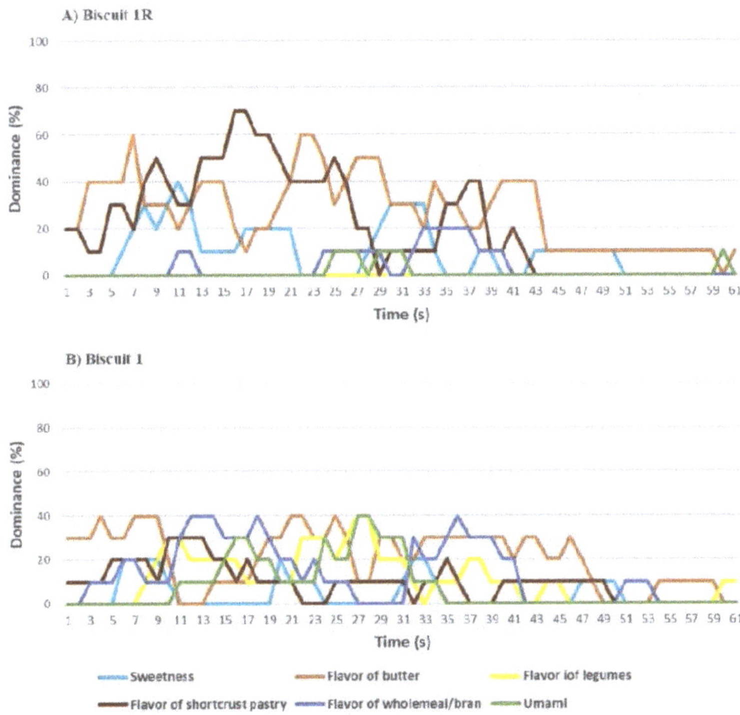


**Figure 7.** Temporal dominance of sensations (TDS) of (**A**) Biscuits 1R (reference, without common bean flour) and (**B**) Biscuit 1 (with 29% common bean flour).

Biscuit 2, as compared to its reference, had a higher mouth coating, while the umami taste was perceived more, as well as the flavor of legumes. The Biscuit 2R showed a higher biscuit odor, consistency and crunchiness; it was perceived to be sweeter than Biscuit 2 and with a more intense flavor of butter and biscuit (Figure 8). A significant difference was recorded for overall liking: Biscuit 2R recorded a score of 6.1, superior to the "Like Slightly" level. On the other hand, Biscuit 2 was rated between "Dislike Slightly" and "Neither Like not Dislike", despite recording an overall liking of 4.8, corresponding to 79% of the reference biscuit score.

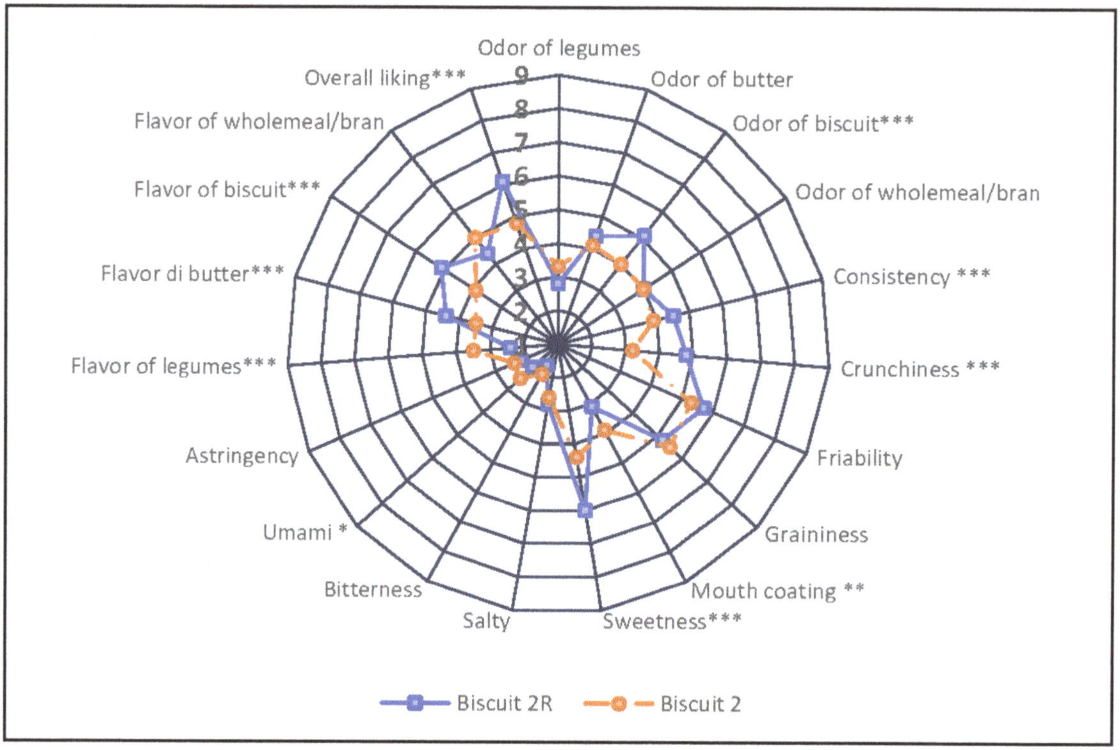


**Figure 8.** Sensorial profiles of type 2 Biscuits, without (2R) and with common bean flour (2). Data analyzed by ANOVA and post hoc test (Tukey's HSD); *, **, *** indicate significant differences between the tested products at $p < 0.1$, $p < 0.5$ and $p < 0.01$, respectively.

The TDS of biscuit 2R with 0% bean flour (Figure 9A) showed a prevalence of flavor of shortcrust pastry and butter during the whole tasting. In the biscuit with 20% bean flour (Figure 9B), the flavor of wholemeal/bran and butter alternated during the tasting (the latter with a lower dominance than in the biscuit without bean flour). Umami and flavor of legumes increased from the beginning to the central segment of tasting where they were the most perceived.

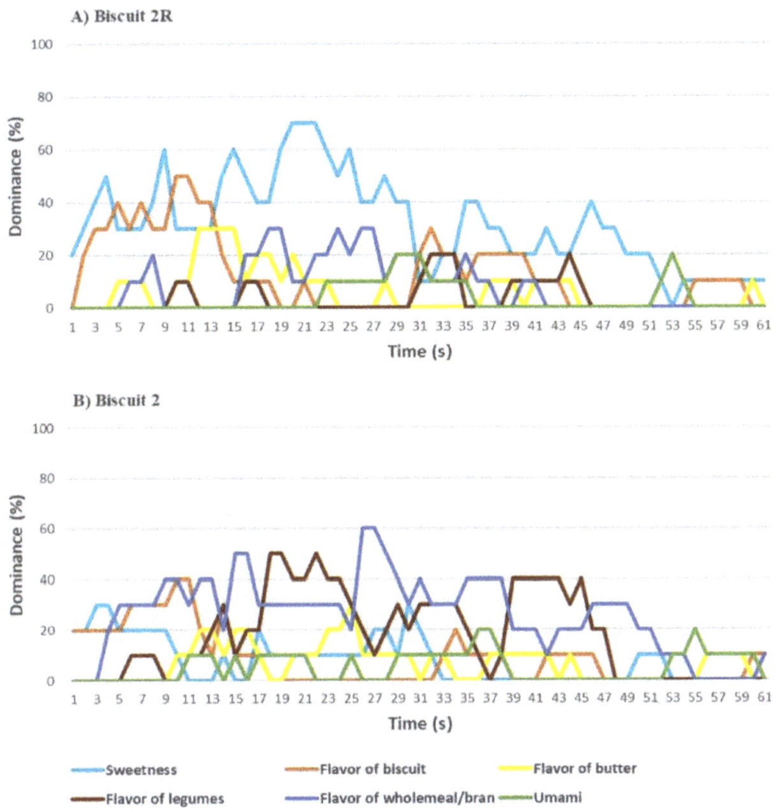


**Figure 9.** Temporal dominance of sensations (TDS) of (**A**) Biscuits 2R (reference, without common bean flour) and (**B**) Biscuit 2 (with 14% common bean flour).

Regarding the TDS flavor of Biscuit 2 (Figure 9A), the sweetness dominated during almost the whole tasting, with the flavor of biscuits being more perceived mainly in theinitial phase. In Biscuit 2 sweetness was dominant at the very beginning of tasting; the wholemeal/bran and legumes flavor alternated in dominance during the remaining tasting time. A slight umami peak was recorded at the end (55–60 s).

Regarding the two cream preparations it was possible to observe several significant differences. In particular, the bean flour Cream had higher odor and flavor of legumes and flavor of almond. Moreover, it was rated as grainier and more sapid (i.e., high umami perception) (Figure 10). The Cream R preparations differed also, for more perceived cream odor and flavor of eggs, cream and lemon. Creaminess and sweetness were higher in Cream R as well as astringency. The overall liking of Cream R was between "Neither Like not Dislike", and "Like Slightly", while bean Cream was below the "Dislike Slightly" point of acceptability, recording an overall liking of 3.9, corresponding to 71% of the reference biscuit score.

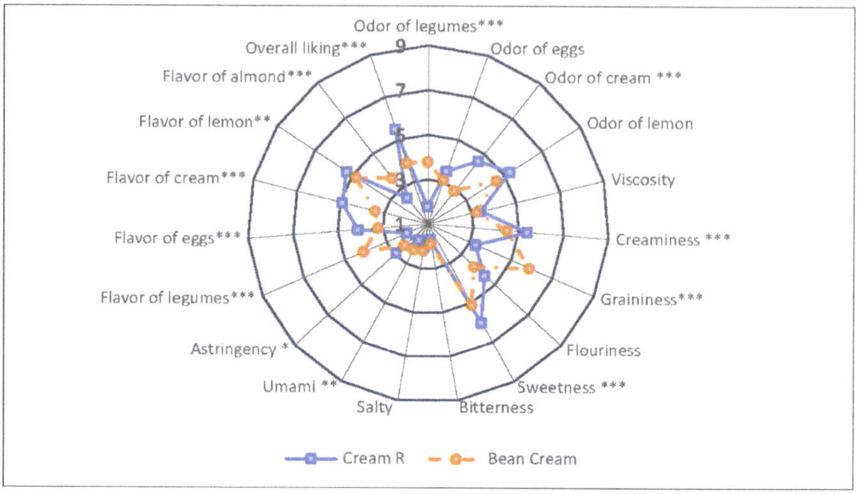


**Figure 10.** Sensorial profile of cream without (Cream R, reference) and with common bean flour (Bean Cream). Data analyzed by ANOVA and post hoc test (Tukey's HSD; *, **, *** indicate significant differences between the tested products at $p < 0.1$, $p < 0.5$ and $p < 0.01$, respectively.

The TDS taste of Cream R (Figure 11A) showed a dominant peak of lemon flavor at the beginning of the tasting (around 11 s), followed by the flavor of cream and eggs, after that the impact of lemon was noteworthy, since it was the sensation that dominated during the remaining tasting time. The bean cream tasting was characterized by lemon flavor and sweetness at the beginning, which were substituted by the flavor of legumes and, later, the flavor of almond (Figure 11B). The last part of tasting was characterized by the flavor of lemon and almond.

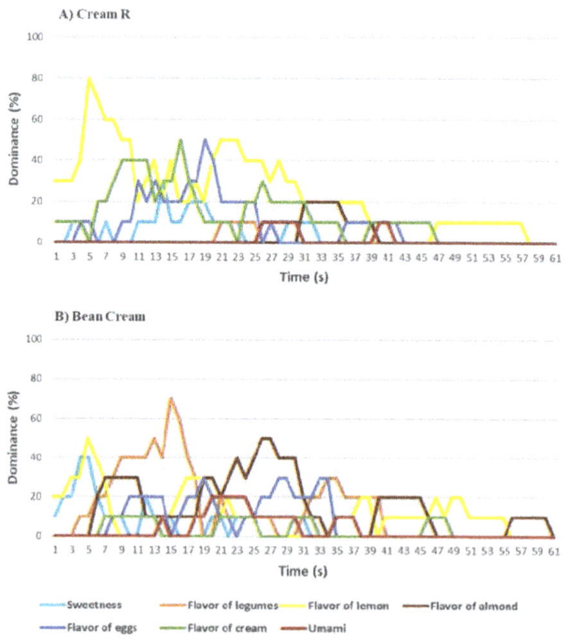


**Figure 11.** Temporal dominance of sensations (TDS) of (**A**) Cream R (reference, without common bean flour) and (**B**) Bean Cream (with 9% common bean flour).

## 4. Discussion

The nutritional quality of traditional snacks is often considered low due to their high sugar and fat content and their protein deficiency [61]. However, for many people they constitute a considerable fraction of daily calories, representing a fast and convenient food source that can be consumed without preparation [62]. In this scenario, healthier food snacks are required to respond to an increasing necessity by consumers to consume healthier snacks. Crackers and biscuits are among the most widely available baked goods used as snacks as they combine nutrition, long shelf-life and practicality. Following the footsteps of our previous study [12], with the present work, we contributed with new knowledge and genetic materials useful to the development of nutritionally improved and healthier snack foods (e.g., crackers, biscuits and a cream) based on a biofortified and low antinutrient common bean flour. Furthermore, data on the sensorial characteristics of these products provided indications to better identify target consumers as well as to develop bean-based products better adapted to consumer needs. From a nutritional standpoint, the use of composite flours containing common bean flour led, as expected and reported in similar studies [7,12,15,63,64], to products with higher protein and total dietary fiber content and lower amounts of starch and total carbohydrates. Depending on the type of pulse-based snacks (e.g., bars, biscuits, chips, bread, extruded snacks) reported protein concentration has ranged between 8.79% and 29.8%, with extruded snacks having the highest protein content, as they mostly or exclusively contain legume flour [7,64]. The direct comparison between the two types of biscuits produced in the current study with other bean-based biscuits previously described [12,15] showed that the protein content of Biscuit 1 and Biscuit 2 was higher (by on average 11%) than that (ranging between 8% and 10%) reported for biscuits containing bean flour ranging from 12% to 29%, indicating that the composition of the composite flour is also important if other flours beside common bean and wheat are used. In fact, the use of maize flour depleted the protein content of the products [12,15], while the addition of buckwheat flour, together with the presence of eggs, increased it in Biscuit 2 (12% of bean flour) compared to Biscuit 1 (20% bean flour). This was also reflected in starch content which was lower in Biscuit 1 and Biscuit 2 (19.3% and 14.8%, respectively) compared to that of biscuits containing from 12% to 29% bean flour (31.1% and 26.4%, respectively) and control biscuits (41.9%) [12].

If an increase in protein and total dietary fiber and a decrease in starch contents in the bean-based products could be expected, the major element of novelty of the present study relates to the fact that the bean-based products were obtained using flours of biofortified and low antinutrient (*lpa lec*$^-$) bean genotypes. Common approaches to increase essential mineral bioavailability and to reduce antinutrients are mainly based on a combination of technological processing, such as soaking, dehulling, heating, cooking, fermentation, germination and extrusion [7]. These treatments, although necessary to reduce or eliminate the antinutrients, are costly and time consuming and may reduce the nutritional quality of the product (e.g., loss of minerals in brine). In contrast, bean genotypes carrying the *lpa* and *lec*$^-$ mutations do not need specific treatments, such as soaking, to reduce phytic acid or heating of the flour to inactivate the lectins, as demonstrated and discussed in this study and in a previous one [12].

Iron deficiency is a serious issue in many areas of the world. Nearly two billion people are currently iron-deficient, especially resource-poor women, infants, and children in developing countries [65]. The common bean is a strategic crop for biofortification and it has been included in breeding programs in the HarvestPlus international research program [66] supporting the research and development of biofortified crops. Although most of the research has been aimed at increasing the amount of iron in the seed, now it is quite clear that the most powerful approach should also consider the reduction of phytic acid, to ensure iron bioavailability [67]. Indeed, a number of studies have demonstrated that the *lpa* bean mutant is able to provide more bioavailable iron both by using a Caco2-cells model system [60] and by performing clinical studies on volunteer women [27,44]. However, no data are available on the evaluation of the *lpa* trait on the iron bioavailability inside processed snacks. Here, it was shown that biscuits made with the *lpa lec*$^-$ bean flour allow iron to be more bioavailable.

Lectins, and in particular common bean PHA, may exert a very toxic action on the consumer if not properly heat inactivated, as shown by Petry et al. (2016) who observed unexpected adverse gastrointestinal symptoms in Rwandese women with low iron status participating in a clinical trial aimed at evaluating the biofortification potential of *lpa* beans. In a recent work, Cominelli et al. [45] demonstrated that these unwanted effects were due to the role played by the *lpa* mutation on the thermal stability of PHA-L, while no significant effects were observed on the thermal stability of PHA-E or PHA-E and PHA-L oligomers. The role of baking on PHA activity was assessed by Sparvoli et al. [12], which demonstrated that, after baking, sample extracts were still partially able to agglutinate erythrocytes, but this activity was significantly decreased with only three minutes of overbaking. Their biscuits retained about 5–10% PHA activity. Our results confirmed the previous finding, showing that no agglutination activity was detected in *lpa lec*$^-$ bean flour and its derivative products. At the same time, a very low residual agglutinating activity was found in bean-based baked products containing active PHA. This activity was higher in the *lpa* genotype for both types of biscuits (about 20–25%) than in the wt genotype Biscuit 1 (about 10–12%). In both Biscuits 1 and 2, the *lpa* genotype retained a more pronounced activity that correlated with higher agglutinating activity of the seed flour. Due to the more efficient heat penetration that allowed complete PHA inactivation, the crackers did not show any residual agglutination activity for all the used common bean genotypes, confirming that proper heat treatment is necessary to obtain a significant reduction or complete inactivation of the toxic protein fraction in PHA-containing genotypes.

In contrast to what was observed for PHA, when the residual α-AI was assayed, all samples containing the *lpa lec*$^-$ bean flour showed higher activity than that measured in samples made with wt bean flour. This finding was not completely unexpected and confirmed the findings of a previous study in which it was observed that B14 biscuits made with Lady Joy bean flour (the parent used to introgress the *lec*$^-$ and α-AI traits in the *lpa lec*$^-$ genotype) showed higher α-AI activity than that of B14 biscuits made with wt bean flours of Taylor and Billò genotypes [12]. Since the α-AI activity measured in the *lpa lec*$^-$ flour was lower than that of the wt, these data, together with the above observations, strongly suggest that the α-AI associated to the *lec*$^-$ trait should be particularly resistant to thermal denaturation and deserves further study to understand the basis for its behavior.

Legume flours are highly recommended to decrease the glycemic index as they are rich in resistant starch and have a high dietary fiber content. For instance, lupin-based biscuits snacks offered to Type 2 diabetes mellitus (T2DM) patients suggested they may improve both their glycemic control and satiety and a number of clinical trials have been conducted, determining the postprandial glycemic response and the pGI of snacks enriched with legume flours [64,68]. Theoretically, foods can be categorized into low (<55), medium (55–69), and high glycemic index (>70) [69]. Based on their pGI values, the two biscuit types (pGI 41.3 and 41.7) and Cracker 2 (pGI 58.5) can be classified as low and medium pGI snacks, respectively, while Cracker 1 (pGI 74), which contained a smaller amount of bean flour, can be classified as high glycemic index (Table 5). These pGI values are, on average, lower than those reported for other bean biscuits, which ranged between 60 and 80 [12], and are in line with those reported for legume-enriched biscuits (reviewed by Binou et al. [64], indicating that the new formulations proposed in this study have an improved nutritional impact (indeed, they also have better protein and starch content). Interestingly, all the products made with the *lpa lec*$^-$ bean flour showed pGI reductions of about 5 units compared to those of corresponding products made with wt flour. The presence of active α-AI, the activity of which was particularly pronounced in baked products made with *lpa lec*$^-$ bean flour, is most likely the reason for the observed reduced pGI values. Indeed, B14 biscuits made with bean flour devoid of α-AI showed a pGI about 4–6 units higher than that of the same product made with a wt bean flour [12]. In a very recent study, a bean-based biscuit containing 25% of common bean flour, and with a composition very similar to that of the

B24 biscuit of Sparvoli et al. [12], has been shown to be able to reduce the glycemic response and to increase satiety perception [15]. Thus, it can be argued that the low pGIs observed for the *lpa lec*$^-$ bean products here described should be predictive of hypoglycemic functional properties. A first step toward the expansion of the legume-based snacks market to more specific consumers' sectors is to ensure their sensorial appreciation by target consumers. The incorporation of legumes into various products usually results in "beany" flavor/back notes, which are considered "unpleasant" by some people. Furthermore, fortification of cereal-based products with legume flours affects the texture of the end-product [64]. In this study, the Biscuit 1 acceptance score was 93% of that of the control and it also received a better score for all the textural attributes, including crunchiness, consistency and friability. The other bean-based products were less appreciated than traditional ones (reference, R), having an acceptance score ranging between "Neither Like nor Dislike" and "Dislike Slightly", corresponding to a range between 70–79% that of the control. Crackers 1 and 2 were the least appreciated due their lack of friability and crunchiness, typical of traditional crackers, but also due to the presence of excessive humidity, adhesiveness and chewiness. Even in the case of the creams, the one containing bean flour was characterized by a worse texture, consisting of a higher graininess and poor creaminess. In contrast, among the bean-based products, the biscuits seemed to have the best textural attributes and a consistency that was comparable to that of traditional ones. Moreover, judges agreed in underlining the dominant flavor and taste of legumes in bean-based crackers (especially in Cracker 2), which was intense, even if to a lesser extent than in biscuits and creams. This property may be not appreciated by typical consumers, because it was also accompanied by a pronounced astringency and bitterness. These results are consistent with those of other studies. In Sparvoli et al. [12] liking scores of bean-based biscuits decreased with increase in the bean flour content. The sensory evaluation of gluten-free biscuits, based on different formulations of cereals with the addition of chickpea or lentil flour, showed that the control biscuit had the best liking score, and the other pulse-based products ranged between "Slightly Pleasant" and "Slightly Unpleasant". Despite the liking scores, they were all considered satisfying as only one sample received the "Slightly Unpleasant" evaluation [70]. Moderate to low texture scores were recorded in a study based on the sensorial evaluation of bean-based, nutrient-enriched, puffed snacks by [71], according to which this characteristic could be attributed to the protein-starch interactions or the starch-fiber interactions that tend to limit expansion of extrudates. Regarding flavor attributes, [72,73] low flavor scores were also associated with the inherent beany flavors, not desired by consumers.

Interestingly, all the bean-based products demonstrated an intense umami taste. Umami, which is also described as the fifth taste, stimulates salivation, improves hypogeusia by enhancing the gustatory-salivary reflex, enhances appetite and satiety, ameliorates eating disorders, and increases the peristaltic reflex and pellet propulsion through the distal colon [74]. This is regarded as a positive attribute, particularly for older consumers, whose senses of taste and smell may have decreased over time; umami could enhance appetite and, as a result, health of tissues like bones and muscles, thus contributing to the maintenance of good health in the elderly [75]. Furthermore, it has been shown that umami compounds significantly attenuate the sensitivity to sucrose, due to competition for binding to the common receptor subunit T1R3 [76] and this might explain the finding that reduced sweetness was perceived for all the bean-based sweet products (Biscuits 1 and 2 and Cream).

These results, although encouraging, also underlie the need to modify the product formulations (especially that of crackers, which require significant enhancement) and to improve their technological, functional and sensorial properties. To this end, sourdough fermentation (by lactic acid bacteria) technology looks very promising as it may improve texture and flavor characteristics, as well as nutritional quality. The use of this type of fermentation on *lec*$^-$ bean flour drastically reduced the phytic acid content, increased antioxidant capacity and, most interestingly, did not affect $\alpha$-AI activity [77]. Therefore, it would be interesting, in future, to apply such technology to produce bean flours to be used for bakery products. Another possibility to improve sensorial properties is to use flours obtained from cooked legumes. In fact, in a study in which sensory analysis of gluten-free rice and common bean biscuits were performed, it has been shown that the most disliked biscuits were those made with raw bean flour, while those made with cooked bean flour received a better acceptance score [78]. Furthermore, it is well known that heat treatment of legume flours may control off-flavors, and hence improve the sensory quality of baked products [79].

The acceptance of bean-based products by the consumer is important to embrace the possibility of extending their market to a more diverse and wider group of people. Beyond their interesting use in feeding of children, bean-based products can be offered to elderly people, as well as successfully employed as efficient substitutes in diets for people suffering from celiac disease, diabetes and obesity. It is quite common that products characterized by the presence of unusual ingredients, including whole wheat, buckwheat and bean flours, tend to have a lower degree of acceptability (due to e.g., high level food neophobia of some consumers) [80]. However, usually their perception from consumers changes when these products are related to a detailed description of their health effects. People are more likely to appreciate uncommon foods when they are sure to receive a positive effect on the body from their regular consumption. For this reason, it would be useful to perform a deeper analysis in which the evaluation also considers this aspect, to understand the real willingness to buy these products by target consumers, considering too that children have a different approach to products from adults. Even teenagers have a different perception of product quality from their parents [81] and this will ultimately lead to different hedonic evaluations and different optimal product formulation.

## 5. Conclusions

In the present study, it was demonstrated that the use of a biofortified and lectin free bean flour (*lpa lec*$^{-}$), instead of that from a wt bean genotype, further improved the nutritional properties of different products. The *lpa lec*$^{-}$ bean flour guaranteed the absence of any possible poisoning due to lectins and provided more bioavailable iron, as demonstrated by using a Caco-2 cells model system. All the products made with the *lpa lec*$^{-}$ bean flour retained higher $\alpha$-AI activities and had lower pGIs. It would be worthwhile in the future to evaluate the impact of such snacks, particularly those with the lowest pGI and the highest $\alpha$-AI activity, on the glycemic response, weight gain, and appetite control using an animal model system, thus providing evidence of the health benefits of such products as has been performed for other types of snacks [82].

These products may meet the nutritional needs of children and elderly people well, as they provide more bioavailable iron, are more protein-rich, have lower pGIs, and their umami taste may enhance appetite and satiety, ameliorating eating disorders.

Although these results are encouraging, there remains a need to improve the technological, functional and sensorial properties of the products and, most importantly, to evaluate the extent of these improvements in the light of how consumer perception may change with information about the nutritional and functional advantages derived from consuming the bean-based products.

**Author Contributions:** F.S., E.C., E.A., G.G., M.R., S.P. conceived and designed the research experiments; S.G., D.L., E.G., M.C., G.M.D. performed the experiments; F.S., E.C., G.G., M.R., S.P. analyzed the data; F.S., S.G., E.C., G.G., M.R., S.P. wrote, reviewed, and edited the paper. All authors have read and agreed to the published version of the manuscript.

**Funding:** This work was partially supported by sPATIALS3 to F.S. and E.C. funded by the European Regional Development Fund under the ROP of the Lombardy Region ERDF 2014–2020—Axis I "Strengthen technological research, development and innovation"—Action 1.b.1.3 "Support for co-operative R&D activities to develop new sustainable technologies, products and services"—Call Hub; CNR-DISBA project NutrAge (project nr. 7022) to F.S. and E.C.; POR FESR 2014/2020 Azione I.1.b.1.2 Project FASTBIOS to E.A. and F.S.

**Institutional Review Board Statement:** Ethical review and approval were waived for this study, since only judges expert in sensory evaluation, accurately trained and informed, and regularly participating to sensory analyses sessions were involved.

**Informed Consent Statement:** Informed consent was obtained from all subjects involved in the study.

**Data Availability Statement:** Not applicable.

**Conflicts of Interest:** The authors declare no conflict of interest.

## References


1. Bielefeld, D.; Grafenauer, S.; Rangan, A. The Effects of Legume Consumption on Markers of Glycaemic Control in Individuals with and without Diabetes Mellitus: A Systematic Literature Review of Randomised Controlled Trials. *Nutrients* **2020**, *12*, 2123. [CrossRef]
2. Clark, S.; Duncan, A.M. The role of pulses in satiety, food intake and body weight management. *J. Funct. Foods* **2017**, *38*, 612–623. [CrossRef]
3. Ferreira, H.; Vasconcelos, M.; Gil, A.M.; Pinto, E. Benefits of pulse consumption on metabolism and health: A systematic review of randomized controlled trials. *Crit. Rev. Food Sci. Nutr.* **2021**, *61*, 85–96. [CrossRef] [PubMed]
4. Padhi, E.M.T.; Ramdath, D.D. A review of the relationship between pulse consumption and reduction of cardiovascular disease risk factors. *J. Funct. Foods* **2017**, *38*, 635–643. [CrossRef]
5. Sievenpiper, J.L.; Kendall, C.W.C.; Esfahani, A.; Wong, J.M.W.; Carleton, A.J.; Jiang, H.Y.R.; Bazinet, P.; Vidgen, E.; Jenkins, D.J.A. Effect of non-oil-seed pulses on glycaemic control: A systematic review and meta-analysis of randomised controlled experimental trials in people with and without diabetes. *Diabetologia* **2009**, *52*, 1479. [CrossRef] [PubMed]
6. Rawal, V.; Navarro, D.K. *The Global Economy of Pulses*; Food and Agriculture Organization: Rome, Italy, 2019.
7. Escobedo, A.; Mojica, L. Pulse-based snacks as functional foods: Processing challenges and biological potential. *Compr. Rev. Food Sci. Food Saf.* **2021**, *20*, 4678–4702. [CrossRef]
8. Dhurandhar, E.J.; Dawson, J.; Alcorn, A.; Larsen, L.H.; Thomas, E.A.; Cardel, M.; Bourland, A.C.; Astrup, A.; St-Onge, M.P.; Hill, J.O.; et al. The effectiveness of Breakfast recommendations on weight loss: A randomized controlled trial. *Am. J. Clin. Nutr.* **2014**, *100*, 507–513. [CrossRef] [PubMed]
9. Popkin, B.M.; Adair, L.S.; Ng, S.W. Global nutrition transition and the pandemic of obesity in developing countries. *Nutr. Rev.* **2012**, *70*, 3–21. [CrossRef]
10. Kerr, M.A.; Rennie, K.L.; Mccaffrey, T.A.; Wallace, J.M.W.; Hannon-Fletcher, M.P.; Livingstone, M.B.E. Snacking patterns among adolescents: A comparison of type, frequency and portion size between Britain in 1997 and Northern Ireland in 2005. *Br. J. Nutr.* **2009**, *101*, 122–131. [CrossRef]
11. World Health Organization. Obesity and Overweight. Available online: https://www.who.int/news-room/fact-sheets/detail/obesity-and-overweight (accessed on 3 November 2021).
12. Sparvoli, F.; Laureati, M.; Pilu, R.; Pagliarini, E.; Toschi, I.; Giuberti, G.; Fortunati, P.; Daminati, M.G.; Cominelli, E.; Bollini, R. Exploitation of common bean flours with low antinutrient content for making nutritionally enhanced biscuits. *Front. Plant Sci.* **2016**, *7*, 928. [CrossRef]
13. Arribas, C.; Cabellos, B.; Cuadrado, C.; Guillamón, E.; Pedrosa, M.M. Bioactive compounds, antioxidant activity, and sensory analysis of rice-based extruded snacks-like fortified with bean and carob fruit flours. *Foods* **2019**, *8*, 381. [CrossRef] [PubMed]
14. Maia, L.C.; Nano, R.M.W.; dos Santos, W.P.C.; de Oliveira, F.S.; Barros, C.O.; de Souza Miranda, K.E. Evaluation of the nutritional quality of cereal bars made with pulse flours using desirability functions. *Food Sci. Technol. Int.* **2021**, *27*, 702–711. [CrossRef] [PubMed]
15. Mecha, E.; Correia, V.; da Silva, A.B.; Ferreira, A.; Sepodes, B.; Figueira, M.E.; Patto, M.C.V.; Rosário Bronze, M. Improvement of wheat cookies' nutritional quality, by partial substitution with common bean and maize flours, sustained human glycemia and enhanced satiety perception. *Cereal Chem.* **2021**, *98*, 1123–1134. [CrossRef]
16. Anton, A.A.; Fulcher, R.G.; Arntfield, S.D. Physical and nutritional impact of fortification of corn starch-based extruded snacks with common bean (Phaseolus vulgaris L.) flour: Effects of bean addition and extrusion cooking. *Food Chem.* **2009**, *113*, 989–996. [CrossRef]
17. Di Cairano, M.; Condelli, N.; Galgano, F.; Caruso, M.C. Experimental gluten-free biscuits with underexploited flours versus commercial products: Preference pattern and sensory characterisation by Check All That Apply Questionnaire. *Int. J. Food Sci. Technol.* **2021**, 1–9. [CrossRef]
18. Simons, C.W.; Hall, C. Consumer acceptability of gluten-free cookies containing raw cooked and germinated pinto bean flours. *Food Sci. Nutr.* **2018**, *6*, 77–84. [CrossRef]
19. Saeed, S.M.G.; Ali, S.A.; Ali, R.; Naz, S.; Sayeed, S.A.; Mobin, L.; Ahmed, R. Utilization of *Vigna mungo* flour as fat mimetic in biscuits: Its impact on antioxidant profile, polyphenolic content, storage stability, and quality attributes. *Legum. Sci.* **2020**, *2*, e58. [CrossRef]
20. Granito, M.; Valero, Y.; Zambrano, R. Baked product development based fermented legumes and cereals for school children snack. *Arch. Latinoam. Nutr.* **2010**, *60*, 85–92.

21. Bogue, J.; Collins, O.; Troy, A.J. Market analysis and concept development of functional foods. In *Developing New Functional Food and Nutraceutical Products*, 1st ed.; Bagch, D., Nair, S., Eds.; Academic Press: San Diego, CA, USA, 2017; pp. 29–45. [CrossRef]
22. Hayat, I.; Ahmad, A.; Masud, T.; Ahmed, A.; Bashir, S. Nutritional and Health Perspectives of Beans (Phaseolus vulgaris L.): An Overview. *Crit. Rev. Food Sci. Nutr.* **2014**, *54*, 580–592. [CrossRef]
23. Petry, N.; Boy, E.; Wirth, J.P.; Hurrell, R.F. Review: The potential of the common bean (*Phaseolus vulgaris*) as a vehicle for iron biofortification. *Nutrients* **2015**, *7*, 1144–1173. [CrossRef]
24. Cominelli, E.; Rodiño, A.P.; De Ron, A.M.; Sparvoli, F. Genetic approaches to improve common bean nutritional quality: Current knowledge and future perspectives. In *Quality Breeding in Field Crops*; Qureshi, A.M.I., Dar, Z.A., Wani, S.H., Eds.; Springer: Cham, Switzerland, 2019; ISBN 978-3-030-04608-8.
25. Noah, N.D.; Bender, A.E.; Reaidi, G.B.; Gilbert, R.J. Food poisoning from raw red kidney beans. *Br. Med. J.* **1980**, *281*, 236–237. [PubMed]
26. Rodhouse, J.C.; Haugh, C.A.; Roberts, D.; Gilbert, R.J. Red kidney bean poisoning in the UK: An analysis of 50 suspected incidents between 1976 and 1989. *Epidemiol. Infect.* **1990**, *105*, 485–491. [CrossRef] [PubMed]
27. Petry, N.; Rohner, F.; Gahutu, J.B.; Campion, B.; Boy, E.; Tugirimana, P.L.; Zimmerman, M.B.; Zwahlen, C.; Wirth, J.P.; Moretti, D. In rwandese women with low iron status, iron absorption from low-phytic acid beans and biofortified beans is comparable, but low-phytic acid beans cause adverse gastrointestinal symptoms. *J. Nutr.* **2016**, *146*, 970–975. [CrossRef]
28. Kumar, S.; Verma, A.K.; Das, M.; Jain, S.K.; Dwivedi, P.D. Clinical complications of kidney bean (*Phaseolus vulgaris* L.) consumption. *Nutrition* **2013**, *29*, 821–827. [CrossRef] [PubMed]
29. Schlemmer, U.; Frølich, W.; Prieto, R.M.; Grases, F. Phytate in foods and significance for humans: Food sources, intake, processing, bioavailability, protective role and analysis. *Mol. Nutr. Food Res.* **2009**, *53*, S330–S375. [CrossRef]
30. Alonso, R.; Aguirre, A.; Marzo, F. Effects of extrusion and traditional processing methods on antinutrients and in vitro digestibility of protein and starch in faba and kidney beans. *Food Chem.* **2000**, *68*, 159–165. [CrossRef]
31. El-Adawy, T.A. Nutritional composition and antinutritional factors of chickpeas (*Cicer arietinum* L.) undergoing different cooking methods and germination. *Plant Foods Hum. Nutr.* **2002**, *57*, 83–97. [CrossRef]
32. Arribas, C.; Cabellos, B.; Cuadrado, C.; Guillamón, E.; Pedrosa, M.M. The effect of extrusion on the bioactive compounds and antioxidant capacity of novel gluten-free expanded products based on carob fruit, pea and rice blends. *Innov. Food Sci. Emerg. Technol.* **2019**, *52*, 100–107. [CrossRef]
33. Vaz Patto, M.C.; Amarowicz, R.; Aryee, A.N.A.; Boye, J.I.; Chung, H.J.; Martín-Cabrejas, M.A.; Domoney, C. Achievements and Challenges in Improving the Nutritional Quality of Food Legumes. *CRC Crit. Rev. Plant Sci.* **2015**, *34*, 105–143. [CrossRef]
34. Wołosiak, R.; Drużyńska, B.; Piecyk, M.; Majewska, E.; Worobiej, E. Effect of Sterilization Process and Storage on the Antioxidative Properties of Runner Bean. *Molecules* **2018**, *23*, 1409. [CrossRef] [PubMed]
35. Confalonieri, M.; Bollini, R.; Berardo, N.; Vitale, A.; Allavena, A. Influence of Phytohemagglutinin on the Agronomic Performance of Beans (*Phaseolus vulgaris* L.). *Plant Breed.* **1992**, *109*, 329–334. [CrossRef]
36. Bollini, R.; Carnovale, E.; Campion, B. Removal of antinutritional factors from bean (*Phaseolus vulgaris* L.) seeds. *Biotechnol. Agron. Soc. Environ.* **1999**, *3*, 217–219.
37. Campion, B.; Sparvoli, F.; Doria, E.; Tagliabue, G.; Galasso, I.; Fileppi, M.; Bollini, R.; Nielsen, E. Isolation and characterisation of an *lpa* (*low phytic acid*) mutant in common bean (*Phaseolus vulgaris* L.). *Theor. Appl. Genet.* **2009**, *118*, 1211–1221. [CrossRef]
38. Cominelli, E.; Confalonieri, M.; Carlessi, M.; Cortinovis, G.; Daminati, M.G.; Porch, T.G.; Losa, A.; Sparvoli, F. Phytic acid transport in Phaseolus vulgaris: A new low phytic acid mutant in the PvMRP1 gene and study of the PvMRPs promoters in two different plant systems. *Plant Sci.* **2018**, *270*, 1–12. [CrossRef]
39. Lioi, L.; Sparvoli, F.; Galasso, I.; Lanave, C.; Bollini, R. Lectin-related resistance factors against bruchids evolved through a number of duplication events. *Theor. Appl. Genet.* **2003**, *107*, 814–822. [CrossRef]
40. Santimone, M.; Koukiekolo, R.; Moreau, Y.; Le Berre, V.; Rougé, P.; Marchis-Mouren, G.; Desseaux, V. Porcine pancreatic α-amylase inhibition by the kidney bean (*Phaseolus vulgaris*) inhibitor (α-AI1) and structural changes in the α-amylase inhibitor complex. *Biochim. Biophys. Acta-Proteins Proteom.* **2004**, *1696*, 181–190. [CrossRef] [PubMed]
41. Barrett, M.L.; Udani, J.K. A proprietary alpha-amylase inhibitor from white bean (*Phaseolus vulgaris*): A review of clinical studies on weight loss and glycemic control. *Nutr. J.* **2011**, *10*, 24. [CrossRef] [PubMed]
42. Layer, P.; Carlson, G.L.; DiMagno, E.P. Partially purified white bean amylase inhibitor reduces starch digestion in vitro and inactivates intraduodenal amylase in humans. *Gastroenterology* **1985**, *88*, 1895–1902. [CrossRef]
43. Layer, P.; Zinsmeister, A.R.; DiMagno, E.P. Effects of decreasing intraluminal amylase activity on starch digestion and postprandial gastrointestinal function in humans. *Gastroenterology* **1986**, *91*, 41–48. [CrossRef]
44. Petry, N.; Egli, I.; Campion, B.; Nielsen, E.; Hurrell, R. Genetic reduction of phytate in common bean (*Phaseolus vulgaris* L.) seeds increases iron absorption in young women. *J. Nutr.* **2013**, *143*, 1219–1224. [CrossRef]
45. Cominelli, E.; Galimberti, M.; Pongrac, P.; Landoni, M.; Losa, A.; Paolo, D.; Daminati, M.G.; Bollini, R.; Cichy, K.A.; Vogel-Mikuš, K.; et al. Calcium redistribution contributes to the hard-to-cook phenotype and increases PHA-L lectin thermal stability in common bean *low phytic acid 1* mutant seeds. *Food Chem.* **2020**, *321*, 126680. [CrossRef]
46. AOAC. *Official Methods of Analysis*, 17th ed.; The Association of Official Analytical Chemists: Gaithersburg, MD, USA, 2000.
47. Giuberti, G.; Gallo, A.; Cerioli, C.; Fortunati, P.; Masoero, F. Cooking quality and starch digestibility of gluten free pasta using new bean flour. *Food Chem.* **2015**, *175*, 43–49. [CrossRef]

48. Granfeldt, Y. Foods Factors Affecting Metabolic Responses to Cereal Products. Ph.D. Thesis, University of Lund, Lund, Sweden, 1994.
49. Altabella, T.; Chrispeels, M.J. Tobacco plants transformed with the bean αAI gene express an inhibitor of insect α-amylase in their seeds. *Plant Physiol.* **1990**, *93*, 805–810. [CrossRef]
50. Varner, J.E.; Mense, R.M. Characteristics of the Process of Enzyme Release from Secretory Plant Cells. *Plant Physiol.* **1972**, *49*, 187–189. [CrossRef] [PubMed]
51. Marshall, J.J.; Lauda, C.M. Purification and properties of phaseolamin, an inhibitor of α amylase, from the kidney bean, Phaseolus vulgaris. *J. Biol. Chem.* **1975**, *250*, 8030–8037. [CrossRef]
52. Glahn, R.P.; Lee, O.A.; Yeung, A.; Goldman, M.I.; Miller, D.D. Caco-2 cell ferritin formation predicts nonradiolabeled food iron availability in an in vitro digestion/Caco-2 cell culture model. *J. Nutr.* **1998**, *128*, 1555–1561. [CrossRef]
53. Pineau, N.; de Bouillé, A.G.; Lepage, M.; Lenfant, F.; Schlich, P.; Martin, N.; Rytz, A. Temporal Dominance of Sensations: What is a good attribute list? *Food Qual. Prefer.* **2012**, *26*, 159–165. [CrossRef]
54. Le Calvé, B.; Saint-Léger, C.; Gaudreau, N.; Cayeux, I. Capturing key sensory moments during biscuit consumption: Using TDS to evaluate several concurrent sensory modalities. *J. Sens. Stud.* **2019**, *34*, e12529. [CrossRef]
55. Meilgaard, M.C.; Carr, B.T.; Carr, B.T. *Sensory Evaluation Techniques*, 4th ed.; CRC Press: Boca Raton, FL, USA, 2007.
56. International Organization for Standardization. *UNI EN ISO 8589 Sensory Analysis-General Guidance for the Design of Test Rooms*; International Organization for Standardization: Paris, France, 2014.
57. Peryam, D.R.; Pilgrim, F.J. Hedonic scale method of measuring food preferences. *Food Technol.* **1957**, *11*, 9–14.
58. Jaworska, D.; Królak, M.; Jezewska-Zychowicz, M. Reformulation of bread rolls using oat fibre: An acceptable way of dietary fibre enrichment? *Nutr. Bull.* **2020**, *45*, 189–198. [CrossRef]
59. Obiro, W.C.; Zhang, T.; Jiang, B. The nutraceutical role of the Phaseolus vulgaris α-amylase inhibitor. *Br. J. Nutr.* **2008**, *100*, 1–12. [CrossRef]
60. Campion, B.; Glahn, R.P.; Tava, A.; Perrone, D.; Doria, E.; Sparvoli, F.; Cecotti, R.; Dani, V.; Nielsen, E. Genetic reduction of antinutrients in common bean (*Phaseolus vulgaris* L.) seed, increases nutrients and in vitro iron bioavailability without depressing main agronomic traits. *Field Crops Res.* **2013**, *141*, 27–37. [CrossRef]
61. Bellisle, F. Meals and snacking, diet quality and energy balance. *Physiol. Behav.* **2014**, *134*, 38–43. [CrossRef] [PubMed]
62. Ciurzyńska, A.; Cieśluk, P.; Barwińska, M.; Marczak, W.; Ordyniak, A.; Lenart, A.; Janowicz, M. Eating habits and sustainable food production in the development of innovative "healthy" snacks (running title: Innovative and "healthy" snacks). *Sustainability* **2019**, *11*, 2800. [CrossRef]
63. Yadav, A.; Sharma, L.; Singh, P.; Batra, P. Nutritional composition and sensory attributes of gluten-free brownie enriched with kidney beans and functional ingredients. *Ann. Biol.* **2021**, *37*, 262–266.
64. Binou, P.; Yanni, A.E.; Karathanos, V.T. Physical properties, sensory acceptance, postprandial glycemic response, and satiety of cereal based foods enriched with legume flours: A review. *Crit. Rev. Food Sci. Nutr.* **2020**, 1–19. [CrossRef]
65. Lynch, S.R. Why nutritional iron deficiency persists as a worldwide problem. *J. Nutr.* **2011**, *141*, 763S–768S. [CrossRef]
66. Harvest Plus. Available online: https://www.harvestplus.org (accessed on 3 November 2021).
67. Hummel, M.; Talsma, E.F.; Taleon, V.; Londoño, L.; Brychkova, G.; Gallego, S.; Raatz, B.; Spillane, C. Iron, zinc and phytic acid retention of biofortified, low phytic acid, and conventional bean varieties when preparing common household recipes. *Nutrients* **2020**, *12*, 658. [CrossRef] [PubMed]
68. Skalkos, S.; Moschonis, G.; Thomas, C.J.; McMillan, J.; Kouris-Blazos, A. Effect of Lupin-Enriched Biscuits as Substitute Mid-Meal Snacks on Post-Prandial Interstitial Glucose Excursions in Post-Surgical Hospital Patients with Type 2 Diabetes. *Nutrients* **2020**, *12*, 1239. [CrossRef]
69. Foster-Powell, K.; Holt, S.H.A.; Brand-Miller, J.C. International table of glycemic index and glycemic load values: 2002. *Am. J. Clin. Nutr.* **2002**, *76*, 5–56. [CrossRef]
70. Di Cairano, M.; Condelli, N.; Caruso, M.C.; Cela, N.; Tolve, R.; Galgano, F. Use of Underexploited Flours for the Reduction of Glycaemic Index of Gluten-Free Biscuits: Physicochemical and Sensory Characterization. *Food Bioprocess Technol.* **2021**, *14*, 1490–1502. [CrossRef]
71. Natabirwa, H.; Nakimbugwe, D.; Lung'aho, M.; Tumwesigye, K.S.; Muyonga, J.H. Bean-based nutrient-enriched puffed snacks: Formulation design, functional evaluation, and optimization. *Food Sci. Nutr.* **2020**, *8*, 4763–4772. [CrossRef] [PubMed]
72. Nyombaire, G.; Siddiq, M.; Dolan, K.D. Physico-chemical and sensory quality of extruded light red kidney bean (*Phaseolus vulgaris* L.) porridge. *LWT-Food Sci. Technol.* **2011**, *44*, 1597–1602. [CrossRef]
73. Siddiq, M.; Uebersax, M.A. (Eds.) *Dry Beans and Pulses Production, Processing and Nutrition*; Wiley: Hoboken, NJ, USA, 2013; ISBN 9780813823874.
74. Wang, W.; Zhou, X.; Liu, Y. Characterization and evaluation of umami taste: A review. *TrAC Trends Anal. Chem.* **2020**, *127*, 115876. [CrossRef]
75. Sasano, T.; Satoh-Kuriwada, S.; Shoji, N.; Iikubo, M.; Kawai, M.; Uneyama, H.; Sakamoto, M. Important Role of Umami Taste Sensitivity in Oral and Overall Health. *Curr. Pharm. Des.* **2014**, *20*, 2750–2754. [CrossRef] [PubMed]
76. Shim, J.; Son, H.J.; Kim, Y.; Kim, K.H.; Kim, J.T.; Moon, H.; Kim, M.J.; Misaka, T.; Rhyu, M.R. Modulation of sweet taste by umami compounds via sweet taste receptor. *PLoS ONE* **2015**, *10*, e0124030. [CrossRef]

77. Gabriele, M.; Sparvoli, F.; Bollini, R.; Lubrano, V.; Longo, V.; Pucci, L. The Impact of Sourdough Fermentation on Non-Nutritive Compounds and Antioxidant Activities of Flours from Different *Phaseolus vulgaris* L. Genotypes. *J. Food Sci.* **2019**, *84*, 1929–1936. [CrossRef] [PubMed]
78. David Wesley, S.; André, B.H.M.; Clerici, M. Gluten-free rice & bean biscuit: Characterization of a new food product. *Heliyon* **2021**, *7*, e05956. [CrossRef]
79. Roland, W.S.U.; Pouvreau, L.; Curran, J.; van de Velde, F.; de Kok, P.M.T. Flavor Aspects of Pulse Ingredients. *Cereal Chem.* **2017**, *94*, 58–65. [CrossRef]
80. Siegrist, M.; Hartmann, C.; Keller, C. Antecedents of food neophobia and its association with eating behavior and food choices. *Food Qual. Prefer.* **2013**, *30*, 293–298. [CrossRef]
81. Bech-Larsen, T.; Jensen, B.B. Food quality assessment in parent–child dyads–A hall-test of healthier in-between meals for adolescents. *Food Qual. Prefer.* **2011**, *22*, 614–619. [CrossRef]
82. Flores-Silva, P.C.; Tovar, J.; Reynoso-Camacho, R.; Bello-Pérez, L.A. Impact of Chickpea- and Raw Planta in-Based Gluten-Free Snacks on Weight Gain, Serum Lipid Profile, and Insulin Resistance of Rats Fed with a High-Fructose Diet. *Cereal Chem.* **2017**, *94*, 124–127. [CrossRef]

nutrients MDPI

Article

# Innovative Application of Phytochemicals from Fermented Legumes and Spices/Herbs Added in Extruded Snacks

**Krystyna Szymandera-Buszka [1], Małgorzata Gumienna [2], Anna Jędrusek-Golińska [1,*], Katarzyna Waszkowiak [1], Marzanna Hęś [1], Artur Szwengiel [2] and Anna Gramza-Michałowska [1]**

[1] Department of Gastronomy Science and Functional Foods, Faculty of Food Science and Nutrition, Poznan University of Life Sciences, Wojska Polskiego 31, 61-624 Poznan, Poland; krystyna.szymandera_buszka@up.poznan.pl (K.S.-B.); katarzyna.waszkowiak@up.poznan.pl (K.W.); marzanna.hes@up.poznan.pl (M.H.); anna.gramza@up.poznan.pl (A.G.-M.)

[2] Department of Food Technology of Plant Origin, Faculty of Food Science and Nutrition, Poznan University of Life Sciences, Wojska Polskiego 31, 61-624 Poznan, Poland; malgorzata.gumienna@up.poznan.pl (M.G.); artur.szwengiel@up.poznan.pl (A.S.)

* Correspondence: anna.jedrusek-golinska@up.poznan.pl; Tel.: +48-061-848-7372

**Abstract:** A trend related to adding legume seeds to various products has been observed. This work aimed to use fermented red bean/broad bean seeds and their hulls to produce extruded snacks with more beneficial nutritional properties and good sensory quality. Extruded snacks containing fermented ground seeds (50%) or hull (10%) of red bean/broad bean and corn grits with the addition of selected herbs/spices (0.5%) were prepared. The chemical composition, phenolic profile, antioxidant activity, and sensory quality were analysed. The results showed that the protein content ranged from 9 to 22.9 g 100 g$^{-1}$, phenolic compounds ranged from 3.97 to 12.80 mg 100 g$^{-1}$ (with the addition of herbs/spices, even up to 62.88 mg 100 g$^{-1}$), and antioxidant activities ranged from 4.32 to 10.23 Trolox g$^{-1}$ (ABTS assay), depending on the type of fermented materials. The addition of ground seeds/hull did not influence the consumer desirability, whereas the addition of selected herbs/spices, particularly lovage, increased it. The application of fermented red bean and broad bean seeds and their hulls, as part of the assumptions of the planetary diet, enabled enrichment of extruded corn products, which are often consumed by vegans and vegetarians, with nutritionally valuable ingredients.

**Keywords:** legumes; fermented beans; extruded snacks; spices; sensory analysis; antioxidant activity

**Citation:** Szymandera-Buszka, K.; Gumienna, M.; Jędrusek-Golińska, A.; Waszkowiak, K.; Hęś, M.; Szwengiel, A.; Gramza-Michałowska, A. Innovative Application of Phytochemicals from Fermented Legumes and Spices/Herbs Added in Extruded Snacks. *Nutrients* **2021**, *13*, 4538. https://doi.org/10.3390/nu13124538

Academic Editors: Stuart Johnson and Rewati Bhattarai

Received: 12 November 2021
Accepted: 15 December 2021
Published: 17 December 2021



## 1. Introduction

Legumes are a rich source of nutrients such as protein, low glycemic carbohydrates, and minerals [1–3]. Their recommended intake, which should be approximately 50 g/person/day (0–100 g/person/day), is not achieved in most countries. It is estimated to be approximately 21 g/day worldwide [4], approximately 4 g/person/day in Australia [3], while data on legume consumption in Poland indicate that it is 0.9 kg/year/person, which amounts to approximately 2.5 g/person/day [5].

Meanwhile, in the face of growing problems of excessive burdening of the environment by agricultural production, especially animal production, and the problem of malnutrition, which has not been solved for years, both in quantitative and qualitative terms, an increasing emphasis is put on the consumption of legumes. According to the assumptions of the so-called planetary health diet that is supposed to improve health, including reduced risk of coronary artery disease [6] and weight control [7], environmental benefits and also food security, the Food and Agriculture Organization of the United Nations (FAO), and other authors postulate that by 2050, global consumption of fruits, vegetables, nuts, and legumes will have to double, while the consumption of red meat and sugar must be reduced by over 50% [1,4,8,9]. In addition to the recommendations to increase the consumption of legumes, there is also a trend of adding them to various products. For example, the work on the nutritional and technological assessment of durum wheat flours enriched with faba bean flours and its use in bread production have been under-taken in Morocco. Such bread with a 40% substitution level showed good sensory quality, with a significantly higher amount of ash, proteins, mineral substances, total phenolic compounds and flavonoids, condensed tannins, and antiradical activity values [10]. Bouhlal et al. [11] showed a significant increase of nutritional quality parameters such as ash, proteins, fat, and energy, as well as of the total polyphenols, flavonoid content, and antiradical activity in four wheat–faba bean-enriched flours. In contrast, Ni et al. [12] showed that the bean hull, a valuable source of dietary fibre in bread, could be used to replace up to 21% of wheat flour without significantly impacting bread texture and volume. Moreover, Fendri et al. [13] revealed that fibre obtained from broad bean pods can be used to improve dough growth and to enhance the textural profile of bread.

However, legume seeds have some anti-nutrients, including trypsin and chymotrypsin inhibitors, which may negatively affect consumers, especially in view of the postulated increasing consumption of these products. Therefore, it seems crucial to subject legume seeds (as a whole or their components) to various technological processes, including fermentation, to reduce oligosaccharide content, but also to generate the formation of functional components such as vitamins (especially B2), antioxidants [14]. Back-slopping fermentation was applied, i.e., to mung bean (*Vigna radiata* (L.) Wilczek) flour, which resulted in better improvement of nutritional values than spontaneous methods; it was stated that this method could be encouraged at community levels [15]. A study has also been undertaken on the replacement of soy flour by fermented and unfermented faba bean flour in the production of gluten-free bread. It was shown that flour fermentation resulted in significant increases in Essential Amino Acid and Biological Value indexes compared to bread made from unfermented flour and soy flour without changing sensory properties [16].

Recent studies also indicate that legume flour can be incorporated into extruded snacks instead of cereals, significantly improving their functional properties as well as their nutritional value [17–19]. The cereal raw material most commonly used to make snacks is corn groat. It provides a good base for the development of corn-based food ingredients for the preparation of nutritious food products enriched with health-promoting ingredients [20]. Previous studies showed the possibility of enhancing their nutritional value by adding legumes [21]. It is also important to increase the functional properties of such products by adding the hull of these seeds. In legume seed production, it is still largely a by-product [22,23]. Therefore, using it as an additive for extruded snacks would constitute a use of waste and, at the same time, an increase in the nutritional value of the product [24]. This is especially true for fermented foods. Red beans and broad beans are often recommended in vegetarian diets [25,26]. Consumption of red kidney beans and broad beans has been linked disease reduction such diabetes, cancer, obesity, and coronary heart diseases. These products are characterised by interesting nutritional properties: high levels of complex proteins, carbohydrates, and dietary fibre, coupled with a low content of saturated lipids and the presence of bioactive compounds, such as polyphenols [26–28]. The addition of them to extruded snacks together with popular spices and herbs such as thyme, marjoram, oregano, basil, and lovage can diversify taste of final products and increase the content of phenolic compounds in them. The potential health benefits of their consumption are related to antioxidants activity, strengthening the immune system, antimicrobial activity, and anticancer effects [29–31]. However, their sensory properties still need to be refined to meet the expectations of increasingly demanding consumers [18].

Therefore, the aim of this study was to use fermented red bean and broad bean seeds as well as their hulls to produce extruded snacks in order to obtain their better nutritional values and desired sensory properties. In order to obtain this, in addition to fermented seeds or hulls, we also added selected spices (thyme, oregano, basil, and lovage) to corn crisps, which not only enriched the product with bioactive elements but also improved their taste and aroma.

## 2. Materials and Methods

### 2.1. Material

Broad bean (*Vicia faba* L., cultivar Bachus) or red bean (*Phaseolus vulgaris* L., cultivar Red Kidney) constituted the sources of the legume seeds. Raw materials (red beans, broad beans) were obtained from Polish crops and purchased at the Seed Headquarters in Nochowo, Poland. Dried leaves of thyme (*Thymus vulgaris* L., cultivar Słoneczko), oregano (*Origanum vulgare* L.), basil (*Ocimum basilicum* L., cultivar Kasia), and lovage (*Levisticum officinale*, cultivar Amor) were purchased from a local herb shop.

### 2.2. Methods

#### 2.2.1. The Course of the Fermentation Process

Legume seeds were subjected to the process of bacterial fermentation (*Lactobacillus plantarum* ATCC 8014). The raw material, after grinding and moistening to 55% humidity, was fermented at 37 °C for 18 h [32]. The obtained product was then dried in an air dryer (Memmert, Büchenbach, Germany) at 50 °C until the final moisture content was 15–17%. The semi-finished product prepared in this way was a component for the construction of solid products—extruded (crisps).

#### 2.2.2. Preparation of Extruded Products

On the basis of preliminary trial sensory of texture (Ranking Test among 40 consumers to evaluate the amount of red and broad bean's added—10, 20, 30, 40, 50, and 60%), we adopted the following variants for further research:

- corn grits (49.5%); 50% ground broad bean or ground red bean seeds after fermentation and 0.5% of herbs/species;
- corn grits (89.5%); 10% ground broad bean hull or ground red bean hull after fermentation and 0.5% herbs/species.

The shredded fermented raw material was mixed with spices or herbs (0.5%) and corn grits and then subjected to the extrusion process. The process of lactic fermentation was carried out at 37 °C for 18 h by *Lactobacillus plantarum* ATCC 8014. The obtained intermediate was dried to a final moisture content of 15–17% dm. The semi-finished product prepared in this way was a component for the construction of solid products—snacks. The extrusion was carried out in a single-screw extruder (type S-45-Metalchem, Gliwice, Poland). The process parameters were established in earlier studies by Gumienna et al. [33].

#### 2.2.3. Chemical Composition and Protein Digestibility

Chemical composition of the extruded samples was analysed, which included: dry mater content [34], ash content [35], and total protein content [36], as well as the reducing sugars with DNS (3,5-dinitrosalicylic acid), according to the Miller method [37], and soluble protein content by the Lowry method [38].

Protein digestibility was determined using in vitro pepsin-pancreatin method according to Saunders et al. [39]. Briefly, water and pepsin dissolved in 0.1 N HCl were added to the sample, and it was placed in a thermostat at 37 °C for 3 h. After that, it was neutralised, pancreatin was dissolved in phosphate buffer at pH = 8 containing 0.005 M, sodium azide was added, and the mixture was again thermostated at 37 °C for 24 h. After complete hydrolysis, 45% TCA was added to precipitate proteins, and the mixture was centrifuged at 5500× *g*. The protein content in the decanted liquid was determined [36] and expressed as percent of total protein content (% of protein digestibility).

2.2.4. Preparation of Samples for the Analysis of Biologically Active Compounds and Their Determination

The sample was ground in a laboratory mill (Witko, Łódź, Poland). A total of 0.5 g of the homogeneous sample was weighed into a centrifuge tube and 10 mL of a 70:30 acetone/water extraction mixture [32]. The extraction mixture used in the research was applied on the basis of the research by Remiszewski et al. [40] and Gumienna et al. [41]. It was shaken for 60 min on a rocker shaker and centrifuged at 4125× *g* for 7 min. The supernatant fluid (extract) was decanted and used to perform the following analyses.

Tannin content: Determination of the amount of tannin was carried out using the reaction with a vanillin reagent [40]. The analysis was performed spectrophotometrically at a wavelength of $\lambda = 500$ nm. The content of tannins in the product was expressed as catechin equivalent (mg $g^{-1}$ dm) (Sigma-Aldrich, Munich, Germany).

Compounds that reduce the Folin–Ciocoltau reagent: The determination was carried out according to the spectrophotometric method with Folin–Ciocoltau reagent described by Singelton and Rossi [42] and modified by Remiszewski et al. [40]. The result was expressed as mg gallic acid equivalent (GAE $g^{-1}$ dm) (Sigma-Aldrich, Munich, Germany).

Antiradical activity (ABTS assay): The antiradical activity was determined against the ABTS reagent (2,2′-azinobis-(3-ethylbenzothiazoline-6-sulphonic acid) (Sigma-Aldrich, Munich, Germany), according to the method described by Re et al. [43] as modified by Remiszewski et al. [40]. The analysis was performed spectrophotometrically at the wavelength $\lambda = 735$ nm by using a spectrophotometer (Biogenet, Józefów, Poland). Results of the ABTS assay were expressed as the capability of antioxidants to scavenge ABTS radicals relative to that of Trolox (a water-soluble vitamin E analogue) and reported as mg Trolox $g^{-1}$ dm (Sigma-Aldrich, Munich, Germany).

2.2.5. Quantitative and Qualitative Determination of Phenolic Compounds by HPLC

Quantitative and qualitative determination of phenolic compounds was analysed by HPLC method, as described by Hertog et al. [44] and modified by Remiszewski et al. [40]. For the determination of phenolic compounds, the extract obtained earlier was subjected to acid hydrolysis in the presence of 6 N HCl (Sigma-Aldrich, Munich, Germany); then, the mixture was incubated at 90 °C for 2 h. The samples prepared in this way were filtered through filters with a diameter of 0.45 µm (Millipore) and subjected to chromatographic analysis. Separation of the compound mixture was carried out on an X-Terra C18 RP column (150 × 3.9 mm i.d. 5 µm) from Waters at 30 °C. The injection volume was 10 µL, and the flow rate of the mobile phase was 1 mL/min. The eluted fractions were monitored by UV–VIS detector (Waters Alliance, Taunton, MA, USA). The measurement time was 50 min. Formic acid (>98%), methanol (99%), and water were used as mobile phase (Sigma-Aldrich, Munich, Germany). A linear concentration gradient from 5% to 60% methanol was applied with a constant concentration of 5% formic acid. The recording of the chromatographic spectra was carried out in the wavelength range from 240–520 nm. Compounds were identified on the basis of spectra and retention times comparable to the standards (Sigma-Aldrich, Munich, Germany).

2.2.6. Sensory Analysis

Sensory Analysis was conducted in an appropriately designed and equipped laboratory of sensory analysis [45] at the Department of Gastronomy Science and Functional Foods, Poznan University of Life Sciences, Poland. The samples were coded with three-digit numbers, and the serving order of samples were random (program ANALSENS—v.5.0; Sopot; Poland, was used for coding and arrangement of serving order).

The 30 g samples were served in plastic containers (150 mL), which were covered with lids. Unsweetened black tea (temperature of ≈45 °C) was used as a taste neutraliser between the samples.

The sensory profiling of taste was conducted by a 10-member trained panel [46,47]. A total of 8 descriptors were adapted for taste (essential oil, herbal, sour, salty, starch, broth, bitter, and strange) and 8 descriptors for aroma (essential oil, herbal, sour, starch, broth, bitter, lemon, and strange). The intensity of each score was determined using a 10 cm linear scale with appropriate margin descriptions. For attributes, uniform margin denotations were applied: "undetectable to very intensive". All samples were assessed in two independent replications.

Consumer traits were conducted on a group of people 380 aged 21–56 (who consume corn snacks and legumes at least twice a month). The study was carried out in accordance with The Code of Ethics of the World Medical Association (consent no. 757/13). Women constituted 62% of the population analysed. All subjects gave written informed consent to participate. In consumer examinations, a 10 cm hedonic graphic scale was applied, with the following margin denotations: undesirable to highly desirable. Consumers evaluated desirability of colour, taste, aroma, texture, and overall desirability. All consumers rated all samples in one session (order of administration: 10 samples, 0.5 h interval, 10 samples, 0.5 h interval, and 12 samples).

### 2.3. Statistical Analysis

The results were analysed statistically with the STATISTICATM PL 13.3 software (StatSoft, Tulusa, OK, USA). The data were analysed for statistically significant differences with Tukey's multiple range test ($p = 0.05$).

The active compounds content of the tested samples were analysed in 6 samples (2 independent samples and 3 measurements for each sample). For the overall evaluation of differences and similarities in sensory profiles, consumer analysis and content of active compounds of the tested samples the analysis of main components (PCA—principal component analysis) was used. Hypotheses were tested at $\alpha = 0.01$.

## 3. Results

### 3.1. Chemical Composition and Protein Digestibility of Extruded Products with the Addition of Fermented Bean Hulls or Ground Beans and Selected Herbs/Spices

All variants of extruded products were analysed for dry matter, ash, reducing substances, soluble protein, total protein, and digestibility of protein (Table 1). It was confirmed that total protein contents were higher in products with ground seeds added than hull—for red bean in the range of 60–80% and broad bean, 90–130%. Moreover, the analysis of these components pointed to higher total protein as well as soluble protein amount in the extruded snakes with added broad bean seeds when compared to red been seeds addition (20–34%). Taking into account the analysed spices, we found no effect on the amount of total and soluble protein. The higher amount of ash (for red bean in the range of 140–220% and broad bean to 140%), as well as reducing sugars (for red bean in the range of 117–180% and broad bean 64–130%), in products with added hulls was also confirmed. The results showed that protein digestibility of the tested extruded snakes was over 65 percent.

**Table 1.** Chemical composition and protein digestibility of extruded snacks with the addition of fermented red or broad bean seeds/hulls and with herbs/species (0.5%), and the controls (without herbs/spices addition).

| Products | Dry Matter ($mg\ g^{-1}$) | Ash ($mg\ g^{-1}$) | Reducing Substance ($mg\ g^{-1}$ dm) | Soluble Protein ($mg\ g^{-1}$ dm) | Total Protein ($mg\ g^{-1}$ dm) | Protein Digestibility (%) |
|---|---|---|---|---|---|---|
| Extruded snacks with addition of ground fermented red bean seeds (50%) | | | | | | |
| Control | 955.50 ± 2.37 [d] | 24.56 ± 0.13 [a] | 17.01 ± 0.29 [c] | 6.03 ± 0.55 [a] | 173.81 ± 4.72 [c] | 66.69 ± 3.66 [a] |
| Lovage | 947.44 ± 1.27 [b] | 25.38 ± 1.00 [b] | 16.32 ± 0.11 [b] | 7.75 ± 0.18 [b] | 174.10 ± 3.36 [c] | 65.86 ± 0.58 [a] |
| Basil | 949.50 ± 2.38 [c] | 26.76 ± 1.44 [c] | 15.80 ± 0.31 [a] | 7.74 ± 0.51 [b] | 172.85 ± 2.10 [b] | 71.65 ± 0.99 [b] |
| Oregano | 943.54 ± 1.13 [a] | 26.71 ± 0.52 [c] | 16.95 ± 0.11 [c] | 6.97 ± 0.51 [a] | 175.23 ± 4.64 [d,] | 69.43 ± 1.36 [b] |
| Thyme | 948.54 ± 0.82 [c] | 26.01 ± 0.62 [b] | 15.95 ± 0.11 [a] | 6.27 ± 0.51 [a] | 170.23 ± 4.62 [a] | 69.29 ± 1.86 [b] |
| Extruded snacks with addition of fermented red bean hull (10%) | | | | | | |
| Control | 946.54 ± 2.08 [b] | 10.44 ± 0.01 [b] | 5.97 ± 0.20 [a] | 10.91 ± 0.21 [a] | 103.14 ± 2.41 [d] | 84.75 ± 2.65 [a] |
| Lovage | 947.41 ± 1.15 [b] | 8.31 ± 0.02 [a] | 6.27 ± 0.09 [v] | 11.042 ± 0.14 [a] | 99.82 ± 0.66 [b] | 84.15 ± 3.81 [a] |
| Basil | 945.85 ± 0.51 [b] | 10.61 ± 0.20 [b] | 7.26 ± 0.01 [d] | 12.582 ± 0.25 [c] | 100.55 ± 1.51 [c] | 85.85 ± 3.82 [a] |
| Oregano | 947.45 ± 0.69 [b] | 8.25 ± 0.05 [a] | 6.57 ± 0.16 [c] | 12.00 ± 0.15 [b] | 100.23 ± 2.73 [c] | 83.84 ± 1.40 [a] |
| Thyme | 942.79 ± 0.86 [a] | 8.40 ± 0.10 [a] | 6.61 ± 0.19 [c] | 12.20 ± 0.10 [b] | 93.31 ± 3.72 [a] | 91.18 ± 1.23 [b] |
| Extruded snacks with addition of ground fermented broad bean seeds (50%) | | | | | | |
| Control | 933.94 ± 1.12 [b] | 19.29 ± 1.29 [a] | 13.66 ± 0.11 [b] | 21.91 ± 0.43 [c] | 212.25 ± 2.20 [b] | 78.45 ± 0.69 [c] |
| Lovage | 934.95 ± 0.53 [c] | 19.40 ± 1.71 [a] | 12.72 ± 0.26 [a] | 17.32 ± 0.39 [a] | 219.05 ± 5.94 [d] | 70.77 ± 1.44 [b] |
| Basil | 928.75 ± 0.61 [b] | 24.46 ± 2.06 [b] | 13.15 ± 0.09 [a] | 19.37 ± 1.70 [b] | 217.34 ± 6.96 [c] | 65.82 ± 1.98 [a] |
| Oregano | 931.25 ± 0.51 [b] | 18.15 ± 1.20 [a] | 13.15 ± 0.07 [a] | 21.02 ± 0.50 [c] | 210.19 ± 4.96 [a] | 69.02 ± 1.68 [b] |
| Thyme | 923.40 ± 0.35 [a] | 22.65 ± 1.42 [b] | 15.15 ± 0.04 [c] | 18.04 ± 1.40 [ba] | 229.42 ± 15.99 [c] | 67.73 ± 2.53 [ba] |
| Extruded snacks with addition of fermented broad bean hull (10%) | | | | | | |
| Control | 930.27 ± 0.52 [a] | 13.69 ± 1.18 [c] | 7.20 ± 0.21 [b] | 10.99 ± 0.22 [a] | 92.03 ± 5.58 [a] | 80.52 ± 1.69 [a] |
| Lovage | 939.63 ± 0.73 [c] | 9.30 ± 1.98 [a] | 6.95 ± 0.59 [b] | 15.98 ± 0.69 [c] | 111.96 ± 28.11 [b] | 87.77 ± 1.44 [b] |
| Basil | 940.25 ± 1.71 [c] | 10.50 ± 0.56 [b] | 7.98 ± 0.10 [c] | 14.85 ± 0.90 [b] | 103.19 ± 0.49 [a] | 89.82 ± 1.98 [bc] |
| Oregano | 939.71 ± 0.25 [c] | 18.80 ± 0.66 [b] | 6.71 ± 0.05 [b] | 14.60 ± 0.38 [b] | 114.12 ± 2.38 [c] | 90.03 ± 2.21 [c] |
| Thyme | 936.01 ± 0.22 [a] | 9.76 ± 0.33 [a] | 6.52 ± 0.04 [a] | 14.49 ± 0.30 [b] | 117.41 ± 6.48 [c] | 92.73 ± 2.53 [c] |

Different letters denote a significant difference for means ($n = 6$) for the same lines at a $\alpha \leq 0.05$.

### *3.2. Phenolic Acid and Flavonoid Amount in the Extruded Products with the Addition of Fermented Bean Hulls or Ground Beans and Selected Herbs/Spices*

A qualitative and quantitative profile of phenolic compounds in the extruded products was shown (Table 2). A higher number of phenolic compounds was found in products with ground fermented beans (3.97 and 4.62 mg 100 $g^{-1}$ dm) than in products with fermented bean hulls (11.17–12.8 mg 100 $g^{-1}$ dm). In the extruded products with fermented bean hulls, as well as in those with ground fermented beans (controls), flavonoids such as myricetin, quercetin, luteolin, and kaempferol were identified. The products with ground fermented beans also had apigenin. Moreover, hydroxycinnamic acids were found in the extruded products with ground fermented red bean seeds and broad bean seeds, i.e., caffeic acid and ferulic acid, respectively.

The results of the study showed that the addition of herbs or spices (0.5%) significantly increased the content of phenolic compounds in the extruded products. The highest phenolic compound content was observed in all extruded products with lovage (42.03–62.88 mg 100 $g^{-1}$ dm) (Tables 2 and S1). In particular, the addition of lovage increased the content of the quercetin flavonoid, as well as hydroxycinnamic acids (mainly ferulic acid). The addition of oregano increased the content of phenolic compounds in the extruded products to 32.49–41.7 mg 100 $g^{-1}$ dm. Luteolin and caffeic acid contributed the most to the phenolic compound profile of these products. In the products with 0.5% of basil and thyme, the contents of the determined phenolic compounds were 22.03–34.13 mg 100 $g^{-1}$ dm (caffeic acid was dominant in the profile) and 17.21–19.5 mg 100 $g^{-1}$ dm (luteolin was dominant).

**Table 2.** Phenolic compounds of extruded snacks with the addition of fermented red or broad bean seeds/hulls and with herbs/species (0.5%), and the controls (without herbs/spices addition).

| Products | Phenolic Compounds [mg 100 $g^{-1}$ dm] | | | | | | | | | |
|---|---|---|---|---|---|---|---|---|---|---|
| | NCA | CA | p-CA | FA | MYR | QUE | LUT | KEMP | API | Total |
| Extruded snacks with addition of ground fermented red bean seeds (50%) | | | | | | | | | | |
| Control | nd | nd | nd | nd | 0.02 ± 0.00 $^{a}$ | 2.67 ± 0.2 $^{c}$ | 0.02 ± 0.00 $^{a}$ | 1.93 ± 0.3 $^{c}$ | nd | 4.62 |
| Lovage | 0.60 ± 0.2 $^{a}$ | 1.41 ± 0.4 $^{a}$ | 2.46 ± 0.5 $^{b}$ | 9.21 ± 1.4 $^{c}$ | 0.63 ± 0.2 $^{b}$ | 34.97 ± 8.1 $^{d}$ | 0.13 ± 0.01 $^{a}$ | 4.9 ± 0.8 $^{d}$ | nd | 54.31 |
| Basil | 0.44 ± 0.2 $^{a}$ | 12.02 ± 2.6 $^{c}$ | 2.74 ± 0.6 $^{b}$ | 2.20 ± 0.8 $^{b}$ | 1.11 ± 0.3 $^{b}$ | 1.21 ± 0.1 $^{a}$ | 1.0 ± 0.2 $^{b}$ | 1.41 ± 0.1 $^{b}$ | 0.9 ± 0.3 $^{a}$ | 22.03 |
| Oregano | 0.41 ± 0.25 $^{a}$ | 6.01 ± 0.6 $^{b}$ | 1.91 ± 0.4 $^{b}$ | 0.79 ± 0.1 $^{a}$ | nd | 1.81 ± 0.3 $^{b}$ | 19.33 ± 5.1 $^{d}$ | 1.81 ± 0.2 $^{c}$ | 1.42 ± 0.5 $^{a}$ | 32.49 |
| Thyme | nd | 1.32 ± 0.3 $^{a}$ | 1.12 ± 0.2 $^{a}$ | 2.33 ± 0.7 $^{b}$ | nd | 1.11 ± 0.1 $^{a}$ | 10.23 ± 0.7 $^{c}$ | 1.09 ± 0.1 $^{a}$ | 1.01 ± 0.6 $^{a}$ | 17.21 |
| Extruded snacks with addition of fermented red bean hull (10%) | | | | | | | | | | |
| Control | nd | 0.86 ± 0.03 | nd | nd | 0.05 ± 0.00 $^{a}$ | 3.98 ± 0.9 $^{c}$ | 0.7 ± 0.01 $^{b}$ | 4.43 ± 0.3 $^{d}$ | 1.15 ± 0.07 $^{b}$ | 11.17 |
| Lovage | 0.63 ± 0.2 $^{a}$ | 1.31 ± 0.15 $^{a}$ | 1.46 ± 0.5 $^{b}$ | 9.97 ± 1.4 $^{c}$ | 0.43 ± 0.1 $^{c}$ | 38.77 ± 9.1 $^{d}$ | 0.11 ± 0.00 $^{a}$ | 7.10 ± 1.3 $^{e}$ | nd | 59.78 |
| Basil | 0.46 ± 0.1 $^{a}$ | 20.12 ± 5.2 $^{c}$ | 1.94 ± 0.4 $^{b}$ | 2.79 ± 0.5 $^{b}$ | 0.67 ± 0.1 $^{d}$ | 1.82 ± 0.25 $^{b}$ | 0.21 ± 0.02 $^{a}$ | 1.11 ± 0.21 $^{a}$ | 0.7 ± 0.01 $^{a}$ | 28.82 |
| Oregano | 0.43 ± 0.1 $^{a}$ | 9.11 ± 2.2 $^{b}$ | 1.31 ± 0.3 $^{b}$ | 0.99 ± 0.01 $^{a}$ | 0.31 ± 0.2 $^{c}$ | 2.17 ± 0.15 $^{b}$ | 22.23 ± 4.5 $^{d}$ | 3.08 ± 0.48 $^{c}$ | 1.44 ± 0.1 $^{c}$ | 41.07 |
| Thyme | nd | 1.32 ± 0.2 $^{a}$ | 0.72 ± 0.2 $^{a}$ | 2.93 ± 0.5 $^{b}$ | 0.14 ± 0.01 $^{b}$ | 0.98 ± 0.11 $^{a}$ | 12.04 ± 0.8 $^{c}$ | 2.14 ± 0.1 $^{b}$ | 1.02 ± 0.2 $^{b}$ | 19.29 |
| Extruded snacks with addition of ground fermented broad bean seeds (50%) | | | | | | | | | | |
| Control | nd | 0.82 ± 0.08 $^{a}$ | nd | nd | nd | 2.01 ± 0.25 $^{c}$ | 0.09 ± 0.00 a | 1.05 ± 0.1 $^{b}$ | nd | 3.97 |
| Lovage | 0.3 ± 0.01 $^{a}$ | 2.41 ± 0.2 $^{b}$ | 2.21 ± 0.5 $^{b}$ | 6.31 ± 1.2 $^{c}$ | 0.41 ± 0.1 $^{b}$ | 25.37 ± 6.1 $^{d}$ | 1.12 ± 0.2 $^{b}$ | 3.9 ± 0.4 $^{d}$ | nd | 42.03 |
| Basil | 0.51 ± 0.11 $^{b}$ | 18.01 ± 4.2 $^{d}$ | 2.19 ± 0.3 $^{b}$ | 2.30 ± 0.2 $^{b}$ | 0.56 ± 0.08 $^{b}$ | 1.19 ± 0.1 $^{b}$ | 1.01 ± 0.1 $^{b}$ | 0.91 ± 0.01 $^{a}$ | nd | 26.68 |
| Oregano | 0.49 ± 0.1 $^{b}$ | 8.78 ± 1.2 $^{c}$ | 1.01 ± 0.19 $^{a}$ | 0.81 ± 0.2 $^{a}$ | 0.06 ± 0.00 $^{a}$ | 1.91 ± 0.2 $^{c}$ | 18.08 ± 4.4 $^{d}$ | 1.91 ± 0.2 $^{c}$ | 0.99 ± 0.15 $^{a}$ | 34.04 |
| Thyme | nd | 1.12 ± 0.2 $^{a}$ | 0.72 ± 0.1 $^{a}$ | 3.33 ± 1.0 $^{b}$ | nd | 0.18 ± 0.00 $^{a}$ | 11.09 ± 0.9 $^{c}$ | 1.08 ± 0.11 $^{b}$ | 0.8 ± 0.1 $^{a}$ | 18.23 |
| Extruded snacks with addition of fermented broad bean hull (10%) | | | | | | | | | | |
| Control | nd | 2.13 ± 0.3 $^{b}$ | nd | 1.81 ± 0.8 $^{b}$ | 0.21 ± 0.02 $^{a}$ | 4.01 ± 0.79 $^{c}$ | 0.8 ± 0.02 $^{b}$ | 3.13 ± 0.2 $^{b}$ | 0.71 ± 0.06 $^{a}$ | 12.8 |
| Lovage | 0.42 ± 0.1 $^{a}$ | 1.01 ± 0.2 $^{a}$ | 1.66 ± 0.6 $^{b}$ | 9.98 ± 1.2 $^{c}$ | 0.43 ± 0.1 $^{c}$ | 40.17 ± 9.6 $^{d}$ | 0.09 ± 0.01 $^{a}$ | 9.12 ± 1.1 $^{d}$ | nd | 62.88 |
| Basil | 0.48 ± 0.1 $^{a}$ | 25.67 ± 6.3 $^{d}$ | 2.04 ± 0.35 $^{b}$ | 3.09 ± 0.5 $^{b}$ | 0.67 ± 0.2 $^{c}$ | 0.92 ± 0.17 $^{a}$ | 0.18 ± 0.01 $^{a}$ | 1.08 ± 0.08 $^{a}$ | nd | 34.13 |
| Oregano | 0.39 ± 0.1 $^{a}$ | 10.13 ± 1.3 $^{c}$ | 1.81 ± 0.4 $^{b}$ | 1.09 ± 0.3 $^{a}$ | 0.31 ± 0.02 $^{b}$ | 2.67 ± 0.3 $^{b}$ | 22.29 ± 4.5 $^{d}$ | 3.67 ± 0.2 $^{c}$ | 1.94 ± 0.9 $^{b}$ | 44.3 |
| Thyme | nd | 1.62 ± 0.3 $^{b}$ | 0.62 ± 0.01 $^{a}$ | 2.98 ± 0.6 $^{b}$ | 0.14 ± 0.01 $^{a}$ | 0.88 ± 0.09 $^{a}$ | 11.02 ± 0.3 $^{c}$ | 1.14 ± 0.03 $^{a}$ | 1.12 ± 0.4 $^{b}$ | 19.5 |

NCA—neochlorogenic acid, CA—caffeic acid, p-CA—*p*-coumaric acid, FA—ferulic acid, MYR—myricetin, QUE—quercetin, UT—luteolin, KEMP—kaempferol, API—apigenin, nd—not detected. Different letters denote a significant difference for means ($n$ = 6) for the same lines at a $\alpha \leq 0.05$.

### 3.3. Tannin Content and Antioxidant Activity of the Extruded Products with Added Fermented Bean Hulls or Ground Beans and Selected Herbs/Spices

Tannins are a class of polyphenolic molecules with molecular weights between 500 and 3000 Da. They act as an antinutrient compound of plant origin because they precipitate proteins, inhibit the digestive enzyme, and decrease the bioavailability of vitamins and minerals. However, tannins have also been considered as compounds with anticarcinogenic and antimutagenic potential, as well as antimicrobial properties [48,49]. Previous studies indicated the antioxidant and antiradical activity of tannins, as well as those extracted from legume seeds [50]. Among the control products (without added herbs/spices) tested in this study (Table 3), the lowest content of tannins (0.37 CAE mg $g^{-1}$ dm) was found in the extruded product with fermented broad bean hull (10% added), and the highest content (1.64 CAE mg $g^{-1}$ dm) in the extruded products with fermented ground red bean seeds (50% added). There were no clear trends regarding the effect of the addition of selected herbs/spices on the tannin amount in the extruded products (Table 3). The extruded products with the addition of fermented red bean were found to have more of tannins in the control sample (1.64 and 1.29 CAE mg $g^{-1}$ dm in the extruded snacks with addition of ground seeds and hulls, respectively) than in the products where spices and herbs were added (0.76–1.32 CAE mg $g^{-1}$ dm and 1.00–1.15 CAE mg $g^{-1}$ dm). In the extruded snacks with broad bean added, there were more tannin in the samples with spices and herbs (1.42–1.61 CAE mg $g^{-1}$ dm in the snacks with addition of ground seeds and 0.47–0.69 CAE mg $g^{-1}$ dm in the snacks with hull addition) than in the control samples. The only exception were the snacks with addition of ground fermented broad bean seeds and oregano.

The Folin–Ciocalteu (FC) reaction is an antioxidant assay based on electron transfer, which measures the capacity to reduce FC reagent [51]. This study showed (Table 3) that a higher ability to reduce the Folin–Ciocolteu reagent was displayed by compounds extracted from the extruded products with the addition of fermented ground beans (3.23–3.35 mg GAE $g^{-1}$ dm and 2.85–3.20 mg GAE $g^{-1}$ dm) than from those where fermented hulls were added (2.18–2.83 mg GAE $g^{-1}$ dm and 1.16–1.70 mg GAE $g^{-1}$ dm). However, the addition of selected herbs/spices only slightly increased the results of FC assay.

As with the FC assay results, the ABTS test showed the effect of adding the fermented ground beans or hulls on the antiradical activity of the extruded products (Table 3). Higher antiradical activity was found for the extruded products with the addition of fermented ground beans (8.48–9.28 mg Trolox $g^{-1}$ dm and 8.64–10.23 mg Trolox $g^{-1}$ dm) than for those with the addition of fermented hulls (6.66–7.86 mg Trolox $g^{-1}$ dm and 4.32–5.02 mg Trolox $g^{-1}$ dm). Similarly to the results described before, adding herbs/spices slightly modified the antiradical activity of the extruded products with added fermented bean hulls/ground beans. There were no clear trends associated with the presence of individual spices in the products tested. Statistically ($p > 0.05$), among ground fermented red bean seeds and fermented red bean hull samples, a significantly higher antiradical activity was revealed for samples with basil and oregano, and basil and thyme, respectively, compared to the controls (without herbs/spices). For the extruded products with fermented broad beans (both hull and ground beans), the highest antiradical activity was demonstrated for samples with thyme (Table 3).

**Table 3.** Antioxidant activity and tannin content of extruded snacks with the addition of fermented red or broad beans/hull and with herbs/species (0.5%), and the controls (without herbs/spices addition).

| Products | Folin–Ciocalteu Assay (FC) (mg GAE $g^{-1}$ dm) | Antiradical Activity (ABTS) (mg Trolox $g^{-1}$ dm) | Tannins (CAE mg $g^{-1}$ dm) |
|---|---|---|---|
| Extruded snacks with addition of ground fermented red bean seeds (50%) | | | |
| Control | $3.27 \pm 0.04$ $^{a}$ | $8.75 \pm 0.21$ $^{a}$ | $1.64 \pm 0.03$ $^{e}$ |
| Lovage | $3.23 \pm 0.06$ $^{a}$ | $8.55 \pm 0.11$ $^{a}$ | $1.32 \pm 0.03$ $^{d}$ |
| Basil | $3.33 \pm 0.08$ $^{a}$ | $9.13 \pm 0.06$ $^{b}$ | $1.13 \pm 0.07$ $^{c}$ |
| Oregano | $3.35 \pm 0.03$ $^{a}$ | $9.28 \pm 0.05$ $^{c}$ | $0.94 \pm 0.01$ $^{b}$ |
| Thyme | $3.32 \pm 0.03$ $^{a}$ | $8.48 \pm 0.05$ $^{a}$ | $0.76 \pm 0.05$ $^{a}$ |
| Extruded snacks with addition of fermented red bean hull (10%) | | | |
| Control | $2.18 \pm 0.02$ $^{a}$ | $6.66 \pm 0.11$ $^{a}$ | $1.29 \pm 0.03$ $^{c}$ |
| Lovage | $2.32 \pm 0.03$ $^{b}$ | $6.73 \pm 0.38$ $^{a}$ | $1.12 \pm 0.05$ $^{b}$ |
| Basil | $2.83 \pm 0.12$ $^{e}$ | $7.86 \pm 0.42$ $^{b}$ | $1.01 \pm 0.02$ $^{a}$ |
| Oregano | $2.43 \pm 0.01$ $^{c}$ | $6.83 \pm 0.24$ $^{a}$ | $1.00 \pm 0.02$ $^{a}$ |
| Thyme | $2.55 \pm 0.07$ $^{d}$ | $7.53 \pm 0.02$ $^{b}$ | $1.15 \pm 0.08$ $^{b}$ |
| Extruded snacks with addition of ground fermented broad bean seeds (50%) | | | |
| Control | $2.85 \pm 0.02$ $^{a}$ | $8.64 \pm 0.28$ $^{a}$ | $1.22 \pm 0.05$ $^{b}$ |
| Lovage | $2.99 \pm 0.04$ $^{b}$ | $9.48 \pm 0.12$ $^{b}$ | $1.61 \pm 0.12$ $^{c}$ |
| Basil | $3.01 \pm 0.07$ $^{b}$ | $9.66 \pm 0.07$ $^{b}$ | $1.42 \pm 0.08$ $^{c}$ |
| Oregano | $3.10 \pm 0.04$ $^{b}$ | $9.39 \pm 0.14$ $^{b}$ | $1.00 \pm 0.08$ $^{a}$ |
| Thyme | $3.20 \pm 0.02$ $^{c}$ | $10.23 \pm 0.13$ $^{c}$ | $1.60 \pm 0.09$ $^{c}$ |
| Extruded snacks with addition of fermented broad bean hull (10%) | | | |
| Control | $1.16 \pm 0.01$ $^{a}$ | $4.32 \pm 0.07$ $^{a}$ | $0.37 \pm 0.01$ $^{a}$ |
| Lovage | $1.33 \pm 0.02$ $^{b}$ | $4.89 \pm 0.10$ $^{b}$ | $0.47 \pm 0.03$ $^{b}$ |
| Basil | $1.57 \pm 0.06$ $^{c}$ | $4.74 \pm 0.33$ $^{b}$ | $0.49 \pm 0.03$ $^{b}$ |
| Oregano | $1.62 \pm 0.05$ $^{c}$ | $4.84 \pm 0.34$ $^{b}$ | $0.59 \pm 0.02$ $^{c}$ |
| Thyme | $1.70 \pm 0.01$ $^{d}$ | $5.02 \pm 0.19$ $^{b}$ | $0.69 \pm 0.08$ $^{c}$ |

CAE—catechin equivalent; GAE—gallic acid equivalent. Different letters denote a significant difference for means ($n$ = 6) for the same lines at a $\alpha \leq 0.05$.

### *3.4. Sensory Quality of the Exruded Snacks with Fermented Broad Beans or Red Beans and Herb/Spices*

The snacks tested can be targeted at many consumer groups, especially vegans. They aim to supplement a vegan diet with a product that has a higher nutritional value than traditional corn crisps. However, the success of a food product in the market is related to its sensory quality, especially consumer desirability. Therefore, this part of the presented experiment presented how the sensory characteristics of corn snacks with bean or broad bean hulls and herb/spices affect consumer desirability.

Consumer evaluation results showed that the extruded snacks with fermented broad beans or red beans and spices had various overall desirability (Table S2 and Figure 1). Different taste and aroma desirability was also confirmed. The highest positive correlation was found between changes in overall desirability and taste desirability ($r$ = 0.918). Samples with low taste desirability scores obtained low total desirability scores (Table S2 and Figure 2).

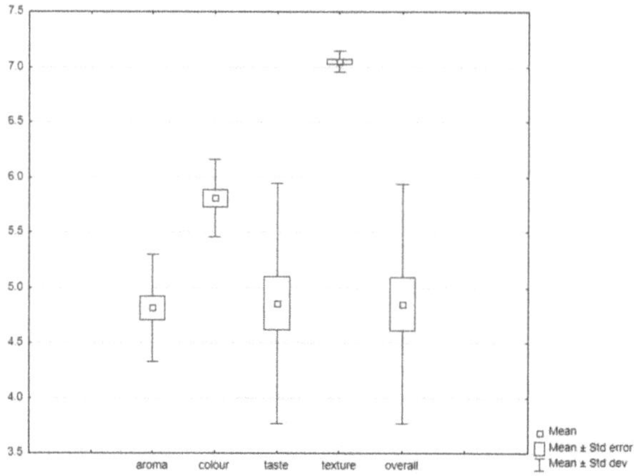


**Figure 1.** Box plot diagram of consumer desirability (colour, taste, aroma, texture, and overall desirability) of extruded snacks depending on the addition of fermented red or broad bean seeds/hulls and herbs/species (0.5%).

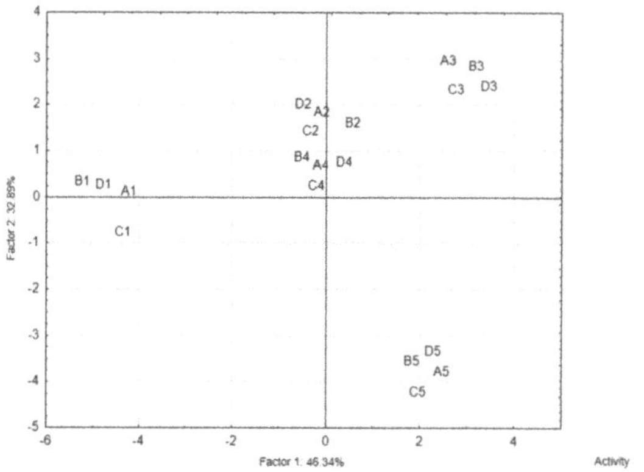


**Figure 2.** Principal component analysis (PCA) plots of data from overall desirability of extruded snacks with the addition of fermented red or broad bean seeds/hull and with herbs/species (0.5%), and without the addition of herbs/spices (Control); Addition of ground fermented red bean seeds (A): Control 1A, Thyme 2A, Oregano 3A, Basil 4A, Lovage 5A; Addition of fermented red bean hull (B): Control 1B, Thyme 2B, Oregano 3B, Basil 4B, Lovage 5B; Addition of ground fermented broad bean seeds (C): Control 1C, Thyme 2C, Oregano 3C, Basil 4C, Lovage 5C; Addition of fermented broad bean hull (D): Control 1D, Thyme 2D, Oregano 3D, Basil 4D; Lovage 5D.

The overall desirability analysis showed that the products were rated between 3.7 and 6.8 points on a 10-point scale (Table S2). Similar rates were confirmed for taste desirability analysis (3.75–7.98).

It should be noted that the type of additive (i.e., fermented seeds or fermented hulls) did not cause differences in taste and overall desirability (Table S2 and Figure 2, $p > 0.05$). No statistically significant differences were identified in consumers' ratings of these factors. However, a statistically significant effect of the herb/spice used on overall desirability was confirmed. The samples were placed in the PCA space (Figure 2), which demonstrated that they mainly focused on the spice variant, or its lack.

The highest taste and overall desirability were identified for the products with lovage (7.75–7.92)—independently of type product. The samples without herb/spices had lower taste and overall desirability. The lowest results, especially taste desirability results, were identified for the samples with basil.

In the sensory profiling of the tested extruded snacks, the perception of the following descriptors was defined and determined: taste (essential oil, herbal, sour, salty, starch, broth, bitter, and strange) and aroma (essential oil, herbal, sour, starch, broth, bitter, lemon, and strange). The results of the sensory profiling of these products are presented in Table S3.

Principal component analysis (PCA, Figure 2) was used to examine the relationship between the characteristic sensory profiles of the extruded snacks with fermented broad bean or red bean and herb/spices, and those without herb/spices —the control sample (consumer analysis variables)—and to identify the derived factors by which these variables can be classified. The PCA was also used to map the variants tested in this experiment (i.e., the samples with selected spices and legume seed variants) on these factors. The PCA showed that the first two factors (F1 and F2) were the most important elements of data variability (79%). The analysis revealed a significant dominant relationship between overall desirability and the taste descriptors (R = 0.91). Therefore, the taste descriptors were chosen to interpret the data. It is worth noting that the F1 factor was responsible for about 70% of taste variance.

The analysis of sensory profiling did not confirm statistically significant differences between sample profilographs (types of descriptors and their intensities) depending on the type of legume variant added. On the other hand, a statistically significant effect of herb/spice on the type of descriptors and their intensity was confirmed. The samples were placed in the PCA space (Figure 3). The samples mainly focused on the spice variant or its lack.

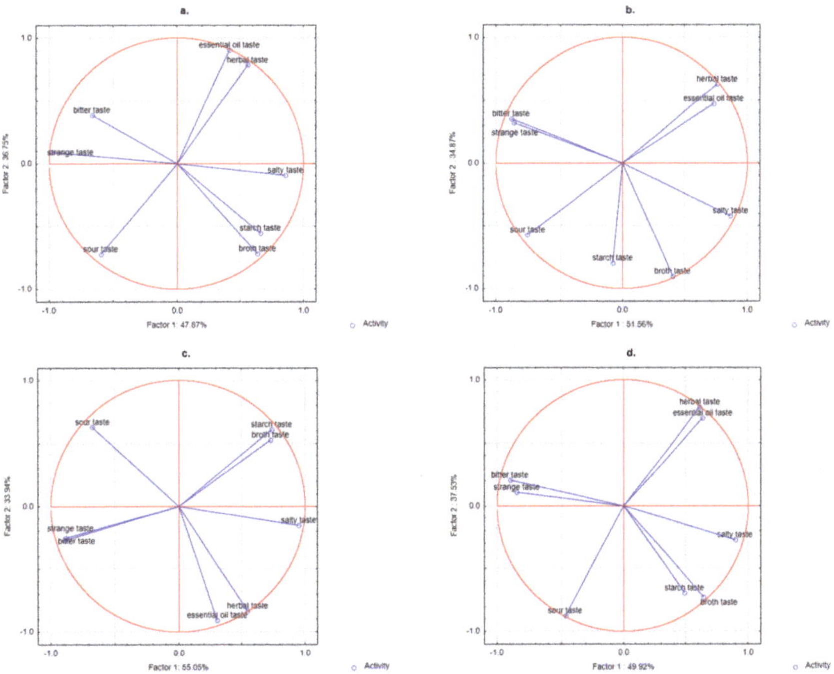


**Figure 3.** Principal component analysis (PCA) plots of data from sensory profiling of extruded snacks with addition of fermented red or broad beans/hull and with herbs/species (0.5%), and without the addition of herbs/spices (Control): (**a**) fermented ground red bean seeds 50%; (**b**) fermented red bean hull 10% panel; (**c**) fermented ground broad bean seeds 50%; (**d**) fermented broad bean hull 10%.

For the taste descriptors, the projection of all variants on the factor-plane F1 × F2 (Figure 3) shows that samples with lovage and oregano were situated on the right of the F1 axis (i.e., they had positive coordinate values for F1). The flavour profile of the extruded snack with lovage, which had a low intensity of bitter and strange taste and a high intensity of broth taste, differed from the others to the largest extent. The samples with oregano showed similarities in profile to both lovage and other samples. A lower intensity of herbal, bitter, and strange taste (but higher than the samples with lovage) was confirmed. The samples with other spices (thyme, basil) and without them (control sample) were situated on the left (negative coordinate values for F1). Their taste profiles were characterised by a higher intensity of bitter and strange taste than the samples with lovage. The samples without spices differed from those with thyme, basil, and oregano in a lower intensity of essential oil and herbal taste.

### 3.5. Effect of Active Compounds of Spices on Sensory Desirability

Taste and overall desirability were found to be positively correlated with the intensity of broth taste ($r$ = 0.923), starch taste ($r$ = 0.714), salty taste ($r$ = 0.758), and broth taste ($r$ = 0.714) in all evaluated products (Table 4). A statistically negative correlation was also confirmed between overall and taste desirability and the intensity of bitter taste ($r$ = −0.838) and strange taste ($r$ = −0.712). The highest intensity of these two descriptors (bitter and strange) was demonstrated for corn crisps with thyme and basil. Previous results indicated that these descriptors and their intensities may be related to the presence of phenolic compounds. Statistical analysis confirmed the relationship between the content of some analysed ingredients and taste and overall desirability. A relationship was found between the intensity of selected taste descriptors and the content of analysed phenolic compounds.

**Table 4.** Correlations coefficients between the intensity of descriptors and consumer analysis parameters of extruded snacks with the addition of fermented red or broad bean seeds/hull and with herbs/species (0.5%), and the controls (without herbs/spices addition).

| | Aroma | Colour | Taste | Texture | Overall |
|---|---|---|---|---|---|
| Essential oil aroma | 0.085 | 0.108 | −0.227 | −0.026 | −0.248 |
| Herbal aroma | 0.181 | 0.165 | −0.110 | −0.066 | −0.132 |
| Starch aroma | −0.438 | −0.322 | −0.658 | 0.021 | −0.671 |
| Lemon aroma | 0.399 | 0.197 | −0.094 | −0.069 | −0.105 |
| Bitter aroma | −0.412 | −0.403 | −0.731 | 0.046 | −0.734 |
| Strange aroma | −0.635 | −0.500 | −0.252 | −0.021 | −0.248 |
| Sour aroma | −0.569 | −0.260 | −0.029 | 0.220 | −0.023 |
| Broth aroma | 0.113 | 0.260 | 0.915 | −0.016 | 0.923 |
| Essential oil taste | 0.046 | 0.085 | −0.063 | −0.064 | −0.092 |
| Herbal taste | 0.203 | 0.233 | −0.035 | −0.043 | −0.057 |
| Sour taste | −0.408 | −0.289 | −0.012 | 0.069 | 0.001 |
| Salty taste | −0.546 | −0.145 | 0.769 | −0.152 | 0.758 |
| Broth taste | 0.112 | 0.244 | 0.922 | −0.017 | 0.932 |
| Starch taste | 0.267 | 0.201 | 0.721 | −0.239 | 0.714 |
| Bitter taste | 0.116 | −0.233 | −0.838 | −0.046 | −0.825 |
| Strange taste | −0.212 | −0.296 | −0.712 | 0.110 | −0.707 |

A high correlation between the products with high overall desirability and with quercetin (QUE) ($r$ = −0.960) was confirmed (Table 5). The analysis showed a high correlation between the amount of this flavonoid and the perceived intensity of broth taste ($r$ = 0.971). There was also a correlation between the amount of QUE and the perceived intensity of salty taste ($r$ = 0.744). The analysis showed a correlation between the amount of luteolin (LUT) and the perceived intensity of essential oil taste ($r$ = 0.840) and herbal taste ($r$ = 0.876). It was also demonstrated that the intensity of bitter taste was negatively correlated with the amount of QUE ($r$ = −0.704). It is worth noting that the lower intensity of bitter taste in the samples with the highest amount of kaempferol (KEMP) was confirmed analytically. The largest content of this flavonoid was found in the samples with lovage. At the same time, these samples had the highest taste desirability and the lowest intensity of bitter taste. Statistical analysis confirmed a negative correlation between the content of this compound and the intensity of bitter taste ($r = -0.636$). These trends were confirmed for strange taste descriptor analysis. It was also confirmed that the higher the content of KEMP, the higher the strange taste reduction. There was a negative correlation concerning the amount of this compound ($r = -0.616$) but also concerning FA (ferulic acid; $r = -0.689$).

**Table 5.** Correlations coefficients between sensory parameters and the content of phenolic compounds of extruded snacks with addition of fermented red bean or broad bean seeds/hulls and with herbs/species (0.5%), and the controls (without herbs/spices addition).

| | NCA | CA | p-CA | FA | MYR | QUE | LUT | KEMP | API | Total |
|---|---|---|---|---|---|---|---|---|---|---|
| Aroma | 0.536 | 0.647 | 0.614 | 0.164 | 0.594 | −0.029 | −0.069 | −0.209 | −0.188 | 0.259 |
| Colour | 0.320 | 0.103 | 0.235 | 0.197 | 0.163 | 0.191 | 0.065 | 0.113 | −0.136 | 0.288 |
| Taste | 0.419 | −0.430 | 0.287 | 0.856 | 0.140 | 0.960 | −0.194 | 0.837 | −0.416 | 0.773 |
| Texture | 0.011 | −0.051 | −0.189 | −0.099 | −0.035 | 0.017 | 0.013 | 0.130 | 0.054 | −0.020 |
| Overall | 0.421 | −0.411 | 0.291 | 0.863 | 0.155 | 0.970 | −0.216 | 0.844 | −0.433 | 0.776 |
| Essential oil aroma | 0.137 | 0.125 | 0.126 | −0.353 | −0.255 | −0.398 | 0.971 | −0.246 | 0.807 | 0.133 |
| Herbal aroma | 0.292 | 0.181 | 0.286 | −0.223 | −0.164 | −0.275 | 0.932 | −0.197 | 0.696 | 0.276 |
| Starch aroma | −0.841 | −0.321 | −0.820 | −0.730 | −0.686 | −0.723 | 0.320 | −0.526 | 0.432 | −0.824 |
| Lemon aroma | 0.500 | 0.424 | 0.561 | 0.006 | 0.219 | −0.150 | 0.688 | −0.092 | 0.612 | 0.435 |
| Bitter aroma | −0.762 | −0.174 | −0.720 | −0.752 | −0.429 | −0.725 | 0.074 | −0.496 | 0.389 | −0.870 |
| Strange aroma | −0.567 | −0.338 | −0.655 | −0.307 | −0.315 | −0.168 | −0.277 | 0.087 | 0.101 | −0.495 |
| Sour aroma | −0.514 | −0.445 | −0.691 | −0.253 | −0.357 | 0.007 | −0.475 | 0.118 | −0.311 | −0.497 |
| Broth aroma | 0.506 | −0.223 | 0.407 | 0.945 | 0.304 | 0.956 | −0.293 | 0.846 | −0.477 | 0.835 |
| Essential oil taste | 0.335 | 0.171 | 0.241 | −0.255 | −0.219 | −0.227 | 0.840 | −0.156 | 0.620 | 0.261 |
| Herbal taste | 0.326 | 0.169 | 0.278 | −0.153 | −0.211 | −0.195 | 0.876 | −0.150 | 0.570 | 0.322 |
| Sour taste | −0.351 | −0.221 | −0.431 | −0.097 | −0.074 | 0.062 | −0.663 | 0.002 | −0.520 | −0.419 |
| Salty taste | −0.089 | −0.718 | −0.126 | 0.630 | −0.204 | 0.744 | −0.203 | 0.719 | −0.230 | 0.385 |
| Broth taste | 0.541 | −0.176 | 0.454 | 0.940 | 0.401 | 0.971 | −0.443 | 0.786 | −0.586 | 0.789 |
| Starch taste | 0.744 | 0.118 | 0.764 | 0.703 | 0.439 | 0.675 | 0.094 | 0.507 | −0.130 | 0.890 |
| Bitter taste | −0.425 | 0.379 | −0.219 | −0.550 | 0.091 | −0.704 | −0.228 | −0.636 | 0.144 | −0.714 |
| Strange taste | −0.479 | 0.132 | −0.543 | −0.689 | −0.199 | −0.616 | −0.199 | −0.404 | 0.141 | −0.745 |

NCA—neochlorogenic acid, CA—caffeic acid, p-CA—*p*-coumaric acid, FA—ferulic acid, MYR—myricetin, QUE—quercetin, LUT—luteolin, KEMP—kaempferol, API—apigenin.

## 4. Discussion

The results confirmed the possibility of using fermented seeds of broad beans or red beans in the production of extruded snacks. They contained flavonoids such as myricetin, quercetin, luteolin, and kaempferol. Products with added fermented hulls also had apigenin. Extruded snacks with fermented hulls had a lower number of phenolic compounds due to their lower content than the content of seeds (10% vs. 50%).

A previous study [52] revealed that raw and cooked beans contain predominant quantities of flavonoids in their phenolic profiles, including quercetin, myricetin, cynidine, procyanidin, naringenin, catechin, hesperidin, and kaempferol, which is consistent with the results of this study. They found out that flavonoids were both in free and bound forms (as glucoside derivatives, mostly apigenin 7-O-glucoside, quercetin 3-O-glucoside, myricetin 3-O-glucoside, naringenin 7-O-glucoside, quercetin 4-O-galactoside, and kaempferol 3-O-glucoside). Variations in the flavonoids and their glucosides were observed depending on bean varieties. Kaempferol and its 3-O-glucosides were primarily found in pinto beans and diglucosides of kaempferol and quercetin in dark red kidney beans. 3-O-glucosides of malvidin, petunidin, and delphinidin were present in black beans; however, a small quantity of quercetin 3-O-glucoside and its malonates were found in light red kidney beans, whereas kaempferol monoglucoside. kaempferol 3-O-glucoside, and kaempferol 3-O-xylosyl glucoside were found in Italian beans. On the other hand, Brazilian beans contained non-glycosylated forms of isoflavonoids, including daidzein (0.82–12.91 mg 100 $g^{-1}$)

and genistein (0.26–0.97 mg 100 $g^{-1}$). Black beans showed the highest concentrations of isoflavonoids among them, with daidzein being the main compound. The study [52] also demonstrated that raw and cooked beans contain phenolic acids, both benzoic acid derivatives (vanillic, *p*-hydroxybenzoic, and gallic acids) and those derived from cinnamic acid (ferulic, *p*-coumaric, and chlorogenic acids). Ferulic acid was the predominant phenolic acid in dry common beans. Studies also revealed demonstrated that cooking common beans at a high temperature does not change the content of phenolic acids. Raw common beans contained *p*-hydroxybenzoic acid (0.45–0.86 mg 100 $g^{-1}$), vanillic acid (0.52–1.66 mg 100 $g^{-1}$), coumaric acid (0.32–0.68 mg 100 $g^{-1}$), and ferulic acid (0.17–0.36 mg 100 $g^{-1}$).

The results of this study showed that adding 0.5% of selected herbs/spices significantly increased the content of flavonoids and phenolic acids in the extruded products. This is particularly true for lovage and oregano samples. The increase in the content of individual phenolic compounds in products with individual herbs/spices resulted from their participation in the phenolic profile of these additives. Slimestad et al. [53] reported that luteolin glucuronide together with rosmarinic acid were the main constituents of thyme. The amount of luteolin in thyme was determined to be 660 mg 100 $g^{-1}$ on average, with the amount of luteolin in fresh herbs being higher (1.489 mg 100 $g^{-1}$). Nieto [54] showed that methanolic extracts of thyme were the sources of flavonols, such as quercetin-7-O-glucoside, and phenolic acids (*p*-coumaric, caffeic, rosmarinic, cinnamic, carnosic, ferulic, quinic, and caeoylquinic acids), as well as flavanones (naringenin) and flavones (apigenin). Using other solvents such as butanol, ethyl acetate, and hexane, they extracted various compounds from thyme, including saponins, steroids, flavonoids, alkaloids, and tannins. In basil, the total amount of phenolic acid was determined to be 52.61 mg of gallic acid per gram of dry extract [55]. Phenolic acids detected in the basil extract were chlorogenic, *p*-hydroxybenzoic, caffeic, vanillic, rosmarinic, and cinnamic acids. The most dominant acids were rosmarinic (0.18 mg $g^{-1}$ of dry extract) and cinnamic (0.18 mg $g^{-1}$ of dry extract). The total flavonoid amount reported by Teofilovic at al. [55] was 0.52 mg of quercetin per gram of dry extract, and among flavonoid components, quercetin (4.77 mg $g^{-1}$ of dry extract) and naringenin (0.18 mg $g^{-1}$ of dry extract). Rosmarinic acid, apigenin, luteolin, quercetin, scutellarein, and their derivatives were identified as the major phenolic compounds in oregano; however. the content and distribution of the phenolic compound in oregano varied depending on cultivar, geographical factors, and environmental factors [56]. Nour et al. [57] analysed the phenolic profile and antiradical activity of culinary herbs (i.e., parsley, dill, lovage, and celery leaves). They found that lovage had the highest total amount of phenolic compound (577.04 mg GAE 100 $g^{-1}$ dm) and flavonoids (298.38 mg QE 100 $g^{-1}$), and the highest antioxidant activity (1462.52 mg Trolox 100 $g^{-1}$) among the selected herbs, demonstrating its outstanding value in terms of antioxidant activity and content of bioactive compounds. In the phenolic profile of lovage, the main compounds were quercetin, myricetin, rutin, ellagic acid, and sinapic acid. In line with the results presented in this study, Kozłowska et al. [58] also identified neochlorogenic acid in the phenolic profile of a lovage extract.

This study showed that a higher ability to reduce the Folin–Ciocolteu reagent was demonstrated by compounds extracted from the extruded products with the addition of fermented ground beans than from those where fermented hulls were added (Table 3). The FC assay has been widely applied in the determination of the total phenol/polyphenol content of plant-derived food and biological samples; however, the method is not specific only for phenolic compounds because it is based on the redox reaction [51,59], and the results can be affected by various interfering substances present in crude plant extracts (e.g., sugars, proteins, and other non-phenolic compounds). This non-specificity of the assay may explain the higher values determined in this study for products with the addition of ground fermented beans compared to those where fermented bean hulls were added, independently of herb/spice additions. Clarification of this problem requires furthers research.

The analysis of the sensory analysis results confirmed a high texture and colour desirability of all samples, regardless of the additive used. Previous data in some cases indicated potential texture deterioration in such products when legumes are added [18]. Similarly, previously concerns about the addition of spices (especially thyme and basil) and colour deterioration in consumer assessment are not confirmed in this study [17]. In contrast, a highly significant impact of taste on overall desirability was confirmed. In this study, lower taste desirability was found, and thus overall desirability among the samples with a high intensity of bitter taste. This is supported by previous studies indicating low acceptance of this descriptor among consumers. Previous studies also revealed that bitter taste can reduce consumer acceptance of food products [60–63]. The highest statistically significant taste and overall desirability were found among the samples with lovage regardless of the addition of legume seeds.

The profile analysis showed the highest intensity of essential oil taste and herbal taste in the samples containing thyme, basil, and oregano. A negative correlation between taste desirability and these descriptors was confirmed. These trends confirm previous data showing restrictions on the use of spices in food production due to the presence of these descriptors [64].

The taste profile analysis also indicated the highest perceived intensity of bitter taste in the samples with thyme and basil, which confirms previous studies on the addition of these spices to meat products [65,66]. Previous literature [65] indicates a correlation between the presence of bitter taste in the product with the presence of kaempferol and quercetin [60,67]. The results presented in this paper did not confirm this trend. The intensity of this taste was found to be the lowest in the samples with the highest concentration of kaempferol and quercetin. This was true for all samples with lovage. Consumer analysis confirmed that taste and overall desirability increased when the intensity of broth taste went up. This may indicate that the broth-like smell and lovage taste, which are due to the presence of terpene-like compounds, mask the bitter taste. A relatively high content of these compounds in lovage noticeably reduced the intensity of the bitter taste. However, this initial hypothesis should be confirmed/verified in subsequent studies.

## 5. Conclusions

This study indicates the possibility of using fermented red bean and broad bean seeds and by-products in the form of their hulls for the production of extruded snacks. Their addition increases the common nutritional value of these snacks, measured by the content of mainly protein and phenolic compounds with proven health-promoting effects in numerous studies. Adding selected herbs, especially lovage, positively affects the sensory profile and consumer desirability of these products. The use of fermented seeds of red beans and broad beans, as well as their hulls, is not only in line with the principles of the planetary health diet but also serves as a nutritionally valuable enrichment of corn products often consumed by vegans and vegetarians.

In future studies, it is worthwhile to further develop this topic by studying for example interactions between corn product components as well as the antioxidant properties of the new products tested to determine their usefulness in the diet of people of today more precisely.

**Supplementary Materials:** The following are available online at https://www.mdpi.com/article/10.3390/nu13124538/s1, Table S1: Phenolic compounds of herb/spices added to extruded snacks. Table S2: Mean scores of consumer desirability of corn snacks with fermented broad bean or bean and herb/spices, and the control sample (without spices addition). Table S3: Mean scores ($n = 17$) of sensory taste and aroma profiling of corn snacks with fermented broad bean or bean and herb/spices, and the control sample (without herb/spices addition).

**Author Contributions:** Conceptualisation, K.S.-B. and M.G.; methodology, K.S.-B., A.J.-G., M.G., K.W., and M.H.; formal analysis, K.S.-B. and M.G.; investigation, K.S.-B., M.G., A.J.-G., K.W., M.H., and A.S.; writing—original draft preparation, K.S.-B., M.G., A.J.-G., K.W., M.H., and A.S.; writing—review and editing, A.G.-M. and A.J.-G.; visualisation, K.S.-B. All authors have read and agreed to the published version of the manuscript.

**Funding:** The research was financed by EU project POIG 01.01.02-00-061/09 "New Bioactive Food with Designed Functional Properties" and statutory funds of Department of Gastronomy Science and Functional Foods, Poznan University of Life Sciences, Poland.

**Institutional Review Board Statement:** The study was conducted according to the guidelines of the Code of Ethics of the World Medical Association POZNAN UNIVERSITY OF MEDICAL SCIENCES (consent no. 757/13).

**Informed Consent Statement:** Informed consent was obtained from all subjects involved in the study.

**Data Availability Statement:** The data are available as Supplementary Data.

**Conflicts of Interest:** The authors declare that they have no conflict of interest.

## References

1. Calles, T.; del Castello, R.; Baratelli, M.; Xipsiti, M.; Navarro, D.K. *The International Year of Pulses—Final Report*; FAO: Rome, Italy, 2019.
2. Didinger, C.; Thompson, H.J. Defining nutritional and functional niches of legumes: A call for clarity to distinguish a future role for pulses in the dietary guidelines for Americans. *Nutrients* **2021**, *13*, 1100. [CrossRef]
3. Bielefeld, D.; Grafenauer, S.; Rangan, A. The effects of legume consumption on markers of glycaemic control in individuals with and without diabetes mellitus: A Systematic Literature Review of Randomised controlled trials. *Nutrients* **2020**, *12*, 2123. [CrossRef]
4. Rawal, V.; Navarro, D.K. *The Global Economy of Pulses*; FAO: Rome, Italy, 2019.
5. Szczebyło, A.; Halicka, E.; Łuczyńska, K. Consumption of pulses in Poland as an element of a sustainable food consumption model. *Pr. Nauk. Akad. im. Jana Długosza w Częstochowie. Pragmata Oikonomias* **2018**, *12*, 35–45. [CrossRef]
6. Bechthold, A.; Boeing, H.; Schwedhelm, C.; Hoffmann, G.; Knuppel, S.; Iqbal, K.; De Henauw, S.; Michels, N.; Devleesschauwer, B.; Schlesinger, S.; et al. Food groups and risk of coronary heart disease, stroke and heart failure: A systematic review and dose-response meta-analysis of prospective studies. *Crit. Rev. Food Sci. Nutr.* **2019**, *59*, 1071–1090. [CrossRef]
7. Kim, S.J.; de Souza, R.J.; Choo, V.L.; Ha, V.; Cozma, A.I.; Chiavaroli, L.; Mirrahimi, A.; Blanco Mejia, S.; Di Buono, M.; Bernstein, A.M.; et al. Effects of dietary pulse consumption on body weight: A systematic review and meta-analysis of randomized controlled trials. *Am. J. Clin. Nutr.* **2016**, *103*, 1213–1223. [CrossRef] [PubMed]
8. Willett, W.; Rockström, J.; Loken, B.; Springmann, M.; Lang, T.; Vermeulen, S.; Garnett, T.; Tilman, D.; DeClerck, F.; Wood, A.; et al. Food in the Anthropocene: The EAT-Lancet Commission on healthy diets from sustainable food systems. *Lancet* **2019**, *393*, 447–492. [CrossRef]
9. Bahl, P.N. Climate change and pulses: Approaches to combat its impact. *Agric. Res.* **2015**, *4*, 103–108. [CrossRef]
10. Benayad, A.; Taghouti, M.; Benali, A.; Aboussaleh, Y.; Benbrahim, N. Nutritional and technological assessment of durum wheat-faba bean enriched flours, and sensory quality of developed composite bread. *Saudi J. Biol. Sci.* **2021**, *28*, 635–641. [CrossRef] [PubMed]
11. Bouhlal, O.; Taghouti, M.; Benbrahim, N.; Benali, A.; Visioni, A.; Benba, J. Wheat-lentil fortified flours: Health benefits, physicochemical, nutritional and technological properties. *J. Mater. Environ. Sci.* **2019**, *10*, 1098–1106.
12. Ni, Q.; Ranawana, V.; Hayes, H.E.; Hayward, N.J.; Stead, D.; Raiko, S.V. Addition of broad bean hull to wheat flour for the development of high-fiber bread: Effects on physical and nutritional properties. *Foods* **2020**, *9*, 1192. [CrossRef] [PubMed]
13. Fendri, L.B.; Chaari, F.; Maaloul, M.; Kallel, F.; Abdelkafi, L.; Ellouz, C.S.; Ghribi-Aydi, D. Wheat bread enrichment by pea and broad bean pods fibers: Effect on dough rheology and bread quality. *LWT* **2016**, *73*, 584–591. [CrossRef]
14. Granito, M.; Flias, J.; Doblado, R.; Guerra, M.; Champ, M. Nutritional improvent of beans (Phaseolus vulgaris) by natural fermentation. *Eur. Food Res. Technol.* **2002**, *214*, 226–231. [CrossRef]
15. Onwurafor, E.U.; Onweluzo, J.C.; Ezeoke, A.M. Effect of fermentation methods on chemical and microbial properties of mung bean (Vigna radiata) flour. *NIFOJ* **2014**, *32*, 89–96. [CrossRef]
16. Sozer, N.; Melama, L.; Silbir, S.; Rizzello, C.G.; Flander, L.; Poutanen, K. Lactic acid fermentation as a pre-treatment process for faba bean flour and its effect on textural, structural and nutritional properties of protein-enriched gluten-free faba bean breads. *Foods* **2019**, *8*, 431. [CrossRef] [PubMed]
17. Anton, A.A.; Gary Fulcher, R.; Arntfield, S.D. Physical and nutritional impact of fortification of corn starch-based extruded snacks with common bean (Phaseolus vulgaris L.) flour: Effects of bean addition and extrusion cooking. *Food Chem.* **2009**, *113*, 989–996. [CrossRef]
18. Tas, A.A.; Shah, A.U. The replacement of cereals by legumes in extruded snack foods: Science, technology and challenges. *Trends Food Sci. Technol.* **2021**, *116*, 701–711. [CrossRef]
19. Pasqualone, A.; Costantini, M.; Coldea, T.E.; Summo, C. Use of Legumes in Extrusion Cooking: A Review. *Foods* **2020**, *9*, 958. [CrossRef] [PubMed]
20. Ai, Y.; Jane, J.L. Macronutrients in Corn and Human Nutrition. *Compr. Rev. Food Sci. Food Saf.* **2016**, *15*, 581–598. [CrossRef]
21. Kulczak, M.; Błasińska, I.; Łuczak, H. Nutritive value and sensory quality of dry food mixes based on prepared 'Red Kidney' beans. *Probl. Hyg. Epidemiol.* **2016**, *97*, 134–139.

22. Preeti, B.; Kalpana, K. Fiber rich snack food products incorporated with green gram husk and their suitability for diabetics. *Asian J. Dairy Food Res.* **2015**, *34*, 300–306. [CrossRef]
23. Ivarsson, E.; Grudén, M.; Södergren, J.; Hultberg, M. Use of faba bean (*Vicia faba* L.) hulls as substrate for Pleurotus ostreatus—Potential for combined mushroom and feed production. *J. Clean. Prod.* **2021**, *313*, 127969. [CrossRef]
24. Renna, M.; De Cillis, F.; Leoni, B.; Acciardi, E.; Santamaria, P. From by-product to unconventional vegetable: Preliminary evaluation of fresh fava hulls highlights richness in L-DOPa and low content of anti-nutritional factor. *Foods* **2020**, *9*, 159. [CrossRef]
25. Kimothi, S.; Dhaliwal, Y.S. Nutritional and health promoting attribute of kidney beans (*Phaseolus vulgaris* L.): A review. *Int. J. Curr. Microbiol. Appl. Sci.* **2020**, *9*, 1201–1209. [CrossRef]
26. Kim, S.; Fenech, M.F.; Kim, P.J. Nutritionally recommended food for semi- to strict vegetarian diets based on large-scale nutrient composition data. *Sci. Rep.* **2018**, *8*, 1–11. [CrossRef]
27. Miedzianka, J.; Styczyńska, M.; Łoźna, K.; Aniołowska, M.; Biernat, J. Nutritional value of raw legumes. *J. Elem.* **2017**, *22*, 643–652. [CrossRef]
28. Mayer Labba, I.C.; Frøkiær, H.; Sandberg, A.S. Nutritional and antinutritional composition of fava bean (*Vicia faba* L., *var. minor*) cultivars. *Food Res. Int.* **2021**, *140*, 110038. [CrossRef] [PubMed]
29. Złotek, U.; Szymanowska, U.; Pecio, U.; Kozachok, S.; Jakubczyk, A. Antioxidative and Potentially Anti-inflammatory Activity of Phenolics from Lovage Leaves Levisticum officinale Koch Elicited with Jasmonic Acid and Yeast Extract. *Molecules* **2019**, *24*, 1441. [CrossRef]
30. Embuscado, M.E. Spices and herbs: Natural sources of antioxidants—A mini review. *J. Funct. Foods* **2015**, *18*, 811–819. [CrossRef]
31. El-Sayed, S.M.; Youssef, A.M. Potential application of herbs and spices and their effects in functional dairy products. *Heliyon* **2019**, *5*, e01989. [CrossRef] [PubMed]
32. Goderska, K.; Gumienna, M.; Czarnecki, Z. Release of phenolic compounds from bean flour, bean-derived chips and black chokeberry juice and changes in their antioxidant activity during digestion in an in vitro gastrointestinal model. *Pol. J. Food Nutr. Sci.* **2008**, *58*, 497–501.
33. Gumienna, M.; Czarnecka, M.; Czarnecki, Z. Zmiany Zawartości Wybranych Składników Żywności W Produktach Otrzymanych Z Nasion Roślin Strączkowych Pod Wpływem Obróbki Biotechnologicznej. *Food. Sci. Technol. Qual.* **2007**, *6*, 159–169.
34. ISO 2483:1973. *Sodium Chloride for Industrial Use—Determination of the Loss of Mass at 110 °C (Reviewed and Confirmed in 2018)*; ISO: Geneva, Switzerland, 2018.
35. AOAC. Ash of Flour (Direct Method), Method 923.03. In *Official Methods of Analsiys*, 18th ed.; AOAC International Publisher: Gaithersburg, MD, USA, 2005.
36. ISO 20483. *Cereals and Pulses—Determination of the Nitrogen Content and Calculation of the Crude Protein Content—Kjeldahl Method*; ISO: Geneva, Switzerland, 2006.
37. Miller, G.L. Use of Dinitrosalicylic Acid Reagent for Determination of Reducing Sugar. *Anal. Chem.* **1959**, *31*, 426–428. [CrossRef]
38. Lowry, O.H.; Rosebrough, N.J.; Farr, A.L.; Randall, R.J. Protein measurement with the Folin phenol reagent. *J. Biol. Chem.* **1951**, *193*, 265–275. [CrossRef]
39. Saunders, R.M.; Connor, M.A.; Booth, A.N.; Bickoff, E.M.; Kohler, G.O. Measurement of digestibility of alfalfa protein concentrates by in vivo and in vitro methods. *J. Nutr.* **1973**, *103*, 530–535. [CrossRef]
40. Remiszewski, M.; Przygoński, K.; Kulczak, M.; Jeżewska, M. The optimization of extractive system and the assessment of antioxidant qualities of seeds of some selected leguminous plants. *Food. Sci. Technol. Qual.* **2006**, *1*, 127–135.
41. Gumienna, M.; Goderska, K.; Nowak, J.; Czarnecki, Z. Changes in the antioxidative activities of bean products and intestinal microflora in the model of the gastrointestinal tract 'in vitro'. *Polish J. Environ. Stud.* **2006**, *15*, 29–32.
42. Singelton, V.L.; Rossi, J.A. Colorymetry of total phenolics with phosphomolybdic-phosphotungstic acid reagents. *Am. J. Enol. Vitic.* **1965**, *16*, 144–158.
43. Re, R.; Pellegirini, N.; Proteggente, A.; Pannala, A.; Yang, M.; Rice-Evans, C. Antioxidant activity applying an improved ABTS radical cation decolorization assay. *Free Radic. Biol. Med.* **1999**, *26*, 1231–1237. [CrossRef]
44. Hertog, M.G.L.; Hollman, P.C.H.; Venema, D.P. Optimization of a quantitative HPLC determination of potentially anticarcinogenic flavonoids in vegetables and fruits. *J. Agric. Food Chem.* **1994**, *40*, 1591–1598. [CrossRef]
45. PN-EN ISO, 8589:2010. *Sensory Analysis. General Guidance for the Design Studio of Sensory Analysis*; Polish Committee for Standardization: Warszawa, Poland, 2010.
46. ISO 13299. *Sensory Analysis—Methodology—General Guidance for Establishing a Sensory Profile*; ISO: Geneva, Switzerland, 2016.
47. ISO 13300-2. *Sensory Analysis—General Guidance for the Staff of a Sensory Evaluation Laboratory—Part 2: Recruitment and Training of Panel Leaders*; ISO: Geneva, Switzerland, 2006.
48. Amarowicz, R. Tannins: The new natural antioxidants? *Eur. J. Lipid Sci. Technol.* **2007**, *109*, 549–551. [CrossRef]
49. Karamać, M.; Kosińska, A.; Rybarczyk, A.; Amarowic, R. Extraction and chromatographic separation of tannin fractions from tannin-rich plant material. *Pol. J. Food Nutr. Sci.* **2007**, *57*, 471–474.
50. Amarowicz, R.; Troszyńska, A.; Baryłko-Pikielna, N.; Shahiid, F. Polyphenolics extracts from legume seeds: Correlation between total antioxidant activity, total phenolics content, tannins content and astringency. *J. Food Lipids* **2004**, *11*, 278–286. [CrossRef]

51. Lamuela-Raventós, R.M. Folin-Ciocalteu method for the measurement of total phenolic content and antioxidant capacity. In *Measurement of Antioxidant Activity & Capacity*; Apak, R., Capanoglu, E., Shahidi, F., Eds.; John Wiley & Sons, Ltd.: Chichester, UK, 2017; pp. 107–115.
52. Ganesan, K.; Xu, B. Polyphenol-rich dry common beans (*Phaseolus vulgaris* L.) and their health benefits. *Int. J. Mol. Sci.* **2017**, *18*, 2331. [CrossRef] [PubMed]
53. Slimestad, R.; Fossen, T.; Brede, C. Flavonoids and other phenolics in herbs commonly used in Norwegian commercial kitchens. *Food Chem.* **2020**, *309*, 125678. [CrossRef] [PubMed]
54. Nieto, G. A review on applications and uses of thymus in the food industry. *Plants* **2020**, *9*, 961. [CrossRef]
55. Teofilović, B.; Tomas, A.; Martić, N.; Stilinović, N.; Popović, M.; Čapo, I.; Grujić, N.; Ilinčić, B.; Rašković, A. Antioxidant and hepatoprotective potential of sweet basil (*Ocimum basilicum* L.) extract in acetaminophen-induced hepatotoxicity in rats. *J. Funct. Foods* **2021**, *87*, 104783. [CrossRef]
56. Gutiérrez-Grijalva, E.; Picos-Salas, M.; Leyva-López, N.; Criollo-Mendoza, M.; Vazquez-Olivo, G.; Heredia, J. Flavonoids and phenolic acids from oregano: Occurrence, biological activity and health benefits. *Plants* **2017**, *7*, 2. [CrossRef]
57. Nour, V.; Trandafir, I.; Cosmulescu, S. Bioactive compounds, antioxidant activity and nutritional quality of different culinary aromatic herbs. *Not. Bot. Horti Agrobot. Cluj-Napoca* **2017**, *45*, 179–184. [CrossRef]
58. Kozłowska, M.; Ścibisz, I.; Przybył, J.; Ziarno, M.; Żbikowska, A.; Majewska, E. Phenolic contents and antioxidant activity of extracts of selected fresh and dried herbal materials. *Pol. J. Food Nutr. Sci.* **2021**, *71*, 269–278. [CrossRef]
59. Prior, R.L.; Wu, X.; Schaich, K. Standarized methods for the determination of antioxidant capacity and phenolics in foods and dietary supplements. *J. Agric. Food Chem.* **2005**, *53*, 4290–4302. [CrossRef]
60. Drewnowski, A.; Gomez-Carneros, C. Bitter taste, phytonutrients, and the consumer: A review. *Am. J. Clin. Nutr.* **2000**, *72*, 1424–1435. [CrossRef]
61. Drewnowski, A. Taste Preferences and food intake. *Annu. Rev. Nutr.* **1997**, *17*, 237–253. [CrossRef] [PubMed]
62. Feng, Y.; Albiol Tapia, M.; Okada, K.; Castaneda Lazo, N.B.; Chapman-Novakofski, K.; Phillips, C.; Lee, S.Y. Consumer Acceptance Comparison Between Seasoned and Unseasoned Vegetables. *J. Food Sci.* **2018**, *83*, 446–453. [CrossRef] [PubMed]
63. Szymandera-Buszka, K.; Jędrusek-Golińska, A.; Waszkowiak, K.; Hęś, M. Sensory sensitivity to sour and bitter taste among people with Crohn's disease and folic acid supplementation. *J. Sens. Stud.* **2020**, *35*, 1–7. [CrossRef]
64. Cardoso-Ugarte, G.A.; Sosa-Morales, M.E. Essential Oils from Herbs and Spices as Natural Antioxidants: Diversity of Promising Food Applications in the past Decade. *Food Rev. Int.* **2021**, 1–31. [CrossRef]
65. Karan, D.; Lukic, M.; Djordjevic, V.; Parunovic, N.; Babic Milijasevic, J.; Jovanovic, J.; Nikolic, A. The influence of the basil on colour, odour and taste of frankfurters. *IOP Conf. Ser. Earth Environ. Sci.* **2019**, *333*, 012069. [CrossRef]
66. Yang, Y.; Pan, D.; Wang, Y.; He, J.; Yue, Y.; Xia, Q.; Zhou, G.; Cao, J. Effect of reconstituted broth on the taste-active metabolites and sensory quality of stewed and roasted pork-hock. *Foods* **2020**, *9*, 513. [CrossRef]
67. Hald, C.; Dawid, C.; Tressel, R.; Hofmann, T. Kaempferol 3- O-(2- O-Sinapoyl-β-sophoroside) Causes the Undesired Bitter Taste of Canola/Rapeseed Protein Isolates. *J. Agric. Food Chem.* **2019**, *67*, 372–378. [CrossRef] [PubMed]

*nutrients*

MDPI

*Review*

# Lupins and Health Outcomes: A Systematic Literature Review

**Lesley Bryant [1], Anna Rangan [1] and Sara Grafenauer [2,3,*]**

[1] Nutrition and Dietetics Group, Sydney Nursing School, Faculty of Medicine and Health, Charles Perkins Centre, The University of Sydney, Camperdown, NSW 2006, Australia; lbry9477@uni.sydney.edu.au (L.B.); anna.rangan@sydney.edu.au (A.R.)
[2] Grains & Legumes Nutrition Council, Mount Street, Sydney, NSW 2060, Australia
[3] School of Medicine and Health, University of New South Wales, Randwick, NSW 2052, Australia
* Correspondence: s.grafenauer@unsw.edu.au

**Abstract:** Lupins have a unique nutrient profile among legumes and may have beneficial health effects when included in the diet. The aim of this systematic review was to investigate the effects of lupin on a range of health outcome measures. Databases included MEDLINE, Embase and CINAHL, and focused on controlled intervention studies on healthy adults and those with chronic disease such as type 2 diabetes, cardiovascular disease and overweight. The Preferred Reporting Items for Systematic Reviews and Meta-Analyses protocol was followed. Investigated intervention diets utilised whole lupin, lupin protein or lupin fibre, and outcomes were measured by markers of chronic disease, body weight and satiety. Quality assessment of results was performed using the Cochrane revised risk of bias tool. Overall, 21 studies with 998 participants were included: 12 using whole lupin, four used lupin protein and five lupin fibre. Beneficial changes were observed in 71% of studies that measured blood pressure, 83% measuring satiety and 64% measuring serum lipids. Unintended weight loss occurred in 25% of studies. Whole lupin demonstrated more consistent beneficial effects for satiety, glycaemic control and blood pressure than lupin protein or lupin fibre. Heterogeneity, low study numbers and a small participant base indicated further studies are required to strengthen current evidence particularly regarding the protein and dietary fibre components of lupin.

**Keywords:** lupin; health outcomes; type 2 diabetes; cardiovascular disease; obesity

**Citation:** Bryant, L.; Rangan, A.; Grafenauer, S. Lupins and Health Outcomes: A Systematic Literature Review. *Nutrients* **2022**, *14*, 327. https://doi.org/10.3390/nu14020327

Academic Editors: Stuart Johnson and Rewati Bhattarai

Received: 17 December 2021
Accepted: 11 January 2022
Published: 13 January 2022



## 1. Introduction

Lupin (*Lupinus*) is a legume of the Fabaceae family that has formed part of the human diet since early civilisations. Legumes such as chickpeas, lentils, peas, beans and pulses continue to be a staple food in many parts of the world. Prized by consumers for their highly nutritious and inexpensive nature, they are becoming increasingly valued by producers for their ecological sustainability. With an increasing awareness of the benefits of regular legume consumption to human health, particularly among people living with chronic disease, lupin may be a useful addition to the food supply. Australia accounts for approximately 85% of the world's lupin production. It is grown predominantly in Western Australia, with some parts of New South Wales, Victoria and South Australia also under cultivation [1,2]. Lupin grows well in poor agricultural conditions, is pest-resistant and requires less water than many other food crops, therefore is ideally suited to Western Australia's climate and sandy soils. Moreover, it helps to improve soil fertility by the nitrogen fixing action of its rhizome, a characteristic shared with all legume plants [3]. The two most common varieties grown are the narrow-leafed lupin, also known as Australian sweet lupin or blue lupin (*Lupinus angustifolius*) and the white or albus lupin (*Lupinus albus*). Species of lupin cultivated globally include *Lupinus mutabilis* and *Lupinus luteus* or yellow lupin.

Macronutrient profiles of the various lupin species differ slightly. Compared with other legumes, Australian sweet lupin has one of the highest combinations of both protein and fibre. While protein can make up as much as 40% and fibre 30% as dry weight, with an additional 5% inulin, its carbohydrate content accounts for less than 10% [4]. As with other legumes, lupins have a high nutrient density, yet their energy density is low. They are rich in minerals such as iron, magnesium, zinc, calcium and potassium; they contain vitamin A, B vitamins and vitamin E; and the fat profile predominantly consists of mono- and polyunsaturated fats, omega 3, 6 and 9 fatty acids [5]. Despite these benefits, Australians include very little lupin or other legumes in their diet, compared with populations in other countries. According to the Grains and Legumes Nutrition Council's 2017 consumption study [5], only 28% of people in Australia eat legumes, a modest 4% increase in three years. Conversely, secondary analysis of the National Nutrition and Physical Activity Survey (NNPAS 2011-12) [6] suggests consumption was around 4 g per day, perhaps indicating use as a snack rather than as a staple food within meals. According to dietary modelling, Australians will need to eat almost five times more legumes to meet the Australian Dietary Guidelines recommendation [7]. Lupin is an excellent source of high-quality plant protein for people who follow a vegetarian or vegan diet. It is gluten-free and provides a more nutrient-dense wheat replacement than other grain and cereal alternatives currently utilised in gluten-free diets. The Australian food industry is beginning to recognise the value of lupin and a range of lupin products is now available, including whole lupin flakes, flour, crumb, meal, kibble and splits. The addition of lupin into other foods enhances their nutritional value and may be an acceptable approach to introduce lupin into the food supply, particularly as an ingredient [8]. Flour made from whole lupin can be easily incorporated into many foods and isolated protein and fibre from the lupin seed may also be of benefit, though separation of the component parts is a more complex process. Interest in lupin as a functional food is increasing among food manufacturers, however it is not clear whether there is a difference in biological effect between whole lupin, lupin protein and lupin fibre when consumed as part of the diet. Sensitivity to legume protein allergens may have an inhibitory effect on lupin consumption in some individuals. In 2017 Food Standards Australia New Zealand included lupin among allergens that must be declared on food products, however, allergic response to lupin is milder and occurs less frequently compared with exposure to peanut and soy [9]. According to Allergy and Anaphylaxis Australia, it is estimated that less than 1% of the Australian population is allergic to lupin [10].

Despite the increasing popularity of plant-based diets, it has been some years since a review of the health benefits of lupin intake was conducted. A 2015 review of the literature on lupins, among a broad range of other legumes, suggested *Lupinus angustifolius* may beneficially effect blood pressure, blood lipids, insulin sensitivity and the gut microbiome [8]. A 2016 review of the nutritional, chemical and health-promoting properties of *Lupinus albus* recognised its potential in the production of functional food [11]. A systematic literature review in 2020 on the effects of legume consumption on markers of glycaemic control in people with and without diabetes excluded studies of less than six weeks' duration, therefore post-meal and short-duration lupin interventions were not captured [12]. The aim of this systematic review was to investigate the effects of lupin on a range of health outcome measures.

## 2. Materials and Methods

This systematic literature review was conducted in accordance with the Preferred Reporting Items for Systematic Reviews and Meta-analyses (PRISMA) guidelines [13]. The study protocol was submitted to the International Prospective Register of Systematic Reviews (PROSPERO) for registration (TBC).

### 2.1. Eligibility and Exclusion Criteria

To research the question 'Is there an effect of lupin consumption on health outcomes in humans?' a framework was developed using the PICO (Population, Intervention, Comparator/Control, Outcome) format (Table S1). Studies were eligible for inclusion in this review if they met the following criteria: (a) controlled intervention trial of any duration and of parallel or cross-over design; (b) populations comprising any adults aged 18 years and above, with or without chronic disease, overweight or obesity; (c) dietary interventions in the form of whole lupin, such as dried, pickled and brined seeds, flakes, flour, crumb, meal, kibble and splits, or components of lupin such as protein and fibre; (d) assessed the effect of lupin consumption on biomarkers of chronic disease such as any related to diabetes, cardiovascular disease, hypertension, hypercholesteremia, hyperlipidaemia, cancer, inflammation, and oxidative stress, or assessed the effect of lupin consumption on anthropometric measurements and perceptions of satiety in relation to overweight and obesity and their association with chronic disease risk. The exclusion criteria applied to the study search were: (a) participants below 18 years of age; (b) assessments of single isolated proteins, peptides or alkaloids from lupin; (c) lupin intake as a supplement in capsule form; (d) non-English language studies; (e) publication dates before 1 January 2000.

### 2.2. Search Strategy

The following databases were searched: EMBASE (via Ovid) MEDLINE (via Ovid), CINHAHL (via EBSCO) from 1 January 2000 until 13 September 2021. Reference lists of eligible studies were scanned and searched manually on PubMed for additional studies.

### 2.3. Study Selection, Data Extraction, and Quality Assessment

Search results were imported into EndNote X9® referencing software (EndNote X9, Clarivate Analytics, Philadelphia, PA, USA) and duplicates were removed. Screening of studies was performed in two stages, first by title and abstract, then by full text. A data extraction form was created in Microsoft® Excel® spreadsheet (Microsoft 365 MSO Version 2109.14430.20306, Redmond, WA, USA) to include study citation, design, and duration; participant numbers and characteristics; intervention and control diet information; outcomes measured, and results obtained. The included studies were assessed for within-study risk of bias using the revised Cochrane risk-of-bias tool (RoB2) for randomised controlled trials [14]. Reviewer L.B. assessed studies to determine whether each study had low, some concerns, or high risk of bias. Areas of uncertainty were resolved in consultation with a second reviewer (S.G.). Assessment domains included risk of bias arising from the randomisation process, period and carryover effects, deviations from intended interventions, missing outcome data, measurement of the outcome, and selection of the reported result [14].

## 3. Results

### 3.1. Search Results and Study Selection

The search was conducted on 13 September 2021, returning a total of 157 records. One additional paper was identified from the reference lists of eligible studies. Following automated removal of duplicates by Endnote X9, 127 studies remained. Screening by title and abstract excluded 94 studies. Principal reasons for exclusion were publication type and study aim, interventions and measured outcomes beyond the scope of this review. A full-text review of the remaining 33 led to the exclusion of a further 12 studies. Reasons for exclusion were: duplications in other journals and/or non-English languages ($n = 4$), brief conference or workshop communications ($n = 2$), lupin intervention administration in capsule or supplement form ($n = 2$), investigation of lupin fractions such as alkaloids or single proteins ($n = 3$) and in vitro study protocol ($n = 1$). A total of 21 journal articles of controlled intervention studies met the inclusion criteria and were included in this qualitative review (Figure 1). Two studies generated three articles that reported on different sets of variables within each. These articles were treated as stand-alone studies and included in the final total.

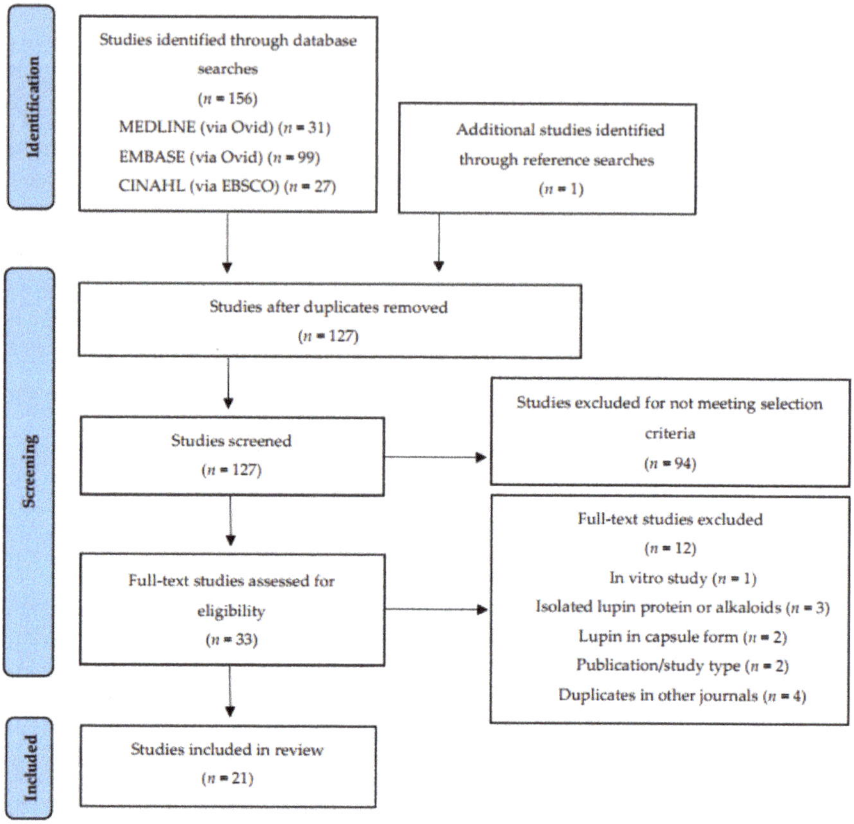


**Figure 1.** Preferred Reporting Items for Systematic Reviews and Meta-Analyses (PRISMA) flow diagram for study selection.

### *3.2. Study Characteristics*

Six of the included studies were randomised controlled trials (RCTs) of parallel design, the remaining 15 were cross-over studies, of which 11 were RCTs, while two were controlled, non-randomised, crossover studies. Research was based predominantly in Australia with smaller number of studies conducted in Germany, Italy and Ecuador. Table 1 lists study locations and the species of lupin under investigation. Five studies recruited healthy men and women [15–19], three recruited healthy men only [20–22], four studies involved men and women with type 2 diabetes [23–26], five with hypercholesterolaemia [27–31] and four involved people who were overweight or obese [32–35]. The number of completing participants ranged from $n = 5$ to $n = 175$ per study. Approximately 25% of the 998 participants across all studies were classified as healthy. Duration of study periods ranged from post-meal studies to 12 months. Categorisation of studies by form of lupin administered during treatment phases identified three distinct groups: whole lupin, lupin protein and lupin fibre. Study characteristics and outcomes of 12 whole lupin studies are summarised in Table 2, four lupin protein studies in Table 3, and five lupin fibre studies in Table 4. Measured outcomes were multiple and varied across studies, with $p < 0.05$ being declared as statistically significant.

**Table 1.** Study location, lupin species and form of lupin consumed in eligible studies investigating the health benefits of lupin consumption.

| Country | Reference | Lupin Species/Common Name (NS = Not Stated) | Whole Seed (W), Protein Isolate (PI) or Fibre Isolate (FI) |
|---|---|---|---|
| Australia | Hall et al., 2005 [18] | *L. angustifolius* | W |
| | Hall et al., 2005 [22] | *L. angustifolius* | FI |
| | Smith et al., 2006 [20] | *L. angustifolius* | FI |
| | Johnson et al., 2006 [21] | *L. angustifolius* | FI |
| | Lee et al., 2006 [16] | NS | W |
| | Lee et al., 2009 [33] | NS | W |
| | Yang et al., 2010 [32] | NS | W |
| | Hodgson et al., 2010 [34] | *L. angustifolius* | W |
| | Dove et al., 2011 [26] | *L. angustifolius* | W |
| | Keogh et al., 2011 [17] | NS | W |
| | Belski et al., 2011 [35] | *L. angustifolius* | W |
| | Skalkos et al., 2020 [24] | Australian Sweet Lupin | W |
| | Ward et al., 2020 [23] | *L. angustifolius* | W |
| Germany | Weiße et al., 2010 [27] | *L. angustifolius* | PI |
| | Bähr et al., 2013 [30] | *L. angustifolius* | PI |
| | Fechner et al., 2013 [19] | *L. angustifolius* and *L. albus* | FI |
| | Fechner et al., 2014 [29] | *L. angustifolius* | FI |
| | Bähr et al., 2015 [31] | *L. angustifolius* | PI |
| | Schopen et al., 2017 [15] | *L. albus* | W |
| Italy | Sirtori et al., 2012 [28] | *L. angustifolius* | PI |
| Ecuador | Fornasini et al., 2019 [25] | *L. mutabilis* | W |

**Table 2.** Characteristics and major outcomes of studies examining whole lupin consumption and health outcomes.

| Reference | Study Type | Subjects (*n*) and Characteristics | Intervention | Control/Comparator | Energy Balance | Main Health Markers | Main Outcomes |
|---|---|---|---|---|---|---|---|
| Hall et al., 2005 [18] | RCT single blind cross-over Post-meal study | *n* = 11 Healthy men (*n* = 9) and women (*n* = 2). Mean age 31.6 years, range 25–45 years | Breakfast including lupin bread with 10% wheat flour replaced with Australian sweet lupin kernel flour | Breakfast including standard recipe white bread | 95kJ difference in lupin breakfast (1338 kJ) and control breakfast (1243 kJ) | 120 min SG, SI 180 min satiety response. Glycaemic index (GI), insulinaemic index (II) and satiety index (WB = 100). Energy intake from *ad libitum* buffet and during remainder of day. | ↓ GI ($p = 0.022$) ↑ II ($p = 0.046$) Trend to lower SG at 30 min. Peak satiety at 10 min WB and 25 min lupin. Below baseline at 160 min for WB. Lupin did not reach zero within 180 min. No difference in SI, satiety response, satiety index, energy intake at *ad libitum* buffet, or energy intake during remainder of day. |
| Lee et al., 2006 [16] | RCT cross-over Study 1: 4 treatments 1 week apart | Study 1: *n* = 16 healthy men (*n* = 8) and women (*n* = 8). Mean age 58.6 ± 7.2 years. Mean BMI 31.3 ± 4.5 kg/m$^2$ | Lupin bread 40% total flour (24% final weight of bread) in 4 treatments: WB-WB/WB-lupin/lupin-WB/lupin-lupin | White bread breakfast and lunch | Isocaloric at breakfast, *ad libitum* lunch | Post breakfast 180 min satiety response. Total energy intake after *ad libitum* lunch. | ↑ Satiety at breakfast for satisfaction and prospective consumption ($p < 0.001$, $p < 0.001$). ↑ Satiety at 180 min for fullness ($p < 0.001$), satisfaction ($p < 0.001$) and prospective consumption ($p < 0.001$). ↓ Energy intake at lunch after lupin breakfast (−488; 95% CI: −798, −178 kJ). ↓ Intrameal energy intake when lupin consumed at lunch(−1028; 95% CI: −1338, −727 kJ). |
| | Study 2: 2 treatments 1 week apart | Study 2: *n* = 17 healthy men (*n* = 11) and women (*n* = 6). Mean age 61.0 ± 5.6 years. Mean BMI 27.2 ± 4.3 kg/m$^2$ | Lupin bread 40% wheat flour replaced with lupin kernel flour | White bread | Isocaloric | Post breakfast 180 min plasma ghrelin, SG and SI | Altered ghrelin response ($p = 0.04$) ↓ 180 min plasma ghrelin ($p = 0.009$). Altered glucose response ($p = 0.01$) ↓ glucose AUC ($p = 0.006$) ↓ insulin AUC ($p = 0.002$). |
| Lee et al., 2009 * [33] | RCT parallel study 16 weeks 2 cohorts | *n* = 74 Overweight and obese men (*n* = 26) and women (*n* = 48) Mean age 59.0 ± 7.4/ 56.8 ± 8.5 years. Mean BMI 30.6 ± 3.6/ 30.5 ± 3.4 kg/m$^2$. | Lupin bread 40% wheat flour replaced with lupin kernel flour (24% final weight bread) to replace usual carbohydrate-rich foods to ~15–20% usual energy intake. | White bread to replace normal bread intake and other carbo-hydrate-rich foods to ~15–20% usual energy intake | Isocaloric | 24-h SBP, DBP, pulse pressure and heart rate | ↓ 24-h SBP ($p = 0.03$) ↓ pulse pressure ($p < 0.001$). No difference in DBP, heart rate |
| Yang et al., 2010 * [32] | Paper refers to the Lee et al., 2009 study above | As above | As above | As above | As above | Plasma and urinary F2-isoprostanes, plasma 20-HETE, plasma and urinary nitrite and nitrate concentrates | No difference between groups |

**Table 2.** *Cont.*

| Reference | Study Type | Subjects ($n$) and Characteristics | Intervention | Control/Comparator | Energy Balance | Main Health Markers | Main Outcomes |
|---|---|---|---|---|---|---|---|
| Hodgson et al. 2010 * [34] | Paper refers to the Lee et al., 2009 study above | As above | As above | As above | As above | BW every 2 weeks<br>Body composition at 16 weeks<br>TC, HDL, TG, LDL, SG, SI, HOMA-IR, plasma leptin and adiponectin, hs-CRP | No difference between groups |
| Dove et al., 2011 [26] | RCT cross-over study of 3 test sessions, 7–14 days apart | $n = 24$<br>type 2 diabetic men ($n = 19$) and women ($n = 5$).<br>Mean age: 57 ± 6.6 years,<br>range 44–66 years.<br>Mean BMI:<br>30.9 ± 4.8 kg/m$^2$ | Beverage of 50 g glucose and 50 g lupin kernel flour | Beverage of 50% glucose (control), beverage of 50% glucose + soya protein and fibre isolates (comparator) | All beverages matched for total volume, carbohydrates and fat content lupin and soya matched for energy, protein and fibre | 240 min SG, SI and C-peptide response | ↓ 240 min SG response ($p < 0.001$),<br>↑ 240 min SI and C-peptide responses ($p < 0.001$)<br>↓ SI response compared with soya ($p = 0.013$).<br>No difference in SG and C-peptide responses between lupin and soya |
| Keogh et al., 2011 [17] | RCT cross-over study | $n = 20$<br>Healthy men ($n = 10$) and women ($n = 10$)<br>Mean age 29.4 years<br>Range: 20.1–44.8 years<br>BMI 21.8 kg/m$^2$<br>Range: 18.4–24.8 kg/m$^2$ | Lupin Bread breakfast | White bread breakfast (control), wholemeal and seeds bread breakfast (comparator) | Isocaloric breakfast, *ad libitum* standardised food and beverage tray 120 min post breakfast | 120 min satiety, PG and insulin response.<br>Food and beverage weight, energy and macronutrient content 120 min post meal | ↑ Fullness response for lupin ($p < 0.01$) and WSB ($p < 0.05$)<br>↓ glucose AUC for lupin and WSB ($p < 0.001$)<br>↓ insulin AUC for lupin and WSB ($p < 0.001$)<br>↓ weight of food and beverage for lupin and WSB ($p < 0.05$)<br>↓ energy and total fat consumed after WSB ($p < 0.05$).<br>No difference in energy and macronutrient intake post lupin meal. |
| Belski et al., 2011 [35] | RCT double blind parallel study 12 months 2 cohorts | $n = 131$<br>$n = 93$ at 12 months<br>Overweight and obese men ($n = 68$) and women ($n = 63$)<br>Mean age 46.5 ± 10.1/ 46.7 ± 9.4.<br>Mean BMI 31.3 ± 2.7/ 31.4 ± 2.8 kg/m$^2$ | Lupin kernel flour in bread, biscuits and pasta | Standard food products without lupin (matched for colour, taste, texture) | Isocaloric | 4 and 12 month BW, body composition SBP, DBP, TC, HDL, LDL, TG, SG and SI, HOMA-IR and hs-CRP. | ↓ 24-h SBP and DBP at 12 months ($p < 0.05$)<br>↓ HDL ($p < 0.05$)<br>↓ SI and HOMA-IR at 4 and 12 months ($p < 0.05$)<br>No difference in TC, LDL, TG, SG, hs-CRP BW or body composition at 4 or 12 months.<br>No difference in maintenance of body weight loss during weight maintenance period<br>(4–12 months) |
| Schopen et al., 2017 [15] | RCT single blind cross-over study 3 test visits 24 h apart | $n = 12$<br>healthy men ($n = 5$) and women ($n = 7$).<br>Mean age men<br>28 ± 3.67 years, women 26.86 ± 3.44 years.<br>Mean BMI men<br>24.72 ± 2.3 kg/m$^2$,<br>women 20.92 ± 1.63 kg/m$^2$ | Sweet lupin flour in lunch meal of pasta and meat sauce (0.94 g lupin flour per kg of participant body weight) | Pasta and meat sauce lunch (reference meal), pasta and meat sauce lunch with whey protein (0.42 g per kg of participant BW) | Standardised breakfast, standardised test lunch. Lupin and whey meals matched for protein. Reference meal ~22% less kJ and ~50% less protein per kg of participant BW. All test meals similar in carbohydrate | 180 min SG<br>180 min SI<br>Post test meal | ↓ SG AUC 0–60 min ($p < 0.001$)<br>↓ SG AUC 0–180 min ($p = 0.030$).<br>No difference in SI |

**Table 2.** *Cont.*

| Reference | Study Type | Subjects (*n*) and Characteristics | Intervention | Control/Comparator | Energy Balance | Main Health Markers | Main Outcomes |
|---|---|---|---|---|---|---|---|
| Fornasini et al., 2019 [25] | Controlled non-randomised single blind cross-over study (28 weeks) (One-group pretest-posttest design with double pretest) | $n = 5$ type 2 diabetic men ($n = 19$) and women ($n = 32$) under conventional non-insulin medication. Mean age 64.1 ± 11.1 years. Mean BMI 30.3 ± 4.5 kg/m$^2$ | Whole *Lupinus mutabilis* 10 g dehydrated lupin snack. One dose per day 7 weeks (following initial 14-week medication only period). 2 doses per day next 7 weeks | Usual diet and medication | N/A | At 14 and 28 weeks BW, SBP, DBP, SG, SI, HbA1c, TC, LDL, HDL, Uric acid, CRP | ↑ SG and SI 0–28 weeks ($p \leq 0.05$) ↑ HDL 0–28 weeks ($p \leq 0.05$) ↓ BW ($p = 0.015$) and BMI ($p = 0.009$) 0–28 weeks ↓ SBP and DBP 0–28 and 14–28 weeks ($p \leq 0.05$). No difference in HbA1c, TC, LDL, uric acid, CRP |
| Skalkos et al., 2020 [24] | Controlled non-randomised cross-over study 3 consecutive days, 1 treatment per day | $n = 20$ Post-surgical hospital patient men ($n = 12$) and women ($n = 8$) with type 2 diabetes. Mean age 74.3 ± 11.7 years. Mean BMI 30.7 ± 4.5 kg/m$^2$ | 4 × lupin biscuit containing 20% lupin flour (2 at morning tea and 2 at afternoon tea) on day 1 | 4 × wholemeal spelt biscuit day 2 4 × Arnott's Marie biscuit (standard hospital option) day 3 | Lupin and spelt biscuits isocaloric (1590 kJ/100 g) and lower than Marie biscuit (1850 kJ/100 g). Higher protein, fat and fibre in lupin and spelt, and lower carbohydrate and sugar than Marie biscuit | CGM interstitial glucose pre- and 5 timepoints post-meal, bowel function (Bristol Stool Chart), hunger and fullness rating | ↓ glucose after dinner following lupin biscuit ($p < 0.001$) No difference in 0–90 min glucose at breakfast, morning tea, lunch, and afternoon tea for all 3 treatments. More patients felt fuller between afternoon tea-dinner following lupin biscuit ($p = 0.018$). No difference in bowel function |
| Ward et al., 2020 [23] | RCT double blind cross-over study 1-week run-in period, 2 × 8-week treatment with 8-week washout period | $n = 22$ $n = 17$ completed, men ($n = 14$) and women ($n = 8$) with moderate-to-well controlled type 2 diabetes (HbA1c < 9%) Mean age 58 ± 6.6 years. Mean BMI 29.9 ± 3.5 kg/m$^2$ | Lupin-enriched foods replacing 20% of daily energy intake. Consumed every breakfast, lunch and at least 3 dinners per week. Average daily intake ~45 g lupin per day (12 g/d protein 10 g/d fibre) | Wheat-based control foods | Isocaloric | SG (at waking, 1 h post breakfast, immediately pre-lunch and 1 h post-lunch), SI, HOMA-IR, BW, BP, TC, LDL, TG, HDL, C-peptide | No difference between treatments. Borderline significant decrease in TG with lupin |

* Part of one study; Abbreviations: Area under the curve (AUC); Body mass index (BMI); Blood pressure (BP); Body weight (BW); Continuous glucose monitor (CGM); Diastolic blood pressure (DBP); Glycated haemoglobin (HbA1c); High density lipoprotein cholesterol (HDL); Homeostasis model assessment of insulin resistance (HOMA-IR); High-sensitivity C-reactive protein (hs-CRP); Low density lipoprotein cholesterol (LDL); Plasma glucose (PG); Systolic blood pressure (SBP); Serum glucose (SG); Serum insulin (SI); Triglycerides (TG); Total cholesterol (TC).

**Table 3.** Characteristics and major outcomes of studies examining lupin protein consumption and health outcomes.

| Reference | Study Type | Subjects ($n$) and Characteristics | Intervention | Control/Comparator | Energy Balance | Main Health Markers | Main Outcomes |
|---|---|---|---|---|---|---|---|
| Weiße et al., 2010 [27] | RCT double blind parallel study 10-day run-in, 6-week treatment | $n$ = 56<br>$n$ = 43 completed, moderately hypercholesterol-aemic (5.7–7.9 mmol/L) men ($n$ = 25) and women ($n$ = 31).<br>Mean age 43.9 ± 11.8 years.<br>Mean BMI 25.9 ± 4.5 kg/m$^2$ | Blue lupin protein isolate, 35 g in 2 snack bars per day | Casein protein (CP), 35 g protein in snack bars per day | Isocaloric | LDL:HDL, TC, LDL, HDL, TG, PG, mRNA SREBP-2, LDL receptor and HMG-CoA reductase | ↓ LDL:HDL for lupin compared to CP ($p < 0.05$)<br>↓ 0–6 week TC and LDL in lupin group<br>↓ 0–6 week TC, HDL, and TG for CP group (all $p < 0.05$)<br>No difference between groups.<br>↓ alanine and glycine after CP<br>↓ methionine after Lupin, and ↓ than CP ($p < 0.05$)<br>↑ SREBP-2 in Lupin group, but not CP<br>↑ LDL receptor and ↓ HMG-CoA reductase in both groups ($p < 0.05$)<br>No difference between groups for all 3 mRNA outcomes |
| Sirtori et al., 2012 [28] | RCT double blind, parallel study 4-week run-in, 4-week treatment | $n$ = 193<br>$n$ = 175 completed, moderately hypercholesterol-aemic (TC > 2200 mg/L) men ($n$ = 82) and women ($n$ = 93).<br>Mean age range 52.7 ± 12.4–55.3 ± 14.6 years.<br>Mean BMI range 24.0 ± 2.0–25.6 ± 3.2 kg/m$^2$ | Blue lupin protein isolate/cellulose fibre combination added to 2 snack bars per day | Control: casein/cellulose.<br>Comparators: lupin/cellulose; pea protein/cellulose; casein/oat fibre; casein/apple pectin; pea protein/oat fibre; pea protein/apple pectin | Isocaloric | TC, LDL, HDL, TG, SG, SI, HOMA-IR, BW, adiponectin, sICAM-1, IL-6, hs-CRP | ↓ TC lupin/cellulose ($p < 0.05$)<br>(Greatest reduction in TC ($p = 0.0098$) and LDL ($p = 0.004$) in pea/apple pectin treatment).<br>No difference lupin/cellulose for LDL, HDL, TG, SG, SI or HOMA-IR<br>(casein/cellulose, casein/apple pectin and pea/oat fibre all decreased SI and HOMA-IR ($p < 0.05$).<br>Pea/oat fibre also decreased SG ($p < 0.05$)).<br>No difference in adiponectin or inflammatory markers |
| Bähr et al., 2013 [30] | RCT double-blind cross-over study 8-week treatment, 4-week washout | $n$ = 33<br>hypercholesterol-aemic (TC ≥ 5.2 mmol/L) men ($n$ = 33) and women ($n$ = 18).<br>Mean age range 49.4 ± 13.9–49.7 ± 12.8 years.<br>Mean BMI range 27.3 ± 5.4–28.8 ± 6.5 kg/m$^2$ | Blue lupin protein isolate (LPI) protein drinks, 25 g LPI per day | Milk protein isolate (MPI) protein drinks, 25 g MPI per day | Isocaloric | TC, LDL, HDL, LDL:HDL, TG, 4 and 8week BW, SBP, DBP, resting pulse, urea, hs-CR | ↑ HDL at week 4 for LPI compared to MPI ($p = 0.036$)<br>No difference between treatments for lipids<br>↓ LDL for both treatments at 4 weeks but not at 8 weeks ($p \leq 0.008$)<br>↓ LDL:HDL for LPI ($p = 0.022$)<br>Both treatments slight<br>↑ BW and body fat from 0–8 weeks ($p \leq 0.045$) No difference between treatments<br>↓ SBP for both ($p \leq 0.014$)<br>↓ DBP and resting pulse for LPI ($p \leq 0.044$)<br>No difference between treatments.<br>No difference between treatments in hs-CRP and urea 0–4 or 0–8 weeks,<br>↑ urea 0–4 weeks for both treatments ($p \leq 0.001$) with smaller increases 0–8 weeks ($p \leq 0.022$) |
| Bähr et al., 2015 [31] | RCT double blind, cross-over 3-phase study 28 days treatment 6-week washout | $n$ = 72<br>$n$ = 68 completed, hypercholesterol-aemic (TC ≥ 5.2 mmol/L) men ($n$ = 28) and women ($n$ = 40).<br>Mean age range 50.4 ± 19.2–59.8. ± 9.3 years.<br>Mean BMI range 24.9 ± 5.0–27.6 ± 4.4 kg/m$^2$ | Blue lupin protein isolate, 25 g consumed daily in 4 food products. | Milk protein (MP) 25 g in 4 food products; MP foods plus 2.5 g/d arginine in capsule form (MPA). Placebo capsules added to LP and MP diets for blindness | Isocaloric | TC, LDL, HDL, LDL:HDL, oxidised LDL, TG, SBP, DBP hs-CRP, urea, uric acid, homocysteine | ↓ LDL after Lupin compared with MP ($p = 0.044$)<br>↓ 0–28 d TC ($p < 0.001$), LDL ($p < 0.01$) and HDL ($p < 0.001$) after lupin and MPA<br>↓ TG ($p < 0.05$) after Lupin<br>Increases in urea were smaller for Lupin ($p = 0.004$) and MP ($p = 0.001$) compared with MPA<br>↓ Uric acid ($p < 0.01$) after lupin<br>↓ homocysteine after lupin compared with MP ($p = 0.001$) and MPA ($p = 0.004$) |

Abbreviations: Body mass index (BMI); Body weight (BW); Diastolic blood pressure (DBP)); High density lipoprotein cholesterol (HDL); Homeostasis model assessment of insulin resistance (HOMA-IR); High-sensitivity C-reactive protein (hs-CRP); Interleukin-6 (IL-6); Low density lipoprotein cholesterol (LDL); Plasma glucose (PG); Systolic blood pressure (SBP); Serum glucose (SG); Serum insulin (SI); soluble intracellular cell adhesion molecule-1 (sICAM-1); Triglycerides (TG); Total cholesterol (TC).

**Table 4.** Characteristics and major outcomes of studies examining lupin fibre consumption and health outcomes.

| Reference | Study Type | Subject ($n$) and Characteristics | Intervention | Control/Comparator | Energy Balance | Main Health Markers | Main Outcomes |
|---|---|---|---|---|---|---|---|
| Hall et al., 2005 * [22] | RCT single blind cross-over study 28 days of treatment 28 days washout period | $n = 44$<br>$n = 38$ completed, healthy men.<br>Mean age 41.0 ± 1.9 years.<br>Mean BMI 26.7 ± 0.5 kg/m$^2$ | Australian sweet lupin kernel fibre in foods within prescribed diet.<br>55 g dietary fibre/day for diets >9 MJ/day, 35 g dietary fibre/day for diets ≤9 MJ/day | Prescribed control diet without added lupin fibre.<br>25 g dietary fibre/day for diets >9 MJ/day, 18 g dietary fibre/day for diets ≤9 MJ/day | Isocaloric | TC, HDL, TG, PG and insulin, HOMA-IR, satiety perception, BW | ↓ TC, LDL, TC:HDL and LDL:HDL for both treatments ($p < 0.05$)<br>↓ TC ($p = 0.001$), LDL ($p = 0.001$) TC:LDL ($p = 0.006$ ) and LDL:HDL ($p = 0.003$) for lupin relative to control.<br>No difference in HDL and TG.<br>No difference in PG for lupin<br>(↓ PG in control ($p = 0.001$))<br>No difference in PG, insulin, HOMA-IR or satiety perception between treatments.<br>No difference in BW for either treatment |
| Smith et al., 2006 * [20] | Paper refers to the Hall 2005 study above | $n = 18$<br>(randomly selected from above study) | As above | As above | As above | Measures of (i) total cells, (ii) total bacteria, (iii) *E. rectale-C. coccoides*, (iv) *Bacteriodes-Prevotella*, (v) *Enterobacteriaceae*, (vi) *C. histolyticum/C. lituseburense* group, (vii) *Lactobacillus-Enterococci*, (viii) *Bifidobacterium*, (ix) *C. ramosum*, *C. spiroforme* and *C. cocleatum* group | ↑ *Bifidobacteria* ($p = 0.001$)<br>↓ *C. ramosum*, *C. spiroforme* and *C. cocleatum* group ($p = 0.039$) in lupin diet.<br>No difference between treatments in total cells, total bacteria or populations of other species.<br>Strong trend ($p = 0.53$) towards decreased *Bacteroides-Prevotella* in lupin diet |
| Johnson et al., 2006 * [21] | Paper refers to the Hall 2005 study above | $n = 38$<br>healthy men.<br>Mean age 41.0 ± 1.9 years.<br>Mean BMI 26.7 ± 0.5 kg/m$^2$ | As above | As above | As above | Frequency and ease of bowel motion, flatulence level, Bristol Stool Form, frequency (events), output, transit time, pH, faecal moisture content SCFA (total, acetate, propionate, isobutyrate, butyrate, isovalerate, valerate) | ↑ Frequency ($p = 0.047$),<br>↑ faecal output ($p = 0.020$),<br>↓ transit time ($p = 0.012$),<br>↑ perception of flatulence level ($p < 0.001$),<br>↓ faecal pH ($p < 0.001$),<br>↑ faecal moisture content ($p = 0.027$),<br>↑ total SCFA concentration ($p = 0.001$) and<br>↑ daily output ($p < 0.001$),<br>↑ acetate concentration ($p < 0.001$) and<br>↑ daily output ($p < 0.001$)<br>↑ butyrate concentration ($p = 0.006$) and output ($p = 0.002$)<br>↑ valerate output ($p = 0.030$) with no difference in concentration.<br>No difference in proprionate, isobutyrate or isovalerate. |

**Table 4.** *Cont.*

| Reference | Study Type | Subject (*n*) and Characteristics | Intervention | Control/Comparator | Energy Balance | Main Health Markers | Main Outcomes |
|---|---|---|---|---|---|---|---|
| Fechner et al., 2013 [19] | RCT double blind cross-over study 4 periods of 2 weeks each: run-in, 2 treatments and washout | $n = 76$ healthy men ($n = 21$) and women ($n = 55$). Mean age 24.4 ± 3.2 years. Mean BMI 21.7 ± 2.4 $kg/m^2$ | Blue lupin kernel fibre and white lupin kernel fibre. Total dietary fibre per treatment 25 g/d in beverages | Citrus fibre as active comparator for 2 lupin and 1 soya fibre treatments | Isocaloric | TC, HDL, LDL, TG, faecal pH, transit time, Bristol Stool Form, faecal SCFAs and bile acids | No change in serum lipids for all treatments, ↓ faecal pH for blue lupin ($p < 0.01$), no difference relative to citrus. ↓ Transit time, ↑ Bristol Stool Form score for blue lupin ($p \leq 0.05$) ↑ Total SCFA, acetate, propionate and *n*-butyrate excretion for blue lupin ($p \leq 0.05$). ↑ Primary bile acid excretion ($p = 0.02$) for blue lupin, ↓ total bile acid excretion for blue lupin relative to citrus. ↓ Total bile acid excretion for white lupin from run-in. ↓ Secondary bile acid excretion for blue and white lupin from run-in ($p \leq 0.05$). |
| Fechner et al., 2014 [29] | RCT double blind cross-over study 3 intervention periods of 4 weeks each, run-in and 2 washout periods of 2 weeks each | $n = 52$ moderately hypercholesterol-aemic (TC >5.2 mmol/L) men ($n = 20$) and women ($n = 32$). Mean age: 46.9 ± 3.2 years. Mean BMI: 26.5 ± 5.9 $kg/m^2$ | Blue lupin kernel fibre 25 g/d | Citrus fibre 25 g/d as active comparator; control diet (CD) with no added fibre | Isocaloric | General excretion markers, faecal concentration or excretion of neutral sterols, bile acids and SCFAs. BW, body composition, BP, TC, HDL, LDL, TG LDL:HDL hs-CRP, satiety score | ↓ Faecal pH from baseline ($p \leq 0.01$) and against CD ($p \leq 0.001$), ↓ transit time against CD ($p \leq 0.05$), no difference in neutral sterols. ↑ Primary bile acids from baseline ($p \leq 0.05$), no difference in total or secondary bile acids. ↑ Formation of total SCFA from baseline ($p \leq 0.001$) and against CD ($p \leq 0.01$), ↑ acetate from baseline and against CD ($p \leq 0.001$), ↑ propionate from baseline ($p \leq 0.001$) and against control ($p \leq 0.05$), ↑ butyrate from baseline ($p \leq 0.01$) and against control ($p \leq 0.05$). ↓ BW, BMI, and WC from baseline ($p \leq 0.001$) and against control ($p \leq 0.01$). ↓ TC (9%), LDL (12%) and TG (10%) for lupin compared with citrus ($p \leq 0.02$), ↓ hs-CRP ($p = 0.02$), SBP ($p = 0.01$) for lupin compared to baseline. ↑ Perception of satiety ($p \leq 0.001$) |

* Part of one study; Abbreviations: Body mass index (BMI); Blood pressure (BP)); Body weight (BW); High density lipoprotein cholesterol (HDL); Homeostasis model assessment of insulin resistance (HOMA-IR); High-sensitivity C-reactive protein (hs-CRP); Low density lipoprotein cholesterol (LDL); Plasma glucose (PG); Short-chain fatty acid (SCFA); Triglycerides (TG); Total cholesterol (TC).

### 3.3. *Risk of Bias*

Each study was assessed according to the criteria outlined in the revised Cochrane RoB2 tool for RCTs. All parallel studies had a low risk of bias. Most cross-over studies had a low risk of bias (Figure 2); exceptions were one study with some concern for risk of bias in Domain1: Randomisation process, and in Domain 5: Selection of the reported result [25]. One other cross-over study had some concern in Domain 1 only [24].

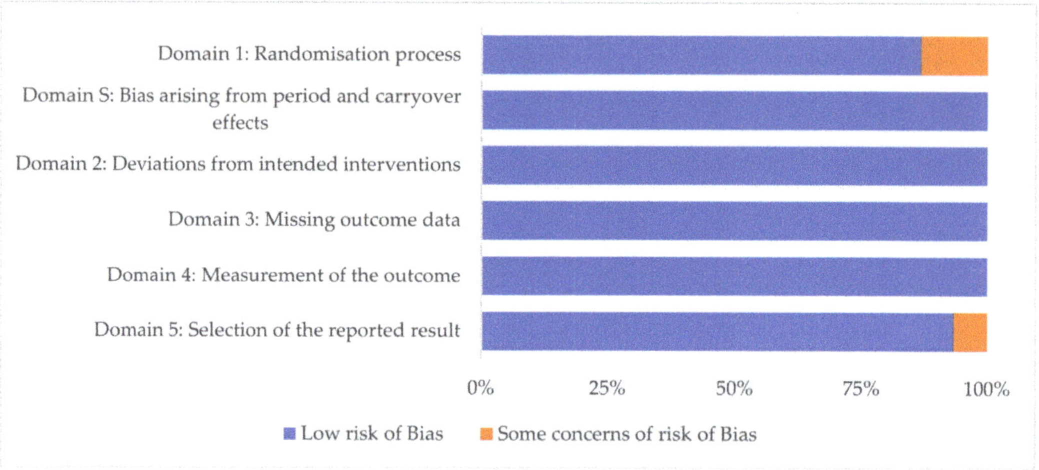


**Figure 2.** Within-study risk of bias assessment using the revised Cochrane risk-of-bias tool (RoB2) in 15 randomised ($n$ = 11) and non-randomised ($n$ = 2) controlled cross-over trials examining health outcomes of lupin consumption.

### 3.4. *Range of Investigated Health Measurements and Their Outcomes*

The five most investigated variables or groups of variables across all studies and the direction of lupin consumption effect are shown in Figure 3. These were:

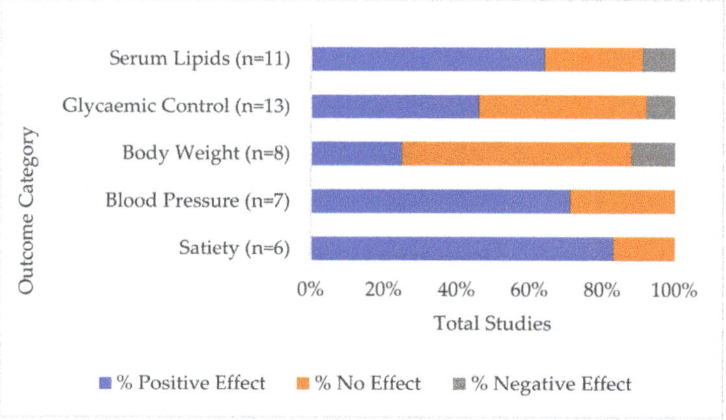


**Figure 3.** Percentage of total studies that reported differences between baseline and/or comparators ($p < 0.05$) by the five most investigated groups of health markers: serum lipids, glycaemic control, body weight, blood pressure and satiety, that had positive (desirable), negative (detrimental) or no effect on health outcomes.

Serum lipids, which included any one or more of total cholesterol, LDL and HDL cholesterol, LDL:HDL ratio and triglycerides. Eleven studies measured serum lipids refs. [19,22,23,25,27–31,34,35], of which 64% had one or more positive outcomes (i.e., showed statistically significant within-study difference(s) from baseline and/or control in a direction considered optimal for good health, depending on the health marker tested). Three studies did not report differences in serum lipids [19,23,34], while one study reported reduced HDL cholesterol with other lipids unchanged [35].

I. Glycaemic control included measurements of post-prandial and long-term glucose and insulin levels, HOMA-IR and HbA1c. These were measured in 13 studies refs. [15–18,22–28,34,35] with positive effects indicated in 46% of these studies refs. [15–17,24,26,35].

II. Body weight measurements were monitored in 8 studies [22,23,25,28–30,34,35], of which 25% recorded unintended weight loss [25,29], while 12.5% (one study) recorded increased body weight [30].

III. Blood pressure was measured in 7 studies [23,25,29–31,33,35], two of which did not find significant differences [23,31] while the remaining five (71%) recorded reductions.

IV. Perceptions of satiety were monitored in 6 studies [16–18,22,24,29]. A significantly higher satiety rating score was given for lupin in 83% of studies, while one study saw no difference to control [22].

Variables measured in five or fewer studies included those relating to inflammation, such as hs-CRP [28–31,35], IL-6 and sICAM-1 [28], oxidative stress [32], and those relating to food intake [16–18,29], appetite, digestion and metabolism, such as concentrations of serum ghrelin [16], leptin [34], adiponectin [28,34], bowel function [19,21,24,29], faecal SCFA concentrations [19,21,29], faecal bile acid concentrations [19,29] and faecal microbiome varieties and populations [20]. No studies found significant differences in inflammation and oxidative stress markers, nor in the gastric hormones and adiponectin. However, bowel function changes were seen in all but one of the studies [24]. All studies that measured faecal SCFA concentrations and the faecal microbiome composition study found significant changes.

### 3.5. *Range of Health Outcomes by Lupin Treatment Category*

#### 3.5.1. Whole Lupin

Seventy-five percent ($n = 9$) of all whole lupin studies ($n = 12$) reported a significant difference in at least one of the health markers under investigation, compared to baseline or to the control group. While most of the directions of change were desirable or positive outcomes for that variable, one study of overweight and obese participants reported decreased levels of HDL cholesterol, thereby reducing its potential heart-protective benefit, while other lipid markers did not change [35]. One study [25] reported increased serum glucose at 14 and 28 weeks and increased insulin levels at 28 weeks following a daily lupin snack in the first phase and two lupin snacks per day in the subsequent phase. Key health outcomes for whole lupin are shown in Figure 4a. All whole lupin studies ($n = 4$) that measured perceptions of satiety reported desirable changes [16–18,24], as well as 75% ($n = 3$) of studies that monitored blood pressure ($n = 4$) [25,33,35] and 67% ($n = 6$) of studies that measured glycaemic control ($n = 10$) [15–17,24,26,35]. Serum lipids moved to healthier levels in 25% ($n = 1$) of studies ($n = 4$) [25], as well as reporting decreased body weight measurements [25].

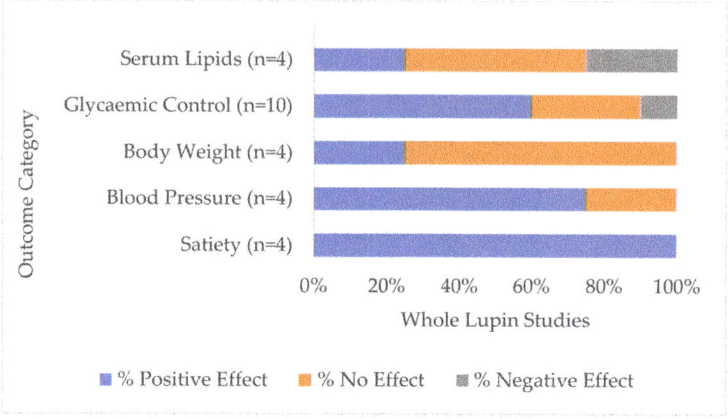


(**a**)

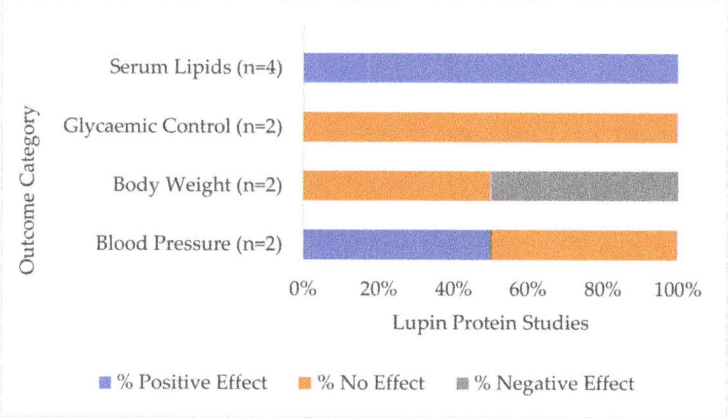


(**b**)

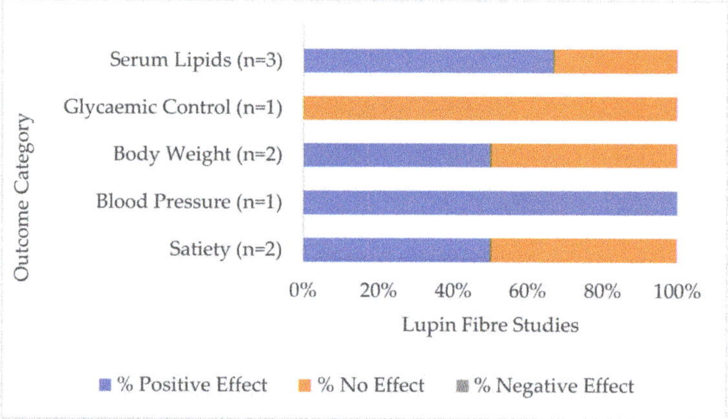


(**c**)

**Figure 4.** Percentage of total studies categorised by (**a**) whole lupin, (**b**) lupin protein and (**c**) lupin fibre treatment that reported positive (potentially beneficial), negative (potentially detrimental) and no significant differences between baseline and/or comparators ($p < 0.05$) in the five most investigated groups of health markers: serum lipids, glycaemic control, body weight, blood pressure and satiety.

3.5.2. Lupin Protein

All lupin protein studies ($n = 4$) reported significant differences in at least one of the targeted health markers. All studies that measured serum lipids [27,28,30,31] and one study [30] of a total of two that measured blood pressure [30,31] reported significantly reduced levels (Figure 4b).

3.5.3. Lupin Fibre

All lupin fibre studies ($n = 5$) reported significant differences in at least one of the measured health markers. Two out of the three lupin fibre studies that measured serum lipids reported significant differences and positive health outcomes [22,29], as well as the one study that measured blood pressure [29] (Figure 4c). All three lupin fibre studies that monitored bowel function reported positive changes [19,21,29].

## 4. Discussion

This systematic review of the evidence for health outcomes from lupin consumption observed a range of results across many biological and anthropometric health markers that variously resulted in no effects, positive effects and negative effects in terms of optimum health. In the 21 studies that met the selection criteria, the strongest evidence related to a lowering effect on total cholesterol, LDL cholesterol and LDL cholesterol:HDL cholesterol ratio; reduction in systolic blood pressure, increased satiety and improvement in post-prandial and glycaemic control. This supports the notion that lupin is equally and possibly more effective among all legumes in protecting long-term health. After categorising studies by form of lupin utilised in each intervention, this review noted potential relationships between lupins in their whole form and increased satiety perception, decreased blood pressure and improved glycaemic control, whereas lupin protein and lupin fibre demonstrated strongest positive results for blood pressure and serum lipids (though from a small study base).

These results correspond with the Kouris-Blazos et al. [8] review which concluded that sweet lupins may favourably effect blood pressure, blood lipids, insulin sensitivity and the gut microbiome. The Prusinski review [11] of white lupin concurred, stating that people who experience health conditions such as diabetes, hypertension, obesity, cardiovascular disease, hyperlipidaemia and colorectal cancer may benefit from the incorporation of this legume in the diet. However, research for the review centred on physiological properties of white lupin rather than on evidence for actual health outcomes. A 2017 review and meta-analyses investigating relationships between mortality and the intake of various food groups [36] found an inverse association between all-cause mortality and increased consumption of legumes, with no further dose response after 150 g/per day.

### *4.1. Whole Lupin*

Nutrients in foods are metabolised in the human body according to the food matrix [37]. Categorisation of studies in this review by type of lupin administration, i.e., whole lupin, its protein and its fibre component, revealed greater health benefits were observed for the consumption of the whole food. Improved health outcomes were consistent for blood pressure [25,33,35], satiety [16–18,24] and glycaemic control markers [15–17,24,26,35] in whole lupin treatment studies, indicating benefits for reducing risk and managing symptoms of hypertension, cardiovascular disease, diabetes and obesity. Although it was noted that evidence for an increased satiety effect was present, evidence was less convincing for weight loss. None of the three whole lupin studies that measured body measurements detected a significant reduction in body weight [23,34,35]. Several reasons may account for this observation. Participants followed *ad libitum* diets that were not intended for weight loss, other lifestyle factors impacting weight, such as physical activity, were not monitored, and treatment duration may have been too short to demonstrate significant change. Hodgson et al. [34] proposed that if the observed trend in weight loss after four months was extended, a significant reduction of 2 kg could be expected within two years.

The study further proposed that while *ad libitum* diets that are high in protein and dietary fibre may result in loss of body weight, the amount of protein and fibre in whole lupin may be a factor. In addition, the mostly insoluble fibre present in lupin may not be as effective as isolated soluble fibre used in many dietary fibre weight loss studies. Nevertheless, the broad health benefits proposed by whole lupin consumption suggest a synergistic interplay of macro- and micronutrient components within the whole food matrix and their influence on multiple biological functions [37], leading to improved long-term health outcomes.

### *4.2. Lupin Protein and Fibre Components*

Protein and dietary fibre components of lupin individually demonstrated consistent evidence for lipid lowering effects [22,27–31]. These benefits were observed in hypercholesterolaemic participants, as well as in one of two studies that recruited healthy participants, yet whose average baseline total cholesterol was above 5 mmol/L [22]. Given that lipid levels did not change significantly in the lupin component study based on healthy men and women with average baseline total cholesterol below 5 mmol/L [19], this suggests lupin protein and fibre had moderating effects above this level. Evidence for blood pressure reduction from lupin component interventions with that of whole lupin does not conclusively favour its protein or fibre alone due to the paucity of studies. Serum lipids were measured by a similar number of studies across all three categories of lupin treatment. While results were variable, mostly positive effects were observed in diets that contained isolated lupin protein and lupin fibre. Substantially higher quantities of protein and fibre were provided in component trial protocols, compared with the amounts obtainable from the whole food in whole lupin studies. This may have contributed to the more consistently positive lipid outcomes in these lupin component studies. While the evidence for any health benefits from isolated lupin component consumption cannot be confirmed from so few studies, the addition of lupin protein to foods and beverages for maximising protein intake may be a useful alternative to soy and whey protein, particularly for consumers avoiding phytoestrogens and animal proteins. Similarly, lupin fibre is a gluten-free alternative for individuals with coeliac disease.

### *4.3. Dose Response*

Indication of a dose response relationship between lupin and health outcomes was not identified due to multiple forms of lupin delivery and study methods among studies. However, one study [25] designed a dose response protocol comprising a doubled intake of whole *Lupinus mutabilis* during one intervention phase from 10 g to 20 g per day. While there was no change in glycaemic response markers between the two doses, blood pressure reduction was greatest after the increased dose phase.

### *4.4. Healthy vs. Unhealthy Participants Health Outcomes*

Substantial evidence for health marker differences between healthy participants and those with type 2 diabetes, hypercholesterolaemia or who were overweight or obese, was unable to be determined because of the heterogeneity among studies. Having undertaken a sub-group analysis, however, one lupin fibre study [22] on generally healthy subjects found no significant effect on serum lipids among normocholesterolaemic participants (baseline total cholesterol < 5.5 mmol/L), while LDL cholesterol was significantly lowered among participants identified by study authors as 'mild to clinically hypercholesterolaemic' (baseline total cholesterol > 5.5 mmol/L). The investigation of whole *Lupinus mutabilis* consumption in type 2 diabetic subjects under conventional non-insulin medication found significantly reduced glycosylated haemoglobin (HbA1c) outcomes among a sub-group with less severe disease (HbA1c maintained at $\leq$ 8%), while no HbA1c effect was found in the remaining group that maintained HbA1C $\geq$ 8% and <10% [25]. Though changes in biomarkers may not be demonstrated consistently in healthy subjects, it can be supposed that lupin consumption offers protective benefits in hypercholesterolaemia and in well-controlled hyperglycaemia, if not in disease of greater severity.

### 4.5. Progression of Lupin and Health Outcomes Knowledge

Reflection on 16 years of lupin and health research revealed a progression from a focus on principal biomarkers for chronic disease, to a broader scope that encompassed other related health markers and possible biological mechanisms of effect. Analysis of faecal SCFA composition after lupin fibre consumption proposed that increased concentration and output of acetate and butyrate may have a protective effect on colorectal cancer risk [21]. A study on lupin fibre that observed a reduction in serum lipids in hypercholesterolaemic individuals proposed that increased bile acid excretion was not the result of bile acids binding to fibre, but a lower environmental pH from the fermentation of lupin fibre in the gut and the subsequent release of SCFAs [29]. Since the single study that focussed on lupin fibre and faecal gut bacteria was in 2006 [20], current interest and greater understanding of the gut microbiome warrants further investigation.

### 4.6. Strengths and Limitations

This systematic literature review was undertaken with the acknowledgement of several strengths and limitations. A major strength was the inclusion of high quality RCTs and non-randomised controlled studies that disclosed valid contextual reasons for non-randomisation. Implementation of the revised Cochrane risk-of-bias tool facilitated recognition and acknowledgement of any limitations within studies. Limitations within this review relate to potential publication bias as only research published in English language journals was targeted. Relatively few studies met the selection criteria and the participant base was limited. Furthermore, the objectives, methods and analyses of the various studies lacked homogeneity, thus precluding a meta-analysis to be performed.

### 4.7. Future Directions

The subject of lupin consumption and health outcomes is a relatively new area of investigation, therefore more research is required to expand the evidence base. This should comprise multiple studies with similar aims, designs and protocols based on adequately sized population groups. Studies should identify the species of lupin and test all lupin forms in quantities that could feasibly be included in a normal diet, preferably in a dose response manner. This would allow for a more accurate assessment of the evidence overall for health benefits and optimum intake. In terms of health outcome measures, those for blood lipids, blood pressure and glycaemic control would be the most useful in identifying the unique nutritional and physiological properties of lupin. Furthermore, studies that involve concurrent investigations on healthy populations and those with different degrees of disease severity will inform whether lupin consumption may be more useful as a risk-reduction strategy or in chronic disease minimisation.

## 5. Conclusions

This is the first systematic review to our knowledge to investigate the range of health outcomes and lupin consumption according to its mode of delivery, either as a whole food or the protein or fibre component. This review found divergent results in the effects of lupin consumption on many health marker outcomes, though greatest indications of benefit were apparent in improved satiety and reductions in blood pressure, and to a lesser extent in reductions in serum lipids and improved glycaemic control. More often, evidence was based on the whole lupin providing a broader range of health benefits than was observed in the smaller number of component studies. While the evidence for lupin's health benefits is promising, more substantial research would be required before health claims could be made. Nevertheless, its unique nutritional and physiological properties, particularly as a whole food, make it an ideal legume to include in a healthy diet.

**Supplementary Materials:** The following are available online at https://www.mdpi.com/article/10.3390/nu14020327/s1, Table S1: PICO (Population, Intervention, Comparator/Control, Outcome) framework to define the search strategy for the question: 'Is there an effect of human lupin consumption on health outcomes?', File S2: Search terms (MEDLINE).

**Author Contributions:** Conceptualization, L.B., S.G. and A.R.; methodology, L.B. and S.G.; formal analysis, L.B.; writing—original draft preparation, L.B.; writing—review and editing, L.B., S.G. and A.R.; supervision, S.G. and A.R. All authors have read and agreed to the published version of the manuscript.

**Funding:** This research received no external funding however was supported by the Grains & Legumes Nutrition Council, a not-for-profit charity organisation.

**Conflicts of Interest:** S.G. was employed by the Grains & Legumes Nutrition Council, a not-for-profit charity organization, during the period when the search was conducted. L.B. and A.R. declare no conflict of interest.

## References

1. Grains & Legumes Nutrition Council (G.L.N.C). Lupins. Available online: https://www.glnc.org.au/legumes/types-of-legumes/lupins/ (accessed on 23 August 2021).
2. NSW Government Industry & Investment. Lupin Growth & Development. Available online: https://www.dpi.nsw.gov.au/__data/assets/pdf_file/0006/516183/Procrop-lupin-growth-and-development.pdf (accessed on 23 August 2021).
3. Foyer, C.H.; Lam, H.-M.; Nguyen, H.T.; Siddique, K.H.M.; Varshney, R.K.; Colmer, T.D.; Cowling, W.; Bramley, H.; Mori, T.A.; Hodgson, J.M.; et al. Neglecting legumes has compromised human health and sustainable food production. *Nat. Plants* **2016**, *2*, 16112. [CrossRef] [PubMed]
4. Food Standards Australia New Zealand (FSANZ). Australian Sweet Lupin. Available online: https://www.foodstandards.gov.au/science/monitoringnutrients/afcd/Documents/Industry%20Report%20-%20Australian%20Sweet%20Lupin.pdf (accessed on 23 August 2021).
5. Grains & Legumes Nutrition Council (G.L.N.C). Consumption of Attitudes Study Results: 2017. Available online: https://www.glnc.org.au/wp-content/uploads/2018/09/GLNC-Consumption-Study-Summary-5.pdf (accessed on 23 August 2021).
6. Figueira, N.; Curtain, F.; Beck, E.; Grafenauer, S. Consumer understanding and culinary use of legumes in Australia. *Nutrients* **2019**, *11*, 1575. [CrossRef] [PubMed]
7. National Health and Medical Research Council (NHMRC). *A Modelling System to Inform the Revision of the Australian Guide to Healthy Eating*; Department of Health and Ageing: Canberra, Australia, 2011; Volume 66.
8. Kouris-Blazos, A.; Belski, R. Health benefits of legumes and pulses with a focus on Australian sweet lupins. *Asia Pac. J. Clin. Nutr.* **2016**, *25*, 1–17. [PubMed]
9. Food Standards Australia New Zealand (FSANZ). Allergen Labelling. Available online: https://www.foodstandards.gov.au/consumer/labelling/Pages/Allergen-labelling.aspx (accessed on 12 November 2021).
10. Allergy and Anaphylaxis Australia. Lupin. Available online: https://allergyfacts.org.au/allergy-anaphylaxis/food-allergens/lupin (accessed on 25 November 2021).
11. Janusz, P. White lupin (*Lupinus albus* L.)—Nutritional and health values in human nutrition—A review. *Czech J. Food Sci.* **2017**, *35*, 95–105. [CrossRef]
12. Bielefeld, D.; Grafenauer, S.; Rangan, A. The effects of legume consumption on markers of glycaemic control in individuals with and without diabetes mellitus: A systematic literature review of randomised controlled trials. *Nutrients* **2020**, *12*, 2123. [CrossRef]
13. Moher, D.; Liberati, A.; Tetzlaff, J.; Altman, D.G. Preferred reporting items for systematic reviews and meta-analyses: The PRISMA statement. *Ann. Intern. Med.* **2009**, *151*, 264–269. [CrossRef]
14. Sterne, J.A.; Savović, J.; Page, M.J.; Elbers, R.G.; Blencowe, N.S.; Boutron, I.; Cates, C.J.; Heng, H.-Y.; Corbett, M.S.; Eldridge, S.M. RoB 2: A revised tool for assessing risk of bias in randomised trials. *BMJ* **2019**, *366*, 14898. [CrossRef]
15. Schopen, K.; Ewald, A.C.; Johannes, B.W.; Bloch, W.; Rittweger, J.; Frings-Meuthen, P. Short-term effects of lupin vs. Whey supplementation on glucose and insulin responses to a standardized meal in a randomized cross-over trial. *Front. Physiol.* **2017**, *8*, 198. [CrossRef]
16. Lee, Y.P.; Mori, T.A.; Sipsas, S.; Barden, A.; Puddey, I.B.; Burke, V.; Hall, R.S.; Hodgson, J.M. Lupin-enriched bread increases satiety and reduces energy intake acutely. *Am. J. Clin. Nutr.* **2006**, *84*, 975–980. [CrossRef]
17. Keogh, J.; Atkinson, F.; Eisenhauer, B.; Inamdar, A.; Brand-Miller, J. Food intake, postprandial glucose, insulin and subjective satiety responses to three different bread-based test meals. *Appetite* **2011**, *57*, 707–710. [CrossRef]
18. Hall, R.S.; Thomas, S.J.; Johnson, S.K. Australian sweet lupin flour addition reduces the glycaemic index of a white bread breakfast without affecting palatability in healthy human volunteers. *Asia Pac. J. Clin. Nutr.* **2005**, *14*, 91–97.
19. Fechner, A.; Fenske, K.; Jahreis, G. Effects of legume kernel fibres and citrus fibre on putative risk factors for colorectal cancer: A randomised, double-blind, crossover human intervention trial. *Nutr. J.* **2013**, *12*, 101. [CrossRef]

20. Smith, S.C.; Choy, R.; Johnson, S.K.; Hall, R.S.; Wildeboer-Veloo, A.C.; Welling, G.W. Lupin kernel fiber consumption modifies fecal microbiota in healthy men as determined by rRNA gene fluorescent in situ hybridization. *Eur. J. Nutr.* **2006**, *45*, 335–341. [CrossRef]
21. Johnson, S.K.; Chua, V.; Hall, R.S.; Baxter, A.L. Lupin kernel fibre foods improve bowel function and beneficially modify some putative faecal risk factors for colon cancer in men. *Br. J. Nutr.* **2006**, *95*, 372–378. [CrossRef]
22. Hall, R.S.; Johnson, S.K.; Baxter, A.L.; Ball, M.J. Lupin kernel fibre-enriched foods beneficially modify serum lipids in men. *Eur. J. Clin. Nutr.* **2005**, *59*, 325–333. [CrossRef]
23. Ward, N.C.; Mori, T.A.; Beilin, L.J.; Johnson, S.; Williams, C.; Gan, S.K.; Puddey, I.B.; Woodman, R.; Phillips, M.; Connolly, E.; et al. The effect of regular consumption of lupin-containing foods on glycaemic control and blood pressure in people with type 2 diabetes mellitus. *Food Funct.* **2020**, *11*, 741–747. [CrossRef]
24. Skalkos, S.; Moschonis, G.; Thomas, C.J.; McMillan, J.; Kouris-Blazos, A. Effect of lupin-enriched biscuits as substitute mid-meal snacks on post-prandial interstitial glucose excursions in post-surgical hospital patients with type 2 diabetes. *Nutrients* **2020**, *12*, 1239. [CrossRef]
25. Fornasini Salvador, M.V.; Abril-Ulloa, S.V.; Beltran Carreno, J.P.; Villacres, E.; Cuadrado-Merino, L.; Robalino, F.; Sanchez, R.; Ricaurte Ortiz, P.S.; Munoz, E.B.; Benitez Loza, N.B.; et al. Efficacy of a Lupinus mutabilis Sweet snack as complement to conventional type 2 diabetes mellitus treatment. *Nutr. Hosp.* **2019**, *36*, 905–911. [CrossRef]
26. Dove, E.R.; Mori, T.A.; Chew, G.T.; Barden, A.E.; Woodman, R.J.; Puddey, I.B.; Sipsas, S.; Hodgson, J.M. Lupin and soya reduce glycaemia acutely in type 2 diabetes. *Br. J. Nutr.* **2011**, *106*, 1045–1051. [CrossRef]
27. Weisse, K.; Brandsch, C.; Zernsdorf, B.; Nkengfack Nembongwe, G.S.; Hofmann, K.; Eder, K.; Stangl, G.I. Lupin protein compared to casein lowers the LDL cholesterol:HDL cholesterol-ratio of hypercholesterolemic adults. *Eur. J. Nutr.* **2010**, *49*, 65–71. [CrossRef]
28. Sirtori, C.R.; Triolo, M.; Bosisio, R.; Bondioli, A.; Calabresi, L.; De Vergori, V.; Gomaraschi, M.; Mombelli, G.; Pazzucconi, F.; Zacherl, C.; et al. Hypocholesterolaemic effects of lupin protein and pea protein/fibre combinations in moderately hypercholesterolaemic individuals. *Br. J. Nutr.* **2012**, *107*, 1176–1183. [CrossRef]
29. Fechner, A.; Kiehntopf, M.; Jahreis, G. The formation of short-chain fatty acids is positively associated with the blood lipid-lowering effect of lupin kernel fiber in moderately hypercholesterolemic adults. *J. Nutr.* **2014**, *144*, 599–607. [CrossRef]
30. Bahr, M.; Fechner, A.; Kramer, J.; Kiehntopf, M.; Jahreis, G. Lupin protein positively affects plasma LDL cholesterol and LDL:HDL cholesterol ratio in hypercholesterolemic adults after four weeks of supplementation: A randomized, controlled crossover study. *Nutr. J.* **2013**, *12*, 107. [CrossRef]
31. Bahr, M.; Fechner, A.; Kiehntopf, M.; Jahreis, G. Consuming a mixed diet enriched with lupin protein beneficially affects plasma lipids in hypercholesterolemic subjects: A randomized controlled trial. *Clin. Nutr.* **2015**, *34*, 7–14. [CrossRef] [PubMed]
32. Yang, X.; Croft, K.D.; Lee, Y.P.; Mori, T.A.; Puddey, I.B.; Sipsas, S.; Barden, A.; Swinny, E.; Hodgson, J.M. The effects of a lupin-enriched diet on oxidative stress and factors influencing vascular function in overweight subjects. *Antioxid. Redox Signal.* **2010**, *13*, 1517–1524. [CrossRef]
33. Lee, Y.P.; Mori, T.A.; Puddey, I.B.; Sipsas, S.; Ackland, T.R.; Beilin, L.J.; Hodgson, J.M. Effects of lupin kernel flour-enriched bread on blood pressure: A controlled intervention study. *Am. J. Clin. Nutr.* **2009**, *89*, 766–772. [CrossRef]
34. Hodgson, J.M.; Lee, Y.P.; Puddey, I.B.; Sipsas, S.; Ackland, T.R.; Beilin, L.J.; Belski, R.; Mori, T.A. Effects of increasing dietary protein and fibre intake with lupin on body weight and composition and blood lipids in overweight men and women. *Int. J. Obes.* **2010**, *34*, 1086–1094. [CrossRef]
35. Belski, R.; Mori, T.A.; Puddey, I.B.; Sipsas, S.; Woodman, R.J.; Ackland, T.R.; Beilin, L.J.; Dove, E.R.; Carlyon, N.B.; Jayaseena, V.; et al. Effects of lupin-enriched foods on body composition and cardiovascular disease risk factors: A 12-month randomized controlled weight loss trial. *Int. J. Obes.* **2011**, *35*, 810–819. [CrossRef] [PubMed]
36. Schwingshackl, L.; Hoffmann, G.; Iqbal, K.; Schwedhelm, C.; Boeing, H. Food groups and intermediate disease markers: A systematic review and network meta-analysis of randomized trials. *Am. J. Clin. Nutr.* **2018**, *108*, 576–586. [CrossRef] [PubMed]
37. Melse-Boonstra, A. Bioavailability of micronutrients from nutrient-dense whole foods: Zooming in on dairy, vegetables, and fruits. *Front. Nutr.* **2020**, *7*, 101. [CrossRef]

*nutrients*

MDPI

*Review*

# Legume Proteins and Peptides as Compounds in Nutraceuticals: A Structural Basis for Dietary Health Effects

**Marina Carbonaro [1,*] and Alessandro Nucara [2]**

[1] Council for Agricultural Research and Economics (CREA), Research Centre for Food and Nutrition, Via Ardeatina 546, 00178 Rome, Italy

[2] Dipartimento di Fisica, Università Sapienza di Roma, Piazzale A. Moro 2, 00185 Rome, Italy; alessandro.nucara@roma1.infn.it

* Correspondence: marina.carbonaro@crea.gov.it

**Abstract:** In the current climate of food security, quality aspects of legume crops have primary market economic and health impact. Legume proteins and peptides have been discovered to have a role far beyond supplying amino acids for growth and maintenance of body tissues. Several proteins (enzymatic inhibitors, lectins, storage globulins) and peptides derived from them (lunasin, hydrophobic peptides) have shown anticarcinogenic, hypocholesterolemic, glucose-lowering, antioxidant, antimicrobial, and immunostimulant properties. Further understanding of how structural features of legume proteins affect in vivo digestion and production of bioactive sequences represents a key step in the valorization of nutraceutical potentiality of legume proteins and peptides derived from them. In this work, the relationship between structure and bioavailability of protein and peptides are reviewed and discussed.

**Keywords:** legume proteins; nutraceuticals; structural properties; bioactive peptides; in vivo digestibility; health effects

**Citation:** Carbonaro, M.; Nucara, A. Legume Proteins and Peptides as Compounds in Nutraceuticals: A Structural Basis for Dietary Health Effects. *Nutrients* **2022**, *14*, 1188. https://doi.org/10.3390/nu14061188

Academic Editors: Stuart Johnson and Rewati Bhattarai

Received: 29 January 2022
Accepted: 9 March 2022
Published: 11 March 2022



## 1. Introduction

Legumes have a potential to add to the nutritional quality of foods and many options have been suggested for their inclusion in novel food preparation with improved nutritional and functional values.

Proteins represent one of the most concentrated nutrients in legumes, and they can be easily used as components in innovative human foods. In addition, legumes have higher protein content than cereals: therefore, they represent a primary source of amino acids for humans. Proteins extracted from legumes are an important font of proteins of plant origin, that can be consumed as an alternative to meat proteins [1].

Legumes, i.e., *Glycine max* (soybean) and *Phaseolus* species (beans), are grown in the tropical and subtropical areas of the world. It has been recognized that legumes have a primary role in the sustainability of agriculture, enhancing soil quality through nitrogen fixation.

Grain legumes (pulses) are included in the traditional diets of many countries. Indeed, dietary guidelines recommend increasing their consumption, especially in developed countries, due to their potential health benefits. They represent main sources of proteins and minerals (iron, zinc, calcium), while having a low amount of lipids, with the exception of soybean, peanut, and lupin (30–35%). Legume seeds contain low amounts of saturated fats. On the other hand, they are rich in carbohydrates (up to 60%), primarily starch, and in many vitamins (thiamine, niacin, biotin, riboflavin, folic acid). Moreover, legumes are a primary source of total dietary fiber (up to 37%) [1]. For this reason, they are a low-glycaemic food [2].

Epidemiological studies have highlighted a correlation between regular intake of legume seeds and maintenance of a good health status in humans. In meta-analyses of prospective observational studies, consumption of legumes has been related with a low risk of coronary heart disease [3,4].

The word "nutraceutical" was coined in the United States in 1989 to define "a food, or components of a food, that provides health benefits, including the prevention and treatment of diseases" [5].

Nutraceutical properties of legumes have been ascribed to non-nutrient compounds —isoflavones, alkaloids, phytates, saponins—and to a number of proteins such as enzymatic (protease and amylase) inhibitors, lectins, storage proteins—as well as to peptides [1]. It is worth reminding that most of these components have originally been considered as antinutrients, because of their adverse effects on nutrient metabolism [1].

## 2. Grain Legume Proteins

Legume proteins have attracted interest from the food industry. Indeed, they have a low cost, and good functional and nutritional attributes [6].

Proteins are accumulated in legume seeds during development inside protein bodies (storage vacuoles) and provide ammonia, carbon, and amino acids during seed development and to proteins. Proteins represent from about 20% in most legumes to 40% in soybean seeds [1]. Storage proteins are prevailing in legume seeds.

According to Osborne classification [7] storage proteins are divided into globulins, albumins, glutelins, and prolamins. Globulins are extracted in salt solutions, albumins are soluble in water, glutelins are soluble in acids or bases, and prolamins are soluble in ethanol [7]. Globulins are predominant in legume seeds, while prolamins and glutelins are prominent in cereals seeds [8]. Legumin, vicilin, and convicilin are the globulins present in legumes. Depending on the source, seeds usually present one or two predominant types of 81 storage proteins. Seed storage proteins also include enzymes, protease 82 inhibitors (trypsin, chymotrypsin, and amylase inhibitors), lectins, defense proteins, and others.

According to their sedimentation coefficient, globulins are divided into 7S and 11S oligomeric proteins. The 7S proteins are called vicilins, the 11S proteins are named legumins. The 7S/11S ratio is variable inside seeds and is dependent on the different legume species.

In *Phaseolus vulgaris* L., the predominant globulin is phaseolin, which accounts for 50% of total protein content, while in *Vicia faba* L. the most abundant globulin is vicilin (30% of total proteins). In *Lens culinaris* L. vicilin and legumin account for 72% and 11% of total proteins, respectively. In *Cicer arietinum* L. legumin and vicilin represent 66.5% and 12% of total protein content, respectively. In addition to the 7S and 11S proteins there are also the 2S proteins (monomeric proteins) [9].

Storage proteins are mostly oligomeric proteins. The 7S globulins are typically trimers (MW about 150 kDa), while the 11S proteins form hexamers (MW about 350–400 kDa), or higher association of subunits, such as the 15–18S globulins found in soybean globulins. From these oligomeric proteins, subunits can be released under dissociating conditions or upon thermal treatment. Reassociation of subunits can result in high-molecular weight aggregates with low susceptibility to digestion [10].

Other proteins present in legume seeds have antinutritional effects. The most important are protease inhibitors (Kunitz and Bowman–Birk inhibitors) and lectins, which have presented some adverse effects on animals and humans. However, the effects of these proteins are lost after processes preceding consumption (cooking, fermentation, germination, or dehulling) [11].

Globulins contain high content of aspartic acid and glutamic acids while albumins are rich in lysine and sulfur-containing amino acids [12]. Sulfur amino acids (methionine and cysteine) and tryptophan are limiting amino acids in legumes. On the other hand, they have a high amount of lysine, a property that makes them complementary to cereal proteins.

Because most of the albumin have higher sulfur amino acid content than globulin, strategies to increase this class of proteins in the seed is relevant. Among these, increasing the proportion of embryo axis to endosperm because the percentage of albumin is higher in the axis than in the endosperm [13]. Among globulins, legumins (11S proteins) have a higher sulfur amino acid content than 7S globulins. Therefore, increasing the legumin to vicilin ratio can result in an increase in sulfur amino acids in the seed. In the complex, plant proteins (legumes and cereals) have a lower nutritional quality than animal proteins.

Besides amino acid composition, studies focused on amino acid availability have pointed out that structural properties of plant proteins, making them resistant to gastrointestinal proteolysis, further lower the nutritional quality, because of limited bioavailability of essential amino acids [14]. Although trypsin inhibitors of many legumes are relatively high sulfur containing proteins, those of *Phaseolus vulgaris* L. and *Glycine max* L. (Bowman–Birk inhibitors) appear not to be readily digested in the rat gut [15].

Legume protein functional properties have been modified by application of several methods, such as thermal treatment, ultrasonication, and high pressure that affect protein structure [16]. In particular, high pressure caused a gradual unfolding of structure with improvement in solubility and emulsifying activity of kidney bean (*Phaseolus vulgaris* L.) isolate, due to the formation of soluble aggregates [17,18].

On the other hand, foaming properties of pea proteins were not improved by thermal treatment. An increase in hydrophobicity as a consequence of changes in structure was observed. Heat treatment induced loss of the oligomeric assembly, subunit denaturation with exposure of hydrophobic and sulfhydryl groups and reassociation into protein aggregates with modified surface properties and limited protein solubility [10,19,20].

Kidney bean and lentil protein isolates showed an improvement in solubility and emulsifying activity with the application of high pressure. However, after high pressure treatment, emulsion properties of kidney bean hydrolysate were impaired, as evidenced by secondary structure modifications (in particular, a shift in amide I and II of the infrared spectrum) [21].

When chickpea protein isolate was subjected to ultrasound treatment, an improvement in solubility, emulsifying, foaming and gel properties was observed [22].

Owing to their biocompatibility, film formation and functional (probiotics) properties, legume proteins have also been used in the encapsulation of several compounds [23]. They are effective for the preservation of probiotic and labile compounds during gastrointestinal digestion [24]. Folate, α-tocopherol, ascorbic acid, and phytase have been incapsulated in protein isolates from pea and chickpea, with 62–100% encapsulation efficiency and good release in the gastrointestinal tract [25,26].

## 3. Nutraceutical Properties of Legume Proteins

The health benefits of consumption of legume seeds have been reported in relation to many diseases, such as cancer, cardiovascular disease, the ageing process, immune response, diabetes, weight control, osteoporosis, digestive tract diseases, and mental health [27–29].

Small peptides, partially digested proteins and intact proteins from soybean, lupin, lentil, chickpea, pea, and the common bean, exert hormone-like activities [30,31]. Specific beneficial effects in humans include cardiovascular protection, anticancer activity, bone protection, control of weight, immune cell action, insulin sensitivity, control of inflammation and of type 2 diabetes mellitus [32–35]. Some examples of bioactive proteins and peptides derived from legume seeds are presented in Table 1.

**Table 1.** Legume proteins and peptides with nutraceutical properties (modified from reference [36]).

| Precursor | Name/Sequence | Nutraceutical Activity |
|---|---|---|
| Soy trypsin/chymotrypsin inhibitor | Kunitz/Bowman Birk inhibitor | Anticancer, anti-inflammatory, weight control |
| Soy amylase inhibitors | α-Amylase inhibitor | Antiobesity, antidiabetic, anticancer |
| Jack bean haemagglutinins | Concanavalin A, Lectins | Anticancer, immunostimolant |
| Bean, soy storage 7S globulins | Phaseolin, conglycinin, 7S protein α′ chain | Hypocholesterolemic |
| Soy storage 11S globulins | Hydrophobic peptides | ACE-inhibitory |
| Soy 2S albumins | Lunasin | Immunostimolant, anticancer, ipotensive |
| Lupin/soy conglutin γ | Conglutin | Hypoglycemic, hypocholesterolemic |
| Soy proteins | YPFVV, LPYPR, IAVPGEVA | ACE-inhibitory, antioxidant, opioid agonist |
| Fermented soybean | LVQGS | Antihypertensive |

Enzyme inhibitors present in legume seeds have been found to be active in the control of proteases, amylases, and glycosidases. These enzymes are involved in the mechanism of defense against exogenous attack (insects and microorganisms) [37].

Once inactivated, proteins known as antinutritional factors, that is protease inhibitors and lectins, have been reported to show health effects. Protease inhibitors are active towards inflammation and cancer [38]. Regular consumption of legumes has been shown to reduce the risk of several cancers, such as colon, prostate, gastric, and pancreatic cancer. Anticarcinogenic properties have been attributed to enzymatic inhibitors, especially Bowman–Birk inhibitors (BBI) [39,40]. Soybean, lentil, and pea BBI have been found to be active in the prevention and suppression of colon, liver, lung, prostate and mammalian cancers induced by chemical and physical agents, soybean BBI being particularly effective [41].

Control of protease activity by protease inhibitors may be responsible for their anticancer power.

α-Amylase inhibitors have shown antidiabetic activity and, therefore, potential applications in the control of obesity. Similarly, α-glucosidase inhibitors have been proposed in the treatment of diabetes. Delay of digestion and absorption of carbohydrates helps in the control of postprandial hyperglycemia in the diabetic patient.

Lectins are blood grouping substances, immunomodulators and tissue markers. One property of lectins is their ability to combine with sugars and glycoconjugates. Lectins have been shown to have a role in the prevention of cancers, in the activation of immune system, and in antimicrobial and insecticidal mechanisms. In addition, they may be used in the control of obesity [42].

Hypocholesterolemic, glucose, and blood pressure-lowering actions have been reported for both proteins and peptides by in vitro and clinical studies. Conglutin γ from lupin has been demonstrated to bind insulin ($K_d = 9 \times 10^{-5}$ M), thus controlling glucose plasma levels [43].

High biological properties of protein extracts of local varieties of *Phaseolus vulgaris* L., such as antiradical, anti α-amylase, and angiotensin converting enzyme-inhibitory activity [44], as well as influence on intestinal permeability, have recently been described [45].

## 4. Structure–Digestibility Relationship of Legume Proteins

The structural properties of legume proteins, by imparting high stability during gastrointestinal digestion, have been reported to play an important role in their in vivo bioactivity and release of bioactive sequences [46,47].

Major structural properties of legume proteins that have been described with nutraceutical activity are reported in Table 2.

**Table 2.** Major structural properties of legume proteins with nutraceutical activity (modified from reference [36]).

| Protein | MW (KDa) | Structure Type [a] | α-Helix (%) | β-Sheet (%) | N° of SS |
|---|---|---|---|---|---|
| Kunitz trypsin inhibitor [b] | 21.5 | Globular, monomeric | 6 | 40–60 | 2 |
| Bowman–Birk inhibitor [b] | 8 | Globular, monomeric | 0 | 60 | 7 |
| α-Amylase inhibitors [c] | 12–60 | Globular, monomeric/dimeric/tetrameric | 15–30 | 25–60 | 2–5 |
| Concanavalin A [d] | 110 | Globular, tetrameric | 0 | 47 | 0 |
| Phaseolin [e] | 150 | Globular, trimeric | 16 | 37 | 0 |
| Glycinin [b] | 340 | Globular, oligomeric | 15 | 36 | 22 |
| Conglycinin [b] | 200 | Globular, oligomeric | 15 | 31 | 2 |
| Conglutin γ [f] | 200 | Globular, tetrameric | 15 | 35 | 24 |

[a] In phosphate-buffered saline, pH 7.0; [b] From soybean; [c] From cereals and legumes; [d] From jack bean; [e] From common bean; [f] From lupin.

The anticarcinogenic effect of BBI towards colon cancer has been related to the native conformation of the inhibitory domain, the inhibitor being found intact in several organs (liver, lung) after ingestion [40]. In particular, trypsin and chymotrypsin inhibitors of the BBI class of both soybean and pea seeds have been found to present anticarcinogenic effects: soybean inhibitors are active toward colon, liver, lung, esophagus, and breast cancers, while pea inhibitors present anti-proliferative activity toward colon cancer [41,42].

Preservation of the conformation of conglutin γ has resulted in being a prerequisite for insulin binding and hypoglycemic activity of the protein, tested in a rat model [43]. Similar properties have been found for the basic 7S globulin, a protein with 64% identity to conglutin γ isolated from soybean seed and built up by two disulfide-bridged subunits of 27 and 16 kDa [43].

Trypsin inhibitors and lectins have been shown to be internalized by the small intestinal villi of rat [48]. These proteins are very stable during processing and gastrointestinal digestion [36].

In addition to stability conferred by disulfide bonds, hydrophobicity is known to affect the physicochemical properties (hydration, gelation, emulsification, foaming, adhesion) of plant proteins, with a consequence on both absorption and nutritional properties. Soybean protein extract showed an average hydrophobicity of 6.44 kJ per residue and control of bitterness of soy hydrolysates in relation to hydrophobicity was successful to increase their functionality.

In oligomeric storage proteins of legume seeds, stabilization conferred by hydrophobic patches between monomers is likely to decrease susceptibility to proteolysis, especially after technological processing [10]. As a consequence, essential amino acids and bioactive peptides may be imprisoned inside stable complexes that are no longer digested [36].

Other adverse effects may include immunological reactions promoted by soluble and stable protein complexes [48]. Major allergens that have been found to be responsible for sensitization are α- and β-conglutins from lupin. Peanut, lentil, and soybean allergens have also been identified. The major lentil allergen is Len c1 (a 48 kDa vicilin), while 33 proteins from soybean (7–71 kDa) have been found to be allergenic. Stability of these proteins during gastrointestinal digestion has been reported as a major cause for their allergenicity [36,49].

Fourier transform infrared spectroscopy (FTIR) has recently been employed to analyze the relationship between structure and bioavailability of food protein by examination of the amide I of the spectrum [46]. FTIR has demonstrated that the secondary structure of several plant proteins, such as legume proteins, is dominated by contributions from β-sheet conformation and, to this respect, it markedly differs from that of animal proteins, characterized by α-helix structure (Figure 1).

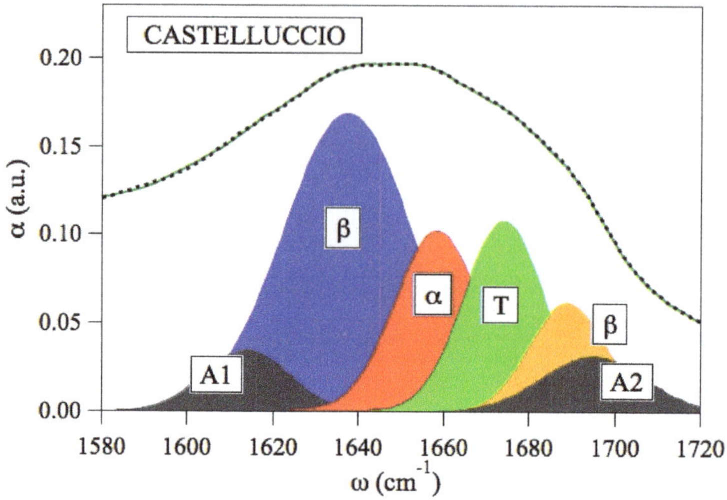


**Figure 1.** Analysis of proteins of Castelluccio lentil PGI (protected geographical indication) by FTIR. The amide I band was deconvolved by gaussian contributes. A1: intermolecular aggregates; β: β-sheet; α: α-helix; T: turns, A2: β-aggregates. (Carbonaro and Nucara, personal communication).

Legume proteins presented quite a lower α-helix to β-sheet ratio than cereal proteins (0.47 and 1.1, respectively) [46]. A different α-helix to β-sheet ratio was found in a recent study on several varieties of *P. vulgaris* coming from different countries, with cannellini and borlotti varieties showing the highest values (form 0.47 to 0.56) (Figure 2).

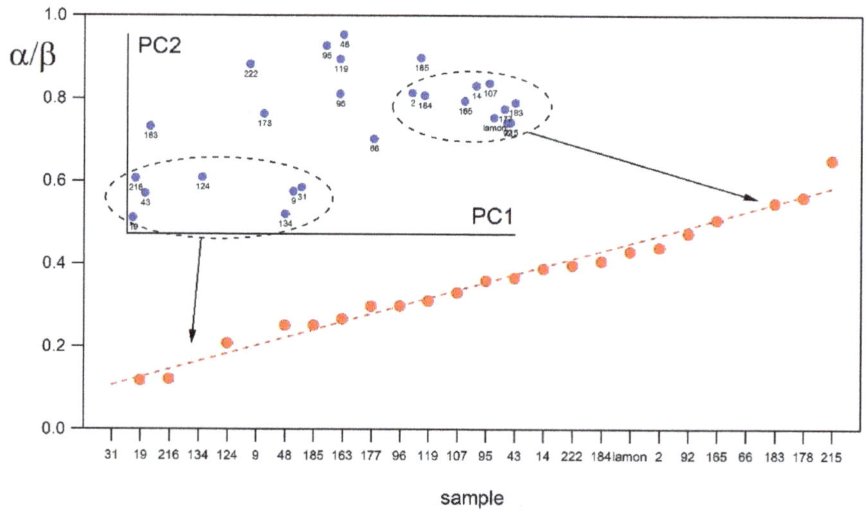


**Figure 2.** Ratio between percentage of α-helix and β-sheet secondary structures for common bean varieties (red points and dashed line). Dashed line is a guide for the eyes. In the inset the score-plot of PC1 and PC2 obtained from a PCA analysis on fit results is reported. (Carbonaro and Nucara, personal communication).

Moreover, β-sheet content of legume proteins has been found to account for the formation of stable intermolecular complexes upon thermal treatment. A high correlation between β-sheet content and protein digestibility has been found for food proteins and for both native and heated legume proteins [46]. Hydrophobic amino acids (alanine, valine, methionine, isoleucine, phenylalanine), together with cysteine, have been found in the small intestinal content of rats fed with legume proteins, further supporting the role of the structural properties of these proteins on the overall nutritional quality [14].

3D structure and surface features of the Bowman–Birk inhibitor are presented in Figures 3a and 3b, respectively. It is evident that this protein is dominated by β-sheet conformation and by large hydrophobic areas on the surface. Stability is conferred by seven disulfide bridges in a small molecular weight protein (8 kDa). Other legume seed proteins (storage globulins, α-amylase inhibitors) have been shown to present similar conformational attributes. These features limit digestibility and digestion rate of the protein in the gastrointestinal tract [46,50].

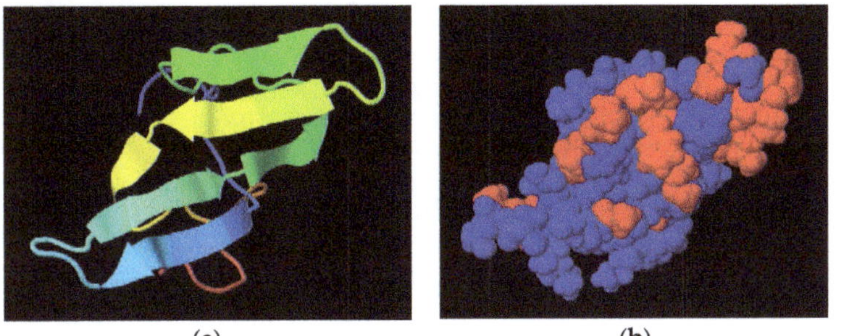

(**a**) (**b**)

**Figure 3.** 3D structure (**a**) and surface properties (**b**) of soybean Bowman–Birk inhibitor as rendered by Jmol software using the coordinates available in the Brookhaven Protein Data Bank (http://www.rcsb.org/pdb) (accessed on 15 December 2021). Surface color mapping: blue for hydrophilic, red for hydrophobic regions.

Exogenous factors may also adversely affect digestibility of legume proteins: these include interaction with other compounds such as carbohydrates, tannins, phytates, lipid, trypsin inhibitors, and lectins [14,36].

Structural properties of legume proteins are likely to also have a role on bioavailability of some micronutrients, such as Fe.

Legume (lentil, chickpea, and pea) seed ferritin concentrates (30–45 mg Fe/100 g) have been investigated for the relationship between structure and resistance to digestion [51]. A correlation between concentration in Asx + Glx of the different legume proteins and iron content was found.

Most of the iron was released by pepsin digestion. Therefore, these iron-rich protein extracts are candidates in the production of functional foods to be used in place of inorganic iron against iron deficiency anemia. The two ferritin polypeptides showed a loose structure, as evidenced by intrinsic fluorescence spectroscopy. This property likely enabled protein degradation and iron release at low pH in the stomach [51].

In another study, determination of Cu, Fe, and protein absorption in the small intestine of rat has been carried out in single-dose experiments [52]. After thermal treatment of legume seeds, most (about 80%) of compounds was extracted in the insoluble fraction. Absorption of proteins, Cu, and Fe in this fraction was low, indicating that insolubilization negatively influences protein, Cu, and Fe absorption from legumes. Increased hydrophobicity of legume proteins after thermal treatment induced protein aggregation and precipitation into insoluble complexes.

These findings suggest that the structure of legume proteins is likely to affect bioavailability not only of essential amino acids, but also that of some micronutrients, such as Cu and Fe.

## 5. Structural Traits of Bioactive Peptides

Similarly to bioactive peptides from animal proteins (milk, meat), those derived from legume proteins are characterized by: (i) a short length (2–20 amino acids); (ii) proline, lysine, arginine, and hydrophobic amino acids; (iii) low susceptibility to digestion [53,54].

Proteolytic digestion of soybean 7S protein produces a pentapeptide (Leu–Leu–Pro–His–His) that has shown antioxidant activity. An Arg–Gly–Asp tripeptide has been found to be the adhesion region of soy lunasin to the cells, adhesion being a prerequisite for its anticarcinogenic properties [55].

Lunasin is a peptide with 44 amino acids and a high content in Asp, extracted from soybean 2S albumins and then isolated from cereal (wheat, barley, rye) proteins [56]. Besides lunasin, Val–Pro–Tyr and $\gamma$-glutamyl peptides from soybean also have anti-inflammatory properties [57]. Hydrophobic peptides from soybean also present anticarcinogenic properties [58].

The antioxidant activity of 28 short-chain peptides attributed to Leu–Leu–Pro–His–His have been examined: the tripeptide Pro–His–His has been shown to be active as metal chelator or radical scavenger, increasing the antioxidant properties of soy protein hydrolysate [59].

Recently, several peptides with antioxidant properties and a high amino acid score have been isolated after hydrolysis of legume proteins [60].

It has also been shown that proteins with low content of Met–Gly and Lys–Arg, i.e., soy and fish proteins, lower cholesterol level [61].

Besides specific residues, charge properties, hydrogen bonding and hydrophobicity are believed to influence both susceptibility to proteolysis and peptide absorption, besides the physiological functions of peptides.

Modern in silico techniques, such as quantitative structure–activity relationship (QSAR) models, consisting in analysis of homology similarity, are available for the screening of the origin of bioactive peptides [62–64]. These approaches have allowed the discovery of bioactive peptides, based on their sequence similarity.

Results from bioinformatic predictions have indicated that fragments with probability to be produced are hydrophilic and, therefore, are present at the external surface of the protein. These regions contain a high percentage of random coil (46%) and low amounts of $\beta$-sheet (17%) [65].

Bioinformatic tools, based on different algorithms, may help in predicting enzymatic hydrolysis of proteins to account for proteolytic process designs.

## 6. Conclusions

The structural properties of legume proteins, as also evidenced by FT-IR analysis, by imparting high stability during processing and gastrointestinal digestion, are likely to affect their bioactivity and production of bioactive peptides.

Further knowledge of the relationships between structure and bioactivity of protein and peptides from legume seeds is required to optimize their use as nutraceuticals, to increase peptide production, and to improve bioavailability of bioactive sequences. Such information may also be useful in planning strategies for eliminating the risk of adverse reactions, such as allergenicity, consequent to consumption of legumes for sensitive population groups, another aspect that has partially been related to high protein stability in the gastrointestinal tract.

**Author Contributions:** Funding acquisition, writing-original paper preparation, writing—review and editing, M.C.; writing—original paper preparation, writing—review and editing, A.N. All authors have read and agreed to the published version of the manuscript.

**Funding:** This work was partially funded by Ministry of Agricultural, Alimentary and Forestry Policies (MiPAAF) "Programma di rilancio leguminose da granella per alimentazione umana—PROLEGU" Project 2019–2023.

**Conflicts of Interest:** Authors declare no conflict of interest.

## References


1. Carbonaro, M. Nutraceutical perspectives of pulses. In *Pulse Foods: Processing, Quality and Nutraceutical Applications*, 2nd ed.; Tiwari, B.K., Gowen, A., Mckenna, B., Eds.; Academic Press: Cambridge, MA, USA; Elsevier: Oxford, UK, 2021; pp. 423–460.
2. Rebello, C.I.; Greenway, F.L.; Finley, J.W. A review of the nutritional value of legumes and their effects on obesity and its related co-morbidities. *Obes. Rev.* **2014**, *15*, 392–407. [CrossRef] [PubMed]
3. Bechthold, A.; Boeing, H.; Schwedhelm, C.; Hoffmann, G.; Knuppel, S.; Iqbal, K.; De Henauw, S.; Michels, N.; Devleesschauwer, B.; Schlesinger, S.; et al. Food groups and risk of coronary heart disease, stroke and heart failure: A systematic review and dose-response meta-analysis of prospective studies. *Crit. Rev. Food Sci. Nutr.* **2019**, *59*, 1071–1090. [CrossRef] [PubMed]
4. Conti, M.V.; Guzzetti, L.; Panzeri, D.; De Giuseppe, R.; Coccetti, P.; Labra, M.; Cena, H. Bioactive compounds in legumes: Implications for sustainable nutrition and health in the elderly population. *Trends Food Sci. Technol.* **2021**, *117*, 139–147. [CrossRef]
5. Diplock, A.T.; Aggett, P.J.; Ashwell, M.; Bornet, F.; Fern, E.B.; Roberfroid, M.B. Scientific concepts of functional foods in Europe: Consensus document. *Br. J. Nutr.* **1999**, *81*, S1–S27. [CrossRef]
6. Prandi, B.; Zurlini, C.; Cicognini, I.M.; Cutroneo, S.; Di Massimo, M.; Bondi, M.; Brutti, A.; Sforza, S.; Tedeschi, T. Targeting the nutritional value of proteins fron legumes by-products through mild extraction technologies. *Front. Nutr.* **2021**, *8*, 695793. [CrossRef]
7. Osborne, T.B. *The Vegetable Proteins*, 2nd ed.; Longmans, Green and Co.: London, UK, 1924.
8. Rajnincová, D.; Spaleková, A.; Galová, Z.; Romanavá, K. The protein profile of cereals, pseudocereals and legumes. *J. Food Sci. Technol.* **2019**, *7*, 49–53.
9. Mittal, R.; Nagi, H.; Sharma, P.; Sharma, S. Effect of processing on chemical composition and antinutritional factors in chickpea flour. *J. Food Sci. Eng.* **2012**, *2*, 180–186. [CrossRef]
10. Carbonaro, M.; Cappelloni, M.; Nicoli, S.; Lucarini, M.; Carnovale, E. Solubility-digestibility relationship of legume proteins. *J. Agric. Food Chem.* **1997**, *45*, 3387–3394. [CrossRef]
11. Champ, M.M.-J. Non-nutrient bioactive substances of pulses. *Br. J. Nutr.* **2002**, *88* (Suppl. S3), 307–319. [CrossRef]
12. Venkidasamy, B.; Selvaraj, D.; Nile, A.S.; Ramalingam, S.; Kai, G.; Nile, S.H. Indian pulses: A review on nutritional, functional and biochemical properties with future perspectives. *Trends Food Sci. Technol.* **2019**, *88*, 228–242. [CrossRef]
13. Ingram, G.; Doll, N. Embryo-endosperm interactions. *Ann. Rev. Plant Biol. Ann. Rev.* **2022**, in press.
14. Carbonaro, M.; Grant, G.; Cappelloni, M.; Pusztai, A. Perspectives into factors limiting in vivo digestion of legume proteins: Antinutritional compounds or storage proteins? *J. Agric. Food Chem.* **2000**, *48*, 742–749. [CrossRef] [PubMed]
15. Moreno, F.J.; Clemente, A. 2S albumin storage proteins: What makes them food allergens? *Open Biochem. J.* **2008**, *2*, 16–28. [CrossRef] [PubMed]
16. Gharibzahedi, S.M.T.; Smith, B. Effect of high hydrostatic pressure on the quality and functionality of protein isolates, concentrates, and hydrolysates derived from pulse legumes: A review. *Trends Food Sci. Technol.* **2021**, *107*, 466–479. [CrossRef]
17. Tang, C.E. Thermal denaturation and gelation of vicilin-rich protein isolates from three Phaseolus legumes: A comparative study. *LWT-Food Sci. Technol.* **2008**, *8*, 1380–1388. [CrossRef]
18. Yin, S.W.; Tang, C.H.; Wen, Q.B.; Li, L. Functional properties and in vitro trypsin digestibility of red kidney bean (*Phaseolus vulgaris* L.) protein isolate: Effect of high-pressure treatment. *Food Chem.* **2008**, *110*, 938–945. [CrossRef]
19. Chao, D.; Aluko, R.E. Modification of the structural, emulsifying, and foaming properties of an isolated pea protein by thermal pretreatment. *Cyta J. Food* **2018**, *16*, 357–366. [CrossRef]
20. Lu, Z.X.; He, J.F.; Zhang, Y.C.; Bing, D.J. Composition, physicochemical properties of pea proteins and its application in functional foods. *Crit. Rev. Food Sci. Nutr.* **2020**, *60*, 2593–2605. [CrossRef]
21. Ahmed, J.; Al-Ruwaih, N.; Mulla, M.; Rahman, M.H. Effect of high pressure treatment on functional, rheological and structural properties of kidney bean protein isolate. *LWT* **2018**, *91*, 191–197. [CrossRef]
22. Wang, Y.; Wang, Y.; Li, K.; Bai, Y.; Li, B.; Xu, W. Effect of high intensity ultrasound on physicochemical, interfacial and gel properties of chickpea protein isolate. *LWT* **2020**, *129*, 109563. [CrossRef]
23. Sharif, H.R.; Williams, P.A.; Sharif, M.K.; Abbas, S.; Majeeh, H.; Masamba, K.G.; Safdar, W.; Zhong, F. Current progress in the utilization of native and modified legume proteins as emulsifiers and encapsulants—A review. *Food Hydrocoll.* **2018**, *76*, 2–16. [CrossRef]
24. Afzaal, M.; Saeed, F.; Aamir, M.; Usman, I.; Ashfaq, I.; Ikram, A.; Hussain, M.; Anjum, F.M.; Waleed, M.; Suleria, H. Encapsulating properties of legume proteins: Recent updates & perspectives. *Int. J. Food Prop.* **2021**, *24*, 1603–1614. [CrossRef]
25. Ariyarathna, I.R.; Karunaratne, D.N. Use of chickpea protein for encapsulation of folate to enhance nutritional potency and stability. *Food Bioprod. Proc.* **2015**, *95*, 76–82. [CrossRef]
26. Gharibzahedi, S.M.T.; Smith, B. Legume proteins are smart carriers to encapsulate hydrophilic and hydrophobic bioactive compounds and probiotic bacteria. A review. *Compr. Rev. Food Sci. Food Saf.* **2021**, *21*, 1250–1279. [CrossRef] [PubMed]
27. Messina, M.J. Legumes and soybeans: Overview of their nutritional profiles and health effects. *Am. J. Clin. Nutr.* **1999**, *70*, 439S–450S. [CrossRef]
28. McCrory, M.A.; Hamaker, B.R.; Lovejoy, J.C.; Eichelsdoerfer, P.E. Pulse consumption, satiety, and weight management. *Adv. Nutr.* **2010**, *1*, 17–30. [CrossRef] [PubMed]
29. Becerra-Thomás, N.; Papandreou, C.; Salas-Salvadó, J. Legume consumption and cardiometabolic health. *Adv. Nutr.* **2019**, *10*, S437–S450. [CrossRef]

30. Moreno-Valdespina, C.A.; Luna-Vital, D.; Camacho-Ruiz, R.M.; Mojica, L. Bioactive proteins and phytochemicals from legumes: Mechanisms of action preventing obesity and type-2 diabetes. *Food Res. Int.* **2020**, *130*, 108905. [CrossRef]
31. Hernandez-Velazquez, I.; Sanchez-Tapia, M.; Ordaz-Nava, G.; Torres, N.; Tovar, A.R.; Galvez, A. Black bean protein concentrates ameliorates hepatic steatosis by decreasing lipogenesis and increasing fatty acid oxidation in rats fed a high fat-sucrose diet. *Food Funct.* **2021**, *12*, 426. [CrossRef]
32. Gonzaléz-Montoya, M.; Cano-Sampedro, E.; Mora-Escobedo, R. Bioactive peptides from legumes as anticancer therapeutic agents. *Int. J. Cancer Clin. Res.* **2017**, *4*, 81. [CrossRef]
33. Jahan-Mihan, A.; Luhovyy, B.L.; Khoury, D.E.L.; Anderson, G.H. Dietary proteins as determinants of metabolic and physiologic functions of the gastrointestinal tract. *Nutrients* **2011**, *3*, 574–603. [CrossRef] [PubMed]
34. Chan-Zapata, I.; Sandoval-Castro, C.; Segura-Campos, M.R. Proteins and peptides from vegetable sources as therapeutic adjuvants for the type 2 diabetes mellitus. *Crit. Rev. Food Sci. Nutr.* **2020**, 1–10. [CrossRef] [PubMed]
35. Jakubczyk, A.; Karas, M.; Zlotek, U.; Szymanowska, U. Identification of potential inhibitory peptides of enzymes involved in the metabolic syndrome obtained by simulated gastrointestinal digestion of fermented bean (Phaseolus vulgaris L.) seeds. *Food Res. Int.* **2017**, *100*, 489–496. [CrossRef] [PubMed]
36. Carbonaro, M.; Maselli, P.; Nucara, A. Structural aspects of legume proteins and nutraceutical properties. *Food Res. Int.* **2015**, *76*, 19–30. [CrossRef]
37. Grosse-Holz, F.M.; van der Hoornm, R.A.L. Juggling jobs: Roles and mechanisms of miltifunctional protease inhibitors in plants. *New Phytol.* **2017**, *210*, 794–807. [CrossRef]
38. Cud-Gallegos, M.S.; Corzo-Ríos, L.J.; Jiménez-Martinez, C.; Sánchez-Chino, X.M. Protease inhibitors from plants as therapeutic agents—A review. *Plant Foods Hum.Nutr.* **2022**. [CrossRef]
39. Kennedy, A.R. Chemopreventive agents: Protease inhibitors. *Pharmacol. Ther.* **1998**, *78*, 167e209. [CrossRef]
40. Clemente, A.; Gee, J.M.; Johnson, I.T.; MacKenzie, D.A.; Domoney, C. Pea (*Pisum sativum* L.) protease inhibitors from the Bowman Birk class influence the growth of human colorectal adenocarcinoma HT29 cells in vitro. *J. Agric. Food Chem.* **2005**, *53*, 8979–8986. [CrossRef]
41. Birk, Y. *Protease Inhibitors as Cancer Chemo Preventive Agents*; Troll, W., Kennedy, A.R., Eds.; Plenum: New York, NY, USA, 1993; pp. 97–106.
42. Roy, F.; Boye, J.I.; Simpson, B.K. Bioactive proteins and peptides in pulse crops: Pea, chickpea and lentil. *Food Res. Int.* **2010**, *43*, 432–442. [CrossRef]
43. Scarafoni, A.; Magni, C.; Duranti, M. Molecular nutraceutics as a mean to investigate the positive effects of legume seed proteins on human health. *Trends Food Sci. Technol.* **2007**, *18*, 454–463. [CrossRef]
44. Mojica, L.; González de Meija, E. Characterization and comparison of protein and peptide profiles and their biological activities of improved common bean cultivars (*Phaseolus vulgaris* L.) from Mexico and Brazil. *Plant Foods Hum. Nutr.* **2015**, *70*, 105–112. [CrossRef] [PubMed]
45. Schisano, C.; Narciso, V.; Maisto, M.; Annunziata, G.; Grieco, P.; Sommella, E.M.; Tenore, G.C.; Novellino, E. In vitro effects of protein fractions from Controne beans (*Phaseolus vulgaris* L. ecotype Controne) on intestinal permeability, ACE and α-amylase activities. *Eur. Food Res. Technol.* **2019**, *254*, 2311–2322. [CrossRef]
46. Carbonaro, M.; Maselli, P.; Nucara, A. Relationship between digestibility and secondary structure of raw and thermally treated legume proteins: A Fourier transform infrared (FT-IR) spectroscopic study. *Amino Acids* **2012**, *43*, 911–921. [CrossRef] [PubMed]
47. Li, Y.; Yu, J. Research progress in structure-activity relationship of bioactive peptides. *J. Med. Food* **2015**, *18*, 147–156. [CrossRef] [PubMed]
48. Pusztai, A.; Grant, G.; Bardocz, S.; Baintner, K.; Gelencser, E.; Ewen, S.W. Both free and complexed trypsin inhibitors stimulate pancreatic secretion and change duodenal enzyme levels. *Am. J. Physiol.* **1997**, *272*, G340–G350. [CrossRef]
49. Rougé, P.; Culerrier, R.; Sabatier, V.; Granier, C.; Rancé, F.; Barre, A. Mapping and conformational analysis of IgE-binding epitopic regions on the molecular surface of the major Ara h 3 legumin allergen of peanut (*Arachis hypogaea*). *Mol. Immunol.* **2009**, *46*, 1067–1075. [CrossRef]
50. Gitlin-Domagalska, A.; Maciejewska, A.; Debowski, D. Bowman-Birk inhibitors: Insights into family of multifunctional proteins and peptides with potential therapeutical applications. *Pharmaceuticals* **2020**, *13*, 421. [CrossRef]
51. Gesinde, F.A.; Udechukwu, M.C.; Aluko, R.E. Structural and functional characterization of legume seed ferritin concentrates. *J. Food Biochem.* **2018**, *42*, e12498. [CrossRef]
52. Carbonaro, M.; Grant, G.; Mattera, M.; Aguzzi, A.; Pusztai, A. Investigation of the mechanisms affecting Cu and Fe bioavailability from legumes. *Biol. Trace Elem. Res.* **2001**, *84*, 181–196. [CrossRef]
53. Chakrabartim, S.; Guha, S.; Maiumder, K. Food-derived bioactive peptides in human health: Challenges and opportunities. *Nutrients* **2018**, *10*, 1738. [CrossRef]
54. Amigo, L.; Hernández-Ledesma, B. Current evidence on the bioavailability of food bioactive peptides. *Molecules* **2020**, *25*, 4479. [CrossRef] [PubMed]
55. Ohashi, Y.; Onuma, R.; Naganuma, T.; Ogawa, T.; Naude, R.; Nokihara, K.; Muramoto, K. Antioxidant properties of tripeptides revealed by a comparison of six different assays. *Food Sci. Technol. Res.* **2015**, *21*, 695–704. [CrossRef]
56. Alaswad, A.A.; Krishnan, H.B. Immunological investigation for the presence of lunasin, a chemopreventive soybean peptide, in the seed of diverse plants. *J. Agric. Food Chem.* **2016**, *64*, 2901–2909. [CrossRef] [PubMed]

57. Guha, S.; Majumder, K. Structural-features of food-derived bioactive peptides with anti-inflammatory activity: A brief review. *J. Food Biochem.* **2019**, *43*, e12531. [CrossRef] [PubMed]
58. Kim, S.E.; Kim, H.H.; Kim, J.Y.; Kang, Y.I.; Woo, H.J.; Lee, S.E. Anticancer activity of hydrophobic peptides from soy proteins. *Biofactors* **2000**, *12*, 151–155. [CrossRef] [PubMed]
59. Shahidi, F.; Zhing, Y. Novel antioxidants in food quality preservation and health promotion. *Eur. J. Lipid Sci. Technol.* **2010**, *112*, 930–940. [CrossRef]
60. Matemu, A.; Nakamura, S.; Katayama, S. Health benefits of antioxidative peptides derived from legume proteins with a high amino acid score. *Antioxidants* **2021**, *10*, 316. [CrossRef] [PubMed]
61. Erdmann, K.; Cheung, B.W.; Schröder, H. The possible roles of food-derived bioactive peptides in reducing the risk of cardiovascular disease. *J. Nutr. Biochem.* **2008**, *19*, 643–654. [CrossRef]
62. Zulvana, A.H.; Andriati, N.; Sri, A.; Widiastuti, S. In silico approach in evaluation of jack bean (Canavalia ensiformis) canavalin protein as precursors of bioactive peptides with dual antioxidant and angiotensin I-converting enzyme inhibitor. *Mater. Sci. Forum* **2019**, *948*, 85–94. [CrossRef]
63. Zou, T.-B.; He, T.-P.; Li, H.-B.; Tang, H.W.; Xia, E.-Q. The structure-activity relationship of the antioxidant peptides from natural proteins. *Molecules* **2016**, *21*, 72. [CrossRef]
64. Wen, C.; Zhang, J.; Zhang, H.; Duan, Y.; Ma, H. Plant protein-derived antioxidant peptides: Isolation, identification, mechanism of action and application in food systems: A review. *Trends Food Sci. Technol.* **2020**, *105*, 308–322. [CrossRef]
65. Dziuba, J.; Niklewicz, M.; Iwaniak, I.; Darewicz, M.; Minkiewicz, P. Bioinformatic-aided prediction for release possibilities of bioactive peptides from plant proteins. *Acta Aliment. Hung.* **2004**, *33*, 227–235. [CrossRef]

nutrients MDPI

Review

# Lupin Kernel Fibre: Nutritional Composition, Processing Methods, Physicochemical Properties, Consumer Acceptability and Health Effects of Its Enriched Products

**Rahil Malekipoor [1], Stuart K. Johnson [1,2] and Rewati R. Bhattarai [1,*]**

[1] School of Molecular and Life Sciences, Faculty of Science and Engineering, Curtin University, Bentley, WA 6102, Australia; rahil.malekipoor@postgrad.curtin.edu.au (R.M.); stuart@ingredientsbydesign.com.au (S.K.J.)
[2] Ingredients by Design Pty Ltd., Lesmurdie, WA 6076, Australia
* Correspondence: r.bhattarai@curtin.edu.au; Tel.: +61-8-9266-5182

**Abstract:** The kernels (dehulled seeds) of lupins (Lupinus spp.) contain far higher dietary fibre levels than other legumes. This fibre is a complex mixture of non-starch polysaccharides making up the thickened cell walls of the kernel. The fibre has properties of both insoluble and soluble fibres. It is a major by-product of the manufacture of lupin protein isolates, which can be dried to produce a purified fibre food ingredient. Such an ingredient possesses a neutral odour and flavour, a smooth texture, and high water-binding and oil-binding properties. These properties allow its incorporation into foods with minimum reduction in their acceptability. The lupin kernel fibre (LKF) has demonstrated beneficial effects in clinical studies on biomarkers for metabolic diseases such as obesity, type 2 diabetes, and cardiovascular disease. It can be described as a "prebiotic fibre" since it improves gut micro-floral balance and the chemical environment within the colon. Thus, LKF is a health-functional ingredient with great opportunity for more widespread use in foods; however, it is evident that more non-thermal methods for the manufacture of lupin kernel fibre should be explored, including their effects on the physicochemical properties of the fibre and the effect on health outcomes in long term clinical trials.

**Keywords:** Lupin; processing; lupin kernel fibre; food ingredient; composition; techno-functionality; health benefits; consumer acceptability

**Citation:** Malekipoor, R.; Johnson, S.K.; Bhattarai, R.R. Lupin Kernel Fibre: Nutritional Composition, Processing Methods, Physicochemical Properties, Consumer Acceptability and Health Effects of Its Enriched Products. *Nutrients* **2022**, *14*, 2845. https://doi.org/10.3390/nu14142845

Academic Editors: Paolo Magni and Anna Gramza-Michałowska

Received: 27 April 2022
Accepted: 8 July 2022
Published: 11 July 2022



## 1. Introduction

The genus Lupinus consists of hundreds of species of legumes, but only a few, including the white lupin (*Lupinus albus*), the narrow-leafed or Australian sweet lupin (*L. angustifolius*), the yellow lupin (*L. luteus*), and the Andean lupin (*L. mutabilis*) are domesticated for seed production. Western Australia harvests most of the world's lupin seed (from the low alkaloid Australian sweet lupin) which is considered to be an important nitrogen-fixing rotation crop favouring sustainable cereal production. Australia's lupin seed is traditionally used for animal feed; however, current interest has shifted towards using lupin seed fractions such as flour protein concentrates/isolates and dietary fibre as human food (Figure 1). This interest may, in part, be due to the high protein and fibre content of lupin flour and the increasing body of evidence from human clinical studies of its potential to reduce the cluster of risk factors that make up metabolic syndrome. In addition, being a gluten-free, non-genetically modified option as well as low in phytoestrogen, lupin may enhance high consumer appeal.

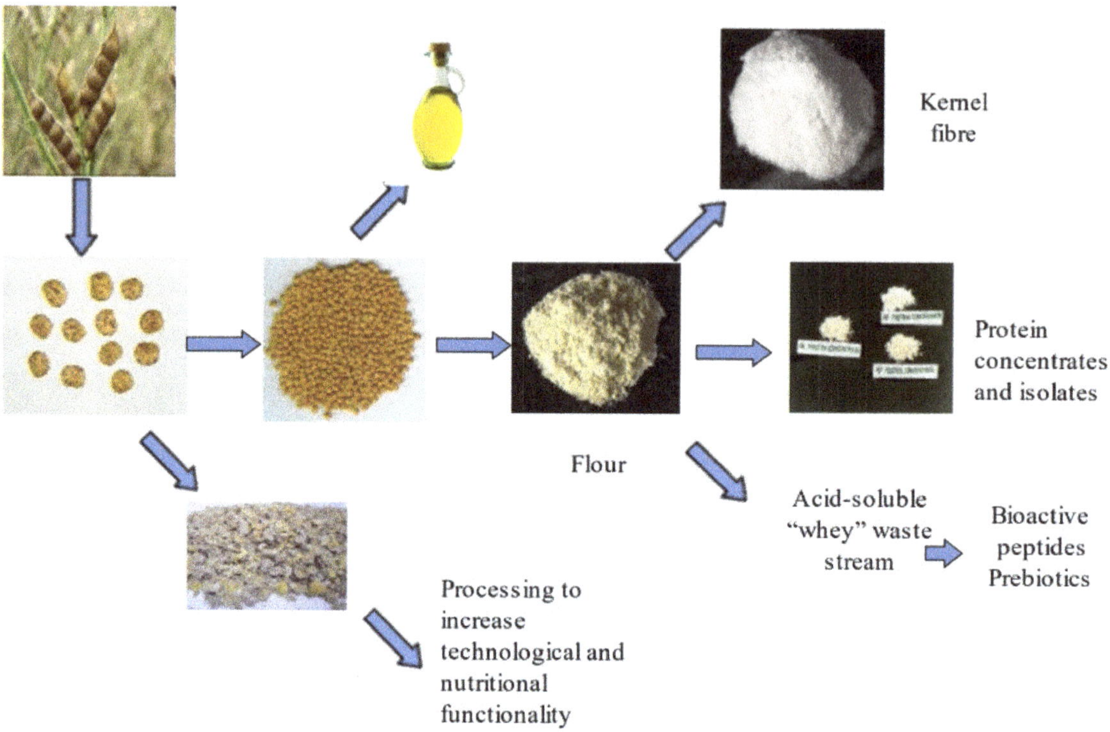


**Figure 1.** End products obtained from lupin seed processing.

The whole lupin seed contains very high levels of dietary fibre in comparison with those of other legumes, including soybean [1], with as much as 50 g/100 g dry basis (db). In lupin seeds, the dietary fibre is located in the protective seed coat (hull) at 80–87 g/100 g db [2], of which half (approx. 40 g/100 g db) is contained within the kernel (endosperm) [3]. This lupin kernel fibre (LKF), in the form of thickened cell walls that serve as storage of carbohydrates for seed germination, was first investigated for its potential as a healthy food ingredient several decades ago, and now its potential commercial value is once again being recognised.

This literature review will focus on the current public domain knowledge of the composition, manufacturing methods, and techno-functional as well as health-functional properties of LKF. The drivers and barriers to its potential commercialisation as a food ingredient will also be discussed.

## 2. Dietary Fibre in Lupin Seeds

Lupin seeds contain both insoluble and soluble dietary fibres. The types of dietary fibres have different chemical structures, especially the level of cellulose vs. non-cellulosic polysaccharides, different physicochemical properties (e.g., water binding, water solubility and viscosity), and different effects on the gut (e.g., fermentability) and metabolism (i.e., health benefits) [4–6]. Insoluble fibres are not water-soluble nor fermented to any significant extent in the human colon; they primarily consist of cellulose, hemicelluloses, and lignin. In contrast, soluble fibres are highly water-soluble and highly fermentable in the colon, while including pectins, gums, and mucilages [7]. Regarding fibre properties and glycemic regulation, Goff et al. [8] reported that, while the viscosity of soluble fibres is well regarded to regulate glucose metabolism, insoluble fibres, despite limited viscosity, lead to improved glycemic control through four different mechanisms—delay in gastric emptying, gut hormonal regulation, reduced activity of small intestinal digestive enzymes, and delayed absorption of sugars.

There are many excellent and comprehensive reviews on the health benefits of dietary fibre for humans, e.g., Anderson et al. [9–11]. In general, insoluble fibres can provide health benefits to the human gastrointestinal tract mainly through their bulking ability which can stimulate healthy bowel movements [10]. However, soluble fibres are generally considered to have more beneficial metabolic effects than those from insoluble fibres [9]. For example, their viscosity can help modulate nutrient digestion and absorption (e.g., blood glucose control). On the other hand, their fermentability can both (a) assist in maintaining a healthy gut microflora and (b) produce a range of metabolically beneficial fermentation products such as short-chain fatty acids that assist with cholesterol control and provide a substrate for healthy colon cell development [10,11].

### *2.1. Dietary Fibre of Lupin Hull*

The seed coat (hull) of lupin represents about 25% of the weight of the seed and consists mainly of structural non-starch polysaccharides that are classified as insoluble dietary fibres combined with low levels of protein and lipids, minerals, and phytochemicals such as polyphenols [12,13]. The non-starch polysaccharide in the hull is primarily cellulose with only low levels of lignins (an anti-nutritional factor) [14]. This hull dietary fibre has been characterised as 96.5% insoluble dietary fibre and 3.5% soluble dietary fibre. Recently, attempts to increase the level of soluble dietary fibre in lupin hull using extrusion cooking reported that a slight increase could be achieved [12]. The lupin seed hull has been used as an ingredient in human food, such as in high-fibre bread and meat products, and as a bulking agent [13]; however, most lupin hull undergoes little if any value addition and is disposed of as waste.

The remainder of this literature review will now focus on LKF.

### *2.2. Dietary Fibre of Lupin Kernel*

In contrast to the hull, the fibre in the lupin kernel contains more of the soluble fraction and a wider range of different classes of polysaccharides, including pectin substances, cellulose, and non-starch non-cellulosic glucans, with an absence of lignin [14]. However, the proportion of insoluble vs. soluble fibre varies widely in the literature. For instance, Naumann et al. [15] reported that the dietary fibre fraction of dehulled lupin seeds was primarily (approx. 90%) insoluble. In contrast, Turnbull et al. [16] found equal levels of insoluble and soluble dietary fibre in a purified LKF ingredient (88 g/100 g as its total dietary fibre). A reason for this could be the form in which the fibre was analysed, i.e., directly using lupin kernel flour for the fibre assay or firstly isolating the fibre before assaying its dietary fibre composition, which may have modified the cell wall construction making the constituent polysaccharides more soluble. In light of this, it is important to directly analyse the insoluble vs. soluble ratio of any lupin kernel dietary fibre ingredient processed in new or modified ways.

The dietary fibre in the kernels is in the form of thickened walls of the mesophyll cells [17,18], which mainly consist of non-starch polysaccharides and raffinose family oligosaccharides such as raffinose, stachyose, and verbascose [19,20]. More specifically, it is composed of the monosaccharides galactose (67.6%), arabinose (11.5%), uronic acids (8.1%), glucose (7.6%), and xylose (2.6%) [15]. The structure of the lupin kernel cell wall fibre is described as *"1-4-linked long-chain galactans and highly branched 1-5-linked arabinans, which are linked to the rhamnosyl residues of a rhamnogalacturonan backbone"* [21]. The presence of galacturonic acid backbone in the primarily insoluble cell wall fibre means that its polysaccharides structure is more similar to pectin, a soluble fibre, than to commonly found insoluble cellulosic dietary fibres [17]. Therefore, upon processing, we hypothesise that the "trapped" pectin within the cell wall changes from insoluble to being released from the cell wall matrix and thus becoming soluble.

## 3. Manufacture of LKF as a Food Ingredient

### *3.1. Processing Approaches*

Lupins, owing to their high protein and dietary fibre contents, have great potential in the manufacture of plant-based food ingredients, which are currently in great demand by the food industry. However, in manufacturing lupin protein concentrates/isolates, the LKF fraction is the main by-product for which value-added commercial utilisation is still in its infancy. The aim of any fractionation is to isolate and quantify fractions of interest and eliminate unwanted components [22]. Potential methods to manufacture LKF can be split into two most common processing types; wet (chemical) processing and dry (physical) processing [23], both to separate the protein and lipid from the dietary fibre using the starting material of lupin kernel flour, flakes, or grits. In addition, enzymatic extractions or combined extraction methods can be used [24,25]. The processing method can greatly affect the composition and properties of the resultant fibre-enriched fraction in food applications and its effect on the human body [25]. This literature review will focus on a comparison between dry and wet processing approaches.

#### 3.1.1. Wet Processing: Methods, Advantages and Disadvantages

The alkaline extraction-isoelectric precipitation method is the most common wet processing technique reported for the separation of protein and fibre of lupin kernels [26]. An example of a wet processing scenario to produce highly dietary-fibre enriched LKF is presented in Figure 2 in which the wet fibre residue is a major by-product of protein concentrate production. First, the protein is extracted from the wet-milled kernel or lupin flour at a high alkaline pH. Centrifugation then results in the protein extract (which is further processed to produce the protein concentrate/isolate) and the fibre residue. This approach provides large volumes of this high-moisture (approx. 80%) paste-like residue that has proven difficult to economically dry to a shelf-stable powder due in part to its very high water-binding properties. Spray drying or freeze-drying to produce the final dry powder ingredient has been reported in the literature [15,27,28]. However, innovation and optimisation for a commercially viable drying method are still required.

There are some drawbacks associated with wet processing for the manufacture of the lupin fractions. The major drawback of this fractionation approach is the requirement of large quantities of water, energy, and chemicals [29]. The high costs of wet processing are also due to extensive losses of solids in the acid-soluble whey (Figure 2) and the need to dry the products as well as recycle the effluents [30]. This method is also time-consuming [25]. In addition, the alkaline extraction and drying steps in wet processing may negatively impact important physico-functional properties of the fibre [26,28].

The techno-functional properties such as solubility and chemical composition of lupin fibre are affected by the extraction conditions; therefore, optimising the condition is crucial [24] to, for instance, improve the fibre yield and maintain or enhance its functionality and reduce processing times. One modified wet milling method involves much less water use and no harmful chemicals for extraction, while producing products with high purity [22].

A novel approach to overcome the issue of drying the LKF residue is to process it directly using high-temperature and high-pressure extrusion cooking that will dry, sterilise, and texturize the fibre in one high throughput and easily up-scalable process. This will give shelf-stable "extrudate pieces" that may be used as a very high fibre ingredient in foods such as breakfast cereal and muesli bars. Alternatively, the extrudate pieces can be milled to give a very high dietary fibre powder with a multitude of application possibilities [25]. The extrusion cooking techniques can improve the colour, flavour, and stability of the fibre fractions [28], as can the total dietary fibre yield [31]. Extrusion cooking, due to its high temperature, pressure, and shear force, can increase the ratio of soluble dietary fibre through the breakdown of bonds of insoluble polysaccharides, converting them into soluble fractions [32,33]. The pectin-like polymers in the cell walls of LKF thus appear as a prime target for solubilisation using extrusion cooking LKF [23]. This potential has recently been reported by Naumann et al. [15,34], who confirmed that extrusion cooking increased the solubility, water binding, and viscosity but decreased the bile acid diffusion, indicating the cholesterol-lowering potential of fibres and thus showing great potential for producing a more health-enhancing dietary fibre ingredient.

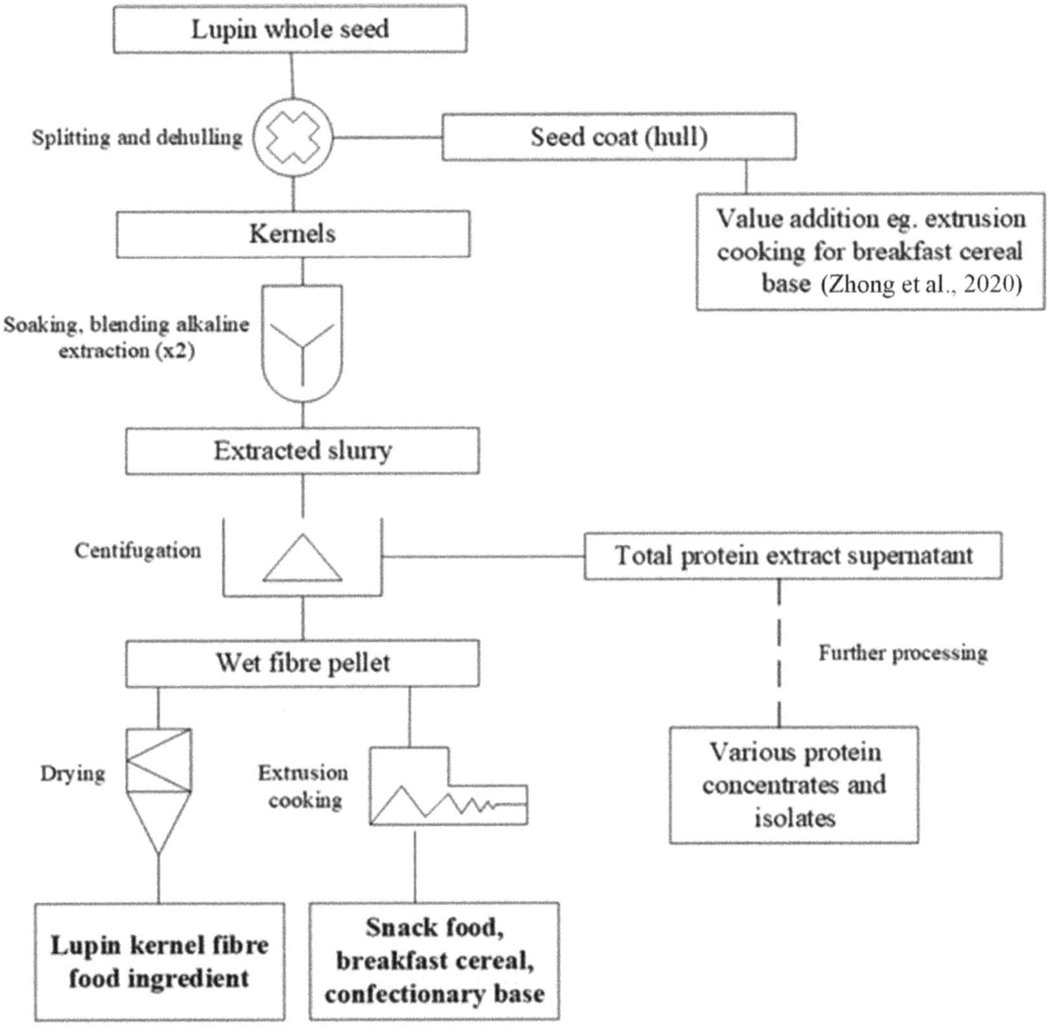


**Figure 2.** Conventional wet processing for the manufacture of LKF illustrating novel direct extrusion cooking [2].

3.1.2. Dry Processing: Methods, Advantages and Disadvantages

Dry processing is used to prepare fibre-enriched fractions from legumes by disintegrating seeds through the process of milling and then air classification into starch, protein, and fibre fractions [22]. For instance, pin-milling of legume seeds results in distinct populations of particles that differ in both size and density. The air classification technique is used to separate the light or fine fraction (containing mostly starches and fibres) from the coarse and relatively heavier fraction (containing mainly proteins and lipids) [35]. Air classification is repeated several times to purify further fractions [25,36]. The advantages of this dry process include: the relatively simple construction of processing plants, no wastewater production, and minimal changes in the structure and functional properties of the components [37]. Therefore, if air classification can be used as an alternative to wet processing techniques for the production of protein and fibre fractions from lupin, it has many advantages, including low capital and labour costs, a less costly effluent disposal system, and minimal sanitation.

The use of dry fractionation to separate the protein from the fibre fraction (lupin has negligible starch content), however, has some drawbacks, including the need for many repetitions of the air classification that can lead to low product recovery [22]. It has been reported that air classification is an efficient method for fractionating legumes that have starch as their main storage material, such as peas and faba beans, but less so for lupin in which the storage of non-starch polysaccharides (cell wall fibre) is more difficult to separate from the protein leading to less purity in the final fractions [25,29]. Gueguen [36] stated that the air classification process technique does not give satisfactory results for lipid-rich seeds such as lupin, wrinkled pea, chickpea, or soybean; however, *L. angustifolius* has a relatively low lipid content compared to some other lupin species and as such the lipids may not be a hindrance to its fractionation. Particle size is critical to efficient separation in air classification; therefore, by decreasing particle size through multiple passes, the purity of the targeted fraction can be improved. The fine fraction purity, such as fibre, can also be increased by lower moisture during air classification [38,39]. Sosulski and Sosulski [30] reported that when using air classification, most of the anti-nutritional factors are recovered in the fibre fraction of legumes; however, sweet varieties of *L. angustifolius* (Australian sweet lupin) are low in alkaloids, and the anti-nutritional factors associated with some other legumes (e.g., trypsin inhibitors) are absent. It is recommended that the alkaloid content of dry fractionated LKF is tested to ensure any concentration effect has not increased its levels above the maximum permitted level of 200 mg/100 g in Australia and New Zealand [40].

Recently, the dry processing method of electrostatic separation has been reviewed for the fractionation of plant materials [41]. This method is based on the different triboelectric charging properties of materials, e.g., different contact electrification when fibre and proteins in lupin are rubbed together, allowing them to be separated between electrodes. This review cited the levels of protein purification from lupin flour by a range of dry processing methods (Table 1) that can indicate the concomitant purification of the fibre fraction [41]. The reason for showing protein purity in Table 1 was that there is no published data on the dry fractionation of lupin kernels that reports the dietary fibre purity of the high fibre fraction. However, the main constituents of the lupin kernel are dietary fibre and protein, and the dry fractionation process gives two main fractions: the "high protein" fraction and the "high dietary fibre" fraction. Therefore, if a process provides the protein with a fraction of high protein purity, it follows that the dietary fibre fraction from that process gives a high dietary fibre purity. It can be seen from Table 1 that using the current technology, electrostatic separation is not superior to air classification; however, technical improvements to the electrostatic separation methods are being researched [41]. The option of pre-concentrating with dry fractionation before the final separation with wet processing is also highlighted as a potential mixed-method approach [41,42]. Figure 3 shows a schematic diagram of how electrostatic separation can be combined with more conventional methods, which may result in increased efficiencies and purities.

**Table 1.** Protein enrichment of lupin by various dry fractionation and mixed (dry + wet) methods. Adapted from [39].

| Method | Protein Purity (g/100 g) | | Reference |
|---|---|---|---|
| | Before | After | |
| Air classification | 40.4 | 59.4 | [43] |
| Electrostatic separation | 40.5 | 57.3 | [44] |
| Recycling electrostatic separation (of protein concentrate) | 57.3 | 65.1 | [44] |
| Air classification + electrostatic separation | 45.1 | 59.3 | [45] |
| Dry separation + Aqueous fractionation | 53.5 | >80 | [42] |

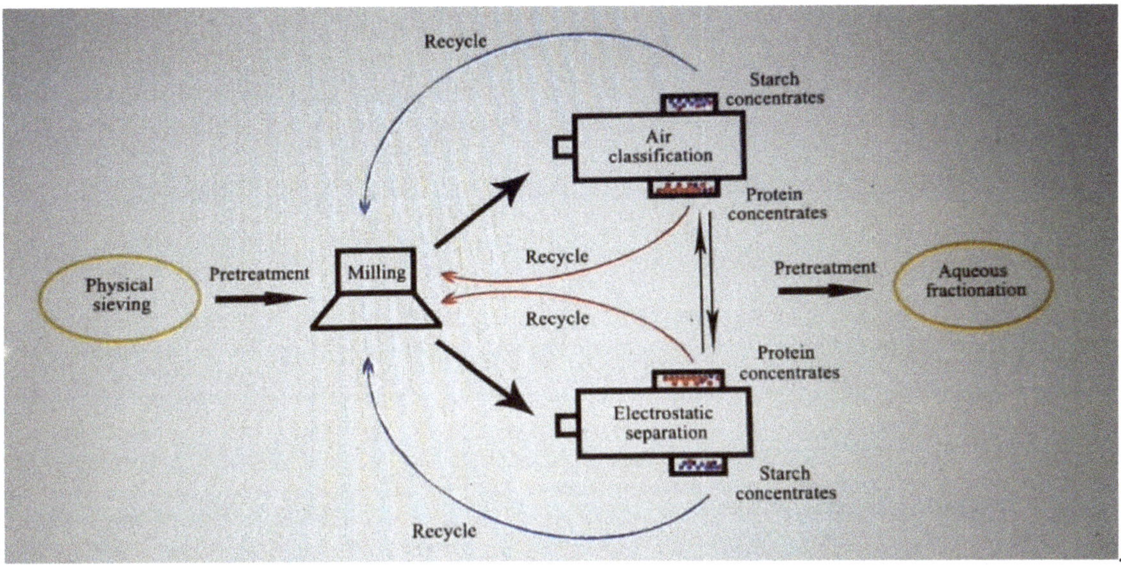


**Figure 3.** Examples of how electrostatic separation can be combined with more conventional separation methods. Reprinted with permission from [41]. 2022, Tong, L.-T.

## 4. Composition and Techno-Functionality of LKF Food Ingredients

### *4.1. Typical Composition of LKF Food Ingredients*

Table 2 presents the composition of lupin (*L. angustifolius*) kernel fibre food ingredients reported in the literature. All these examples were produced using wet processing by alkaline extraction and acid precipitation followed by the drying of the protein precipitate. Full fat lupin was used in all of the studies except that of Fechner et al. [46]. These data show the high level of purification of the dietary fibres, the residual level of protein and to a lesser extent fat, even when the kernels were not defatted. This is because most of the fat will solubilise in the highly alkaline conditions used to dissolve the protein and will not associate with the insoluble fibre residue.

**Table 2.** Proximate and dietary fibre composition of LKF food ingredients reported in the literature. These were produced by alkaline extraction and acid precipitation. [1] Full fat lupin used for manufacture, [2] defatted lupin used for manufacture.

| Energy kj/100 g | Protein g/100 g | Available Carbohydrate g/100 g | Total Dietary Fibre g/100 g | Soluble Dietary Fibre g/100 g | Insoluble Dietary Fibre g/100 g | Fat g/100 g | Ash g/100 g | Reference |
|---|---|---|---|---|---|---|---|---|
| - | 9.0 | - | 80.2 | 48.7 | 31.5 | 1.0 | 1.5 | [46] [1] |
| - | - | - | 77.5 | - | - | - | - | [47] [2] |
| 883 | 5.9 | <0.1 | 88 | 44.8 | 43.2 | 2.1 | - | [48] [1] |
| - | 5.7 | - | 77.1 | 8.5 | 68.6 | 2.5 | 1.2 | [49] [1] |
| - | 11.1 | 3.7 | 83.3 | 3.7 | 79.7 | - | 1.8 | [50] [2] |

### *4.2. Physicochemical Properties of LKF Food Ingredients*

A wide range of physicochemical properties influences the effect of adding dietary fibre ingredients into food products. These include hydration food properties, binding of fat/oil, available surface area and porosity, fibre particle size and bulk volume, and ion exchange capacity [51]. Only a few publications have reported some of these properties for LKF.

#### 4.2.1. Colour, Odour, Flavour and Texture

Colour, odour, flavour, and texture are factors that must be considered in the application of fibre ingredients due to their impact on the sensory characteristics of foods in which they are incorporated. One aim during the manufacture of LKF is to produce a product nearly white in colour, with little odour and a neutral flavour [52]. Thus, its pale colour and low odour and flavour make LKF suitable for fibre enrichment of a wide range of foods such as dairy, baked goods, and meat products; it was dubbed an "invisible fibre" [47]. However, Stephany et al. reported non-enzymatic oxidation in LKF during storage and gave it an unacceptable odour as determined by sensory evaluation. The authors recommended preheat treatment of the lupin seed to reduce the lipoxygenase activity prior to the manufacture of the LKF. In addition, LKF has a smooth texture that makes it an excellent ingredient for fibre enrichment of food formulations [23]. In contrast, lupin kernel flour has some limitations as a fibre enrichment ingredient in foods due to its pale yellow colour and slight beany flavour [53].

#### 4.2.2. Hydration, Water Binding and Viscosity

Hydration properties of the fibre depend on the chemical and physical structures, environmental conditions of the aqueous solution, and the different processing treatments applied to extract the fibre. Hydration terms such as water binding, water holding, and water retention are used interchangeably [51]. Analytically, the water-binding capacity (WBC) of a fibre refers to its ability to bind water and hold it under centrifugal force, with high and stable water-binding properties of fibres preferable for most food product development applications [54]. However, high water binding of lupin kernel fibres can reduce the level of water available for the development of gluten in leavened bread, preventing the full formation of the visco-elastic network in the dough needed to trap gas during fermentation; therefore, for this application, lower water-binding fibres are easier to incorporate [52]. In contrast, the high water-binding capacity of LKF could result in the production of a bread with a slower staling rate through inhibiting moisture migration in the crumb and loss through the crust; however, there is no published evidence of this potentially useful phenomenon. Fibres aid in the modification of food texture through water retention. Various food processing treatments, such as extrusion or grinding, can modify hydration properties and improve functionality [51]. The water-binding capacity of LKF at 11 mL/g dry solids is high compared with other primarily insoluble fibre types, e.g., soy kernel fibre (7 mL/g dry solids), pea hull (5 mL/g dry solids), cellulose (5 mL/g dry solids), and wheat bran fibre (4 mL/d dry solids) [16]. The high water-binding capacity of the lupin kernel results from the high level of pectin-like hydrophilic non-starch polysaccharides embedded in the cell wall structure [54].

In general, the viscosity of primarily insoluble fibre ingredients such as LKF is lower than that of soluble fibre ingredients such as pectins, gums, and β-glucans. However, amongst the primarily insoluble fibres, Turnbull et al. [16] reported a higher viscosity for LKF than for soy kernel fibre, pea hull, cellulose, and wheat bran fibre. Thus, LKF has unusual hydration properties, intermediate between those of soluble and insoluble fibres ingredients [16].

#### 4.2.3. Oil Binding

Fat/oil binding is an important physicochemical property of dietary fibre food ingredients [51]. LKF can interact with oil in a food formulation, and according to McCleary and Prosky [54], this good oil-binding potential plus its water-binding capacity makes it an ideal additive for food preparation such as burgers, as well as a potential fat part-replacer in low-fat processed meat products [55].

#### 4.2.4. Consumer Acceptability of Lupin Kernel Fibre-Enriched Food Products

In the development of fibre-enriched products, the sensory properties, including taste, are expected to be like the equivalent convention product. LKF can fulfil this desire as they have a neutral colour, odour, flavour, and smooth texture that assists in the sensory acceptability of food formulations enriched with this fibre source [48,52,54]. There are only very few published studies on the sensory acceptability of foods containing LKF. As outlined by McCleary and Prosky [54], LKF can improve baking stability without affecting taste and is an ideal choice for baking-stable fillings or toppings and coatings for fried foods due to its texturizing properties and the high and stable water-binding and good oil-binding properties. Foods incorporated with LKF, such as white bread, muffin, pasta, orange juice, and breakfast bar, have demonstrated acceptable palatability in sensory evaluation trials, though some food types of the LKF variant showed lower palatability compared to the control. In this study, the products were enriched with at least 3 g/serving of the lupin kernel fibre. No changes in overall acceptability were observed for fibre enrichment at serving levels of 4.4 and 7.3 g in bread and pasta compared to 5.5, 2.9, and 5.4 g in the muffin, the orange juice, and the breakfast bar, respectively. In all foods enriched with fibre, the flavour significantly affected the overall acceptability. This means the specific food formulation with added lupin kernel fibre requires knowledge of physicochemical properties of fibres and their interactions in the food matrices and processing-induced changes [47]. In a study by Hall et al. [56], the liking of LKF-containing foods (muesli, bread, muffin, chocolate brownie, chocolate milk drink and pasta) was evaluated after repeated consumption in a dietary setting and it was reported that the fibre addition gave no severe effects on product palatability. In contrast to LKF, lupin flour has a beany flavour that can reduce the sensory attribute of products to which it is added; this may have contributed to its slow uptake in baked products due to poor sensory quality [57].

## 5. Commercial Examples of LKF Food Ingredients and Their Use in Foods

There are very few examples of commercially available LKF food ingredients or commercial applications of it in food products. Prolupin GmbH [58] advertises that they make a dietary fibre ingredient "from the innermost parts of the seed", suggesting it may be LKF. The fibre is described as having a smooth mouthfeel and can be used as a "roughage" source and fat substitute, such as in meat products. However, no standard product information form was available on the web page. Prolupin also markets food products containing their ingredients under the brand "Made with LUVE" [59]; however, it was not possible from the web page to determine if LKF was used in any of their products.

## 6. Evidence of Health Benefits of LKF from Human Trials

A summary of the evidence of health benefits of LKF from human (clinical) trials is presented in Table 3; each study is described in more detail and recommendations for further research are presented. There are also some cell model and animal model trials giving supporting evidence to these clinical trials, but a discussion of these is beyond the scope of this review.

**Table 3.** A summary of the findings of clinical studies investigating effects of LKF intake on markers/physiological responses of chronic diseases.

| Reference | Effect on Biomarker/Physiological Responses | | | | | |
|---|---|---|---|---|---|---|
| | Satiety/ Weight Loss | Blood Glucose | Blood Cholesterol | Blood Pressure | Bowel Function | Probiotic |
| | | | *Short-term (post-meal) studies* | | | |
| [49] | ND | 0 | ND | ND | ND | ND |
| [55] | +ve | ND | ND | ND | ND | ND |
| | | | *Longer term (dietary intervention) studies* | | | |
| [48] | ND | 0 | +ve | ND | ND | ND |
| [60] | ND | ND | ND | ND | +ve | ND |

**Table 3.** *Cont.*

| Reference | Effect on Biomarker/Physiological Responses | | | | | |
|---|---|---|---|---|---|---|
| | Satiety/ Weight Loss | Blood Glucose | Blood Cholesterol | Blood Pressure | Bowel Function | Probiotic |
| [61] | ND | ND | ND | ND | ND | +ve |
| [46] | ND | ND | 0 | ND | +ve | ND |
| [62] | ND | ND | +ve | ND | +ve | ND |

+ve the lupin treatment gave significantly improved levels of the biomarker/physiological response compared to the control (non-lupin) treatment; 0 no difference in the levels of the biomarker between the lupin and the control treatment; ND biomarker not assessed, or experimental design not valid for comparison.

### *6.1. Metabolic Syndrome Protection*

Major risk factors for the development of cardiovascular disease are: (i) obesity; (ii) elevated total cholesterol (TC) and triglycerides and low-density lipoprotein cholesterol (LDL-C) concentrations; (iii) insulin resistance, which is usually indicated by high blood glucose and insulin concentrations; and (iv) high blood pressure. The presence of some or all of these risk factors is known as the metabolic syndrome; a condition afflicting up to one quarter of the world's population. The unique properties of LKF—bioactive protein complex (Section 4.2)—could help lower these risk factors as part of a healthy diet and lifestyle.

### *6.2. Appetite and Body Weight Reduction*

Foods that are highly satiating, in that they strongly reduce appetite after eating, may help in longer-term reduction in food intake and therefore assist with maintaining a healthy body weight. One critical study demonstrating the highly satiating effect of LKF in a post-meal setting was reported by Archer et al. [55]. In this study, 38 men consumed breakfast either with a full-fat sausage patty or a reduced-fat patty where some fat was replaced by lupin fibre. The participants reported that the LKF-containing breakfast gave higher ($p < 0.05$) perceptions of "fullness" in the post-meal period and lower ($p < 0.05$) total energy intake over the day than the full-fat breakfast even though it had lower total energy. The authors hypothesised that this potentially beneficial effect was due to a combination of high water-binding properties in the stomach and small intestine [16] and fermentation of the LKF to short-chain fatty acids in the colon [60].

The post-meal study by Archer et al. [55], however, does not provide direct evidence of the ability of LKF to help fight obesity. For this evidence, longer-term (over weeks or months) double-blind trials in overweight and/or obese participants are required in which the body weight of the participants on a diet with kernel fibre-containing foods is compared to one with control foods without the fibre. This study has yet to be reported for LKF but is highly recommended to gain crucial evidence of any "anti-obesity" properties of this fibre.

### *6.3. Cholesterol and Blood Pressure*

High blood pressure and elevated fasting total cholesterol levels, low-density lipoprotein cholesterol and triglycerides, and low levels of high-density lipoprotein cholesterol are biomarkers of increased risk of cardiovascular disease [63]. Several studies have investigated the effect of diets containing LKF on these cardiovascular disease risk biomarkers. Hall et al. [48] found a clinically significant ($p < 0.05$) reduction in LDL-cholesterol and TC in overweight but otherwise healthy men when they consumed LKF (30 g/day) incorporated into foods compared to when they consumed the same food without the fibre addition for 28 days in a randomised cross-over study. The authors hypothesised that this beneficial effect might be due to the LKF inhibiting cholesterol re-absorption and increasing the production of short-chain fatty acids in the colon. They also noted that the residual protein in the LKF ingredient may have played a role; however, this remains unclear.

Fechner et al. [46] performed a similar study to Hall et al. [48] in participants with normal cholesterol levels but found no cholesterol-lowering effects of the LKF-containing diet. However, when they performed the study with a cohort of moderately hypercholesteremic adults, a cholesterol-lowering effect of the LKF-containing diet was found; the authors attributed this effect to increases in short-chain fatty acids butyrate and acetate found in the faeces [62].

No studies have been reported investigating the effect of LKF ingestion on blood pressure; however, several studies have shown beneficial effects on blood pressure from consumption of lupin kernel flour-containing foods in non-diabetic participants [64,65]. The authors hypothesised that the effect was due to the increased protein and fibre in the lupin flour-containing diet. In contrast, a recent similar study but in well-controlled type 2 diabetic participants found no effect of the lupin kernel flour-containing diet on blood pressure [66].

### *6.4. Protection from Type 2 Diabetes*

Loss of control of blood glucose levels after a meal is a key risk factor for type 2 diabetes; therefore, the measurement of the blood glucose response of food after a meal compared with a standard and calculation of the food glycaemic index (GI) has become a popular technique for evaluating the "healthiness" of foods in terms of maintaining good blood glucose control. However, the GI is only applicable for foods with high levels of available carbohydrates (digestible starch and sugars), since the participant is required to inject a test meal containing 50 g of available carbohydrates [67]. Since the level of available carbohydrates in both lupin seed/flour and LKF is very low, their GI cannot be determined [25]. Adding lupin seed/flour of LKF to food can lower its glycemic load (GL), which is the GI multiplied by its available carbohydrate content (g/per serving) divided by 100. Inclusion of low GL foods is recommended for diets to protect against type 2 diabetes [68]. There is potential that LKF, when incorporated into foods such as wheat bread, can interact with 50 g of available carbohydrates from the wheat to lower blood glucose response, and thus the GI of the bread, through its high water-binding capacity that could slow stomach emptying and inhibit starch digestion and glucose absorption in the small intestine [27]. One study has reported the post-meal effects on the glycaemic response of food with added LKF. There was no difference in the post-meal blood glucose response in 21 healthy adults with the LKF compared to the control standard white bread [49]. The lack of effect of the LKF bread may be related to its low inclusion rate (control bread 2.7% total dietary fibre; LKF bread 8.5% TDF) in order to maintain good acceptability of the LKF bread by the participants. It is recommended that further optimisation of LKF bread manufacture is performed in order to allow a higher rate of LKF inclusion whilst maintaining palatability and to then test the glycaemic response/GI of this bread. A nutrient content labelling claim for low glycaemic index and glycaemic load can be made for foods under conditions specified by the FSANZ Food Standards Code, Schedule 4 Nutrition, health, and related claims [69].

More substantial evidence for the type 2 diabetes protective effects of LKF would be its potential to reduce fasting blood glucose and/or insulin and glycosylated haemoglobin (a marker of chronic high blood sugar levels) after consuming food containing the fibre compared to control foods without the fibre for weeks or months. One study by Hall et al. [48] reported no reduction ($p > 0.05$) in fasting glucose and insulin in 38 overweight men who consumed diets containing food with and without LKF in a cross-over design. Further studies on the effect of LKF ingestion in participants with elevated fasting blood glucose (pre-diabetic) are warranted.

### *6.5. Potential Gastrointestinal Health Benefits*

The great importance of the microbiological ecology of the colon and the role of colonic bacterial fermentation of the dietary fibre for health is now fully recognised. This has recently been comprehensively reviewed with respect to the effects of whole grains and their fibre fractions consumption by Seal et al. [70]. There is good evidence from several independent studies that LKF consumption can beneficially affect microbiological (i.e., the balance of "good" probiotics and "bad" potential pathogens) and chemical markers of good bowel health and function, and thus be classified as a 'prebiotic' food ingredient.

Several studies have provided important evidence that LKF can act as a prebiotic ingredient and promote the growth of desirable gut bacteria while supporting digestive system function. For example, a study by Smith et al. [61] reported reduced faecal levels of Clostridia bacteria (potential pathogens) and increased levels of Bifidobacterium (beneficial probiotics) in 38 overweight but otherwise healthy men after consuming a 28-day diet incorporating food containing LKF (approx. 28 g of LKF per day) compared with a diet containing control foods with no LKF.

Improved markers of healthy bowel function (e.g., reduced transit time) and increased levels of health-protective compounds in the faeces (e.g., increased levels of the short-chain fatty acid butyrate, a substrate for healthy colonic cell development) are noted [60]. Similarly, Fechner et al. [46,62] also reported that adding LKF to the diet improved bowel function and faecal chemistry makers.

## 7. Conclusions

Dietary fibre is a major fraction of the lupin kernel, so its utilisation is critical when manufacturing food ingredients from lupin. As a food ingredient with a light colour and little flavour and aroma, LKF can act as an "invisible" fibre source to boost the fibre levels in foods, thus allowing dietary fibre nutrient content claims. Sensory evaluation trials performed with foods incorporated with LKF (at least 3 g/serving), such as white bread and pasta, demonstrated a higher acceptable palatability. In contrast, the muffin, orange juice, and breakfast bar showed lower palatability than the control. This implies that the fibre physicochemical properties and their interactions with the food systems become crucial for product formulations. In addition, the high water-binding properties of the fibre are a commercial barrier to drying it to a stable powder food ingredient. Therefore, investigations into an alternative drying approach such as extrusion cooking of the wet fibre into high fibre extrudates are suggested.

Current evidence suggests that LKF, with its unique non-starch polysaccharide structure, may have potential health benefits in the human diet, related to satiety, blood cholesterol levels and prebiotic activity; however, further independent research studies are required before health claims for this fibre can be made. LKF appears to have commercial potential as a new dietary fibre, but there are currently few, if any, commercially available LKF food ingredients.

**Author Contributions:** Writing—original paper preparation, writing—reviewing and editing, R.M.; conceptualisation, writing—original paper preparation, writing—reviewing and editing, S.K.J.; funding acquisition, project administration, conceptualisation, writing—reviewing and editing, R.R.B. All authors have read and agreed to the published version of the manuscript.

**Funding:** This literature review work was supported by Wide Open Agriculture Ltd., Western Australia, Project number RES64098, 2021.

**Conflicts of Interest:** The authors declare no conflict of interest.

## References

1. Písaříková, B.; Zralý, Z. Dietary fibre content in lupine (*Lupinus albus* L.) and soya (*Glycine max* L.) seeds. *Acta Vet. Brno* **2010**, *79*, 211–216. [CrossRef]
2. Zhong, L.; Ali, H.; Fang, Z.; Wahlqvist, M.L.; Hodgson, J.M.; Johnson, S.K. Lupin seed coat as a promising food ingredient: Physicochemical, nutritional, antioxidant properties, and effect of genotype and environment. *Int. J. Food Sci. Technol.* **2020**, *55*, 1816–1824. [CrossRef]
3. Villarino, C.B.; Jayasena, V.; Coorey, R.; Chakrabarti-Bell, S.; Johnson, S. The effects of bread-making process factors on Australian sweet lupin-wheat bread quality characteristics. *Int. J. Food Sci. Technol.* **2014**, *49*, 2373–2381. [CrossRef]

4. Augustin, L.S.; Aas, A.-M.; Astrup, A.; Atkinson, F.S.; Baer-Sinnott, S.; Barclay, A.W.; Brand-Miller, J.C.; Brighenti, F.; Bullo, M.; Buyken, A.E. Dietary fibre consensus from the international carbohydrate quality consortium (ICQC). *Nutrients* **2020**, *12*, 2553. [CrossRef] [PubMed]
5. Bhattarai, R.R.; Dhital, S.; Williams, B.A.; Yang, H.J.; Mikkelsen, D.; Flanagan, B.M.; Gidley, M.J. In vitro fermentation of legume cells and components: Effects of cell encapsulation and starch/protein interactions. *Food Hydrocoll.* **2021**, *113*, 106538. [CrossRef]
6. Stephen, A.M.; Champ, M.M.-J.; Cloran, S.J.; Fleith, M.; Van Lieshout, L.; Mejborn, H.; Burley, V.J. Dietary fibre in Europe: Current state of knowledge on definitions, sources, recommendations, intakes and relationships to health. *Nutr. Res. Rev.* **2017**, *30*, 149–190. [CrossRef] [PubMed]
7. Dhingra, D.; Michael, M.; Rajput, H.; Patil, R. Dietary fibre in foods: A review. *J. Food Sci. Technol.* **2012**, *49*, 255–266. [CrossRef] [PubMed]
8. Goff, H.D.; Repin, N.; Fabek, H.; El Khoury, D.; Gidley, M.J. Dietary fibre for glycaemia control: Towards a mechanistic understanding. *Bioact. Carbohydr. Diet. Fibre.* **2018**, *14*, 39–53. [CrossRef]
9. Anderson, J.W.; Baird, P.; Davis, R.H.; Ferreri, S.; Knudtson, M.; Koraym, A.; Waters, V.; Williams, C.L. Health benefits of dietary fiber. *Nutr. Rev.* **2009**, *67*, 188–205. [CrossRef]
10. Slavin, J.L. Position of the American Dietetic Association: Health implications of dietary fiber. *J. Am. Diet. Assoc.* **2008**, *108*, 1716–1731.
11. Slavin, J. Fiber and prebiotics: Mechanisms and health benefits. *Nutrients* **2013**, *5*, 1417–1435. [CrossRef] [PubMed]
12. Zhong, L.; Fang, Z.; Wahlqvist, M.L.; Hodgson, J.M.; Johnson, S.K. Multi-response surface optimisation of extrusion cooking to increase soluble dietary fibre and polyphenols in lupin seed coat. *LWT* **2021**, *140*, 110767. [CrossRef]
13. Zhong, L.; Fang, Z.; Wahlqvist, M.L.; Wu, G.; Hodgson, J.M.; Johnson, S.K. Seed coats of pulses as a food ingredient: Characterisation, processing, and applications. *Trends Food Sci. Technol.* **2018**, *80*, 35–42. [CrossRef]
14. Mohamed, A.; Rayas-Duarte, P. Nonstarchy polysaccharide analysis of cotyledon and hull of *Lupinus albus. Cereal Chem.* **1995**, *72*, 648–651.
15. Naumann, S.; Schweiggert-Weisz, U.; Haller, D.; Eisner, P. Retention of primary bile acids by lupin cell wall polysaccharides under in vitro digestion conditions. *Nutrients* **2019**, *11*, 2117. [CrossRef]
16. Turnbull, C.M.; Baxter, A.L.; Johnson, S.K. Water-binding capacity and viscosity of Australian sweet lupin kernel fibre under in vitro conditions simulating the human upper gastrointestinal tract. *Int. J. Food Sci. Nutr.* **2005**, *56*, 87–94. [CrossRef] [PubMed]
17. Brillouet, J.M.; Riochet, D. Cell wall polysaccharides and lignin in cotyledons and hulls of seeds from various lupin (*Lupinus* L.) species. *J. Sci. Food Agric.* **1983**, *34*, 861–868. [CrossRef]
18. Crawshaw, L.A.; Reid, J.G. Changes in cell-wall polysaccharides in relation to seedling development and the mobilisation of reserves in the cotyledons of *Lupinus angustifolius* cv. Unicrop. *Planta* **1984**, *160*, 449–454. [CrossRef]
19. Evans, A.J.; Cheung, P.C.K.; Cheetham, N.W. The carbohydrate composition of cotyledons and hulls of cultivars *of Lupinus angustifolius* from Western Australia. *J. Sci. Food Agric.* **1993**, *61*, 189–194. [CrossRef]
20. Trugo, L.; Von Baer, D.; Von Baer, E. Lupin. In *Encyclopedia of Food Sciences and Nutrition*; Academic Press: Cambridge, MA, USA, 2003; pp. 3623–3629.
21. Cheetham, N.W.; Cheung, P.C.-K.; Evans, A.J. Structure of the principal non-starch polysaccharide from the cotyledons of *Lupinus angustifolius* (cultivar Gungurru). *Carbohydr. Polym.* **1993**, *22*, 37–47. [CrossRef]
22. Elleuch, M.; Bedigian, D.; Roiseux, O.; Besbes, S.; Blecker, C.; Attia, H. Dietary fibre and fibre-rich by-products of food processing: Characterisation, technological functionality and commercial applications: A review. *Food Chem.* **2011**, *124*, 411–421. [CrossRef]
23. Bader Ul Ain, H.; Saeed, F.; Ahmed, A.; Asif Khan, M.; Niaz, B.; Tufail, T. Improving the physicochemical properties of partially enhanced soluble dietary fiber through innovative techniques: A coherent review. *J. Food Processing Preserv.* **2019**, *43*, e13917. [CrossRef]
24. Tejada-Ortigoza, V.; Garcia-Amezquita, L.E.; Serna-Saldívar, S.O.; Welti-Chanes, J. Advances in the functional characterisation and extraction processes of dietary fiber. *Food Eng. Rev.* **2016**, *8*, 251–271. [CrossRef]
25. Yangilar, F. The application of dietary fibre in food industry: Structural features, effects on health and definition, obtaining and analysis of dietary fibre: A review. *J. Food Nutr. Res.* **2013**, *1*, 13–23. [CrossRef]
26. Chew, P.G.; Casey, A.J.; Johnson, S.K. Protein quality and physico-functionality of Australian sweet lupin (*Lupinus angustifolius* cv. Gungurru) protein concentrates prepared by isoelectric precipitation or ultrafiltration. *Food Chem.* **2003**, *83*, 575–583. [CrossRef]
27. Johnson, S.K.; Clements, J.; Villarino, C.B.J.; Coorey, R. Lupins: Their unique nutritional and health-promoting attributes. In *Gluten-Free Ancient Grains*; Elsevier: Amsterdam, The Netherlands, 2017; pp. 179–221. [CrossRef]
28. Xiaoran, D.; Suyan, G.; Shoumin, J.; Zhang, L.; Ruihuan, D.; Xiang, A.; Biao, Q.; Lei, W. Research Status and Progress on Modification of Dietary Fiber at Home and Abroad. *Agric. Biotechnol.* **2019**, *8*, 131–135.
29. Maphosa, Y.; Jideani, V.A. Dietary fiber extraction for human nutrition—A review. *Food Rev. Int.* **2016**, *32*, 98–115. [CrossRef]
30. Sosulski, F.; Sosulski, K. Composition and functionality of protein, starch, and fiber from wet and dry processing of grain legumes. *Plant Proteins Appl. Biol. Eff. Chem.* **1986**, *312*, 176–189. [CrossRef]
31. Stojceska, V.; Ainsworth, P.; Plunkett, A.; İbanoğlu, Ş. The advantage of using extrusion processing for increasing dietary fibre level in gluten-free products. *Food Chem.* **2010**, *121*, 156–164. [CrossRef]
32. Wolf, B. Polysaccharide functionality through extrusion processing. *Curr. Opin. Colloid Interface Sci.* **2010**, *15*, 50–54. [CrossRef]

33. Yang, Y.-Y.; Ma, S.; Wang, X.-X.; Zheng, X. Modification and application of dietary fiber in foods. *J. Chem.* **2017**, *2017*, 9340427. [CrossRef]
34. Naumann, S.; Schweiggert-Weisz, U.; Martin, A.; Schuster, M.; Eisner, P. Effects of extrusion processing on the physiochemical and functional properties of lupin kernel fibre. *Food Hydrocoll.* **2021**, *111*, 106222. [CrossRef]
35. Wang, N.; Maximiuk, L. Effect of air classification processing variables on yield, composition, and certain antinutrients of air-classified fractions from field peas by response surface methodology. *J. Food Processing Preserv.* **2019**, *43*, e13999. [CrossRef]
36. Gueguen, J. Legume seed protein extraction, processing, and end product characteristics. *Plant Foods Hum. Nutr.* **1983**, *32*, 267–303. [CrossRef]
37. Hedley, C.L. *Carbohydrates in Grain Legume Seeds: Improving Nutritional Quality and Agronomic Characteristics*; CABI: Wallingford, UK, 2001.
38. Arntfield, S.; Maskus, H. Peas and other legume proteins. In *Handbook of Food Proteins*; Elsevier: Amsterdam, The Netherlands, 2011; pp. 233–266. [CrossRef]
39. Tyler, R.; Panchuk, B. Effect of seed moisture content on the air classification of field peas and faba beans. *Cereal Chem.* **1982**, *59*, 31–33.
40. Food Standards Australia New Zealand. Maximum Levels of Contaminants and Natural Toxicants. Australia New Zealand Food Standards-Schedule 19. 2017. Available online: https://www.foodstandards.gov.au/code/Documents/Sched%2019%20Contaminant%20MLs%20v157.pdf (accessed on 21 September 2021).
41. Zhu, H.-G.; Tang, H.-Q.; Cheng, Y.-Q.; Li, Z.-G.; Tong, L.-T. Electrostatic separation technology for obtaining plant protein concentrates: A review. *Trends Food Sci. Technol.* **2021**, *113*, 66–76. [CrossRef]
42. Berghout, J.; Pelgrom, P.; Schutyser, M.; Boom, R.; Van Der Goot, A. Sustainability assessment of oilseed fractionation processes: A case study on lupin seeds. *J. Food Eng.* **2015**, *150*, 117–124. [CrossRef]
43. Silventoinen, P.; Sipponen, M.H.; Holopainen-Mantila, U.; Poutanen, K.; Sozer, N. Use of air classification technology to produce protein-enriched barley ingredients. *J. Food Eng.* **2018**, *222*, 169–177. [CrossRef]
44. Wang, J.; Zhao, J.; De Wit, M.; Boom, R.M.; Schutyser, M.A. Lupine protein enrichment by milling and electrostatic separation. *Innov. Food Sci. Emerg. Technol.* **2016**, *33*, 596–602. [CrossRef]
45. Pelgrom, P.J.; Wang, J.; Boom, R.M.; Schutyser, M.A. Pre-and post-treatment enhance the protein enrichment from milling and air classification of legumes. *J. Food Eng.* **2015**, *155*, 53–61. [CrossRef]
46. Fechner, A.; Fenske, K.; Jahreis, G. Effects of legume kernel fibres and citrus fibre on putative risk factors for colorectal cancer: A randomised, double-blind, cross-over human intervention trial. *Nutr. J.* **2013**, *12*, 101. [CrossRef] [PubMed]
47. Clark, R.; Johnson, S. Sensory acceptability of foods with added lupin (*Lupinus angustifolius*) kernel fiber using pre-set criteria. *J. Food Sci.* **2002**, *67*, 356–362. [CrossRef]
48. Hall, R.; Johnson, S.; Baxter, A.; Ball, M. Lupin kernel fibre-enriched foods beneficially modify serum lipids in men. *Eur. J. Clin. Nutr.* **2005**, *59*, 325–333. [CrossRef] [PubMed]
49. Johnson, S.K.; McQuillan, P.L.; Sin, J.H.; Ball, M.J. Sensory acceptability of white bread with added Australian sweet lupin (*Lupinus angustifolius*) kernel fibre and its glycaemic and insulinaemic responses when eaten as a breakfast. *J. Sci. Food Agric.* **2003**, *83*, 1366–1372. [CrossRef]
50. Naumann, S.; Schweiggert-Weisz, U.; Eisner, P. Characterisation of the molecular interactions between primary bile acids and fractionated lupin cotyledons (*Lupinus angustifolius* L.). *Food Chem.* **2020**, *323*, 126780. [CrossRef] [PubMed]
51. Tosh, S.M.; Yada, S. Dietary fibres in pulse seeds and fractions: Characterisation, functional attributes, and applications. *Food Res. Int.* **2010**, *43*, 450–460. [CrossRef]
52. Tiwari, U.; Cummins, E. Functional and physicochemical properties of legume fibers. *Pulse Foods Processing Qual. Nutraceutical Appl.* **2011**, 121–156. [CrossRef]
53. Villarino, C.; Jayasena, V.; Coorey, R.; Chakrabarti-Bell, S.; Johnson, S.K. Nutritional, health, and technological functionality of lupin flour addition to bread and other baked products: Benefits and challenges. *Crit. Rev. Food Sci. Nutr.* **2016**, *56*, 835–857. [CrossRef]
54. McCleary, B.; Prosky, L. *Advanced Dietary Fibre Technology*; John Wiley & Sons: Hoboken, NJ, USA, 2008.
55. Archer, B.J.; Johnson, S.K.; Devereux, H.M.; Baxter, A.L. Effect of fat replacement by inulin or lupin-kernel fibre on sausage patty acceptability, post-meal perceptions of satiety and food intake in men. *Br. J. Nutr.* **2004**, *91*, 591–599. [CrossRef]
56. Hall, R.S.; Baxter, A.L.; Fryirs, C.; Johnson, S.K. Liking of health-functional foods containing lupin kernel fibre following repeated consumption in a dietary intervention setting. *Appetite* **2010**, *55*, 232–237. [CrossRef]
57. Visvanathan, R.; Madhujith, T.; Gamage, A.; Zhang, N. *Lupin*; Manickavasagan, A., Pulses, P.T., Eds.; Springer: Berlin/Heidelberg, Germany, 2020; pp. 169–203. [CrossRef]
58. Prolupin, Germany. Available online: https://www.prolupin.com/company.html. (accessed on 31 May 2022).
59. Luve Aus Lupinen. Available online: https://madewithluve.de (accessed on 31 May 2022).
60. Johnson, S.K.; Chua, V.; Hall, R.S.; Baxter, A.L. Lupin kernel fibre foods improve bowel function and beneficially modify some putative faecal risk factors for colon cancer in men. *Br. J. Nutr.* **2006**, *95*, 372–378. [CrossRef]
61. Smith, S.C.; Choy, R.; Johnson, S.K.; Hall, R.S.; Wildeboer-Veloo, A.; Welling, G.W. Lupin kernel fiber consumption modifies fecal microbiota in healthy men as determined by rRNA gene fluorescent in situ hybridisation. *Eur. J. Nutr.* **2006**, *45*, 335–341. [CrossRef] [PubMed]

62. Fechner, A.; Kiehntopf, M.; Jahreis, G. The formation of short-chain fatty acids is positively associated with the blood lipid–lowering effect of lupin kernel fiber in moderately hypercholesterolemic adults. *J. Nutr.* **2014**, *144*, 599–607. [CrossRef] [PubMed]
63. Gabb, G.M.; Mangoni, A.A.; Anderson, C.S.; Cowley, D.; Dowden, J.S.; Golledge, J.; Hankey, G.J.; Howes, F.S.; Leckie, L.; Perkovic, V. Guideline for the diagnosis and management of hypertension in adults—2016. *Med. J. Aust.* **2016**, *205*, 85–89. [CrossRef] [PubMed]
64. Belski, R.; Mori, T.A.; Puddey, I.B.; Sipsas, S.; Woodman, R.J.; Ackland, T.R.; Beilin, L.J.; Dove, E.R.; Carlyon, N.; Jayaseena, V. Effects of lupin-enriched foods on body composition and cardiovascular disease risk factors: A 12-month randomised controlled weight loss trial. *Int. J. Obes.* **2011**, *35*, 810–819. [CrossRef]
65. Lee, Y.P.; Mori, T.A.; Puddey, I.B.; Sipsas, S.; Ackland, T.R.; Beilin, L.J.; Hodgson, J.M. Effects of lupin kernel flour–enriched bread on blood pressure: A controlled intervention study. *Am. J. Clin. Nutr.* **2009**, *89*, 766–772. [CrossRef]
66. Ward, N.C.; Mori, T.A.; Beilin, L.J.; Johnson, S.; Williams, C.; Gan, S.K.; Puddey, I.B.; Woodman, R.; Phillips, M.; Connolly, E. The effect of regular consumption of lupin-containing foods on glycaemic control and blood pressure in people with type 2 diabetes mellitus. *Food Funct.* **2020**, *11*, 741–747. [CrossRef]
67. Wolever, T.M.; Jenkins, D.J.; Jenkins, A.L.; Josse, R.G. The glycemic index: Methodology and clinical implications. *Am. J. Clin. Nutr.* **1991**, *54*, 846–854. [CrossRef]
68. Buyken, A.; Mitchell, P.; Ceriello, A.; Brand-Miller, J. Optimal dietary approaches for prevention of type 2 diabetes: A life-course perspective. *Diabetologia* **2010**, *53*, 406–418. [CrossRef]
69. Food Standards Australia New Zealand. Food Standards Code, Schedule 4 Nutrition, Health and Related Claims. 2021. Available online: https://www.foodstandards.gov.au/code/Documents/Sched%204%20Nutrition%20and%20health%20claims%20v159.pdf (accessed on 21 September 2021).
70. Seal, C.J.; Courtin, C.M.; Venema, K.; de Vries, J. Health benefits of whole grain: Effects on dietary carbohydrate quality, the gut microbiome, and consequences of processing. *Compr. Rev. Food Sci. Food Saf.* **2021**, *20*, 2742–2768. [CrossRef]

*nutrients*

MDPI

*Article*

# Legume Intake Is Associated with Potential Savings in Coronary Heart Disease-Related Health Care Costs in Australia

**Mohammad M. H. Abdullah [1,*], Jaimee Hughes [2] and Sara Grafenauer [3]**

[1] Department of Food Science and Nutrition, Kuwait University, Kuwait 10002, Kuwait
[2] Grains & Legumes Nutrition Council, 1 Rivett Rd, North Ryde, Sydney 2113, Australia; j.hughes@glnc.org.au
[3] Faculty of Medicine and Health, School of Health Sciences, University of New South Wales, Randwick 1466, Australia; s.grafenauer@unsw.edu.au
* Correspondence: mohamad.abdullah@ku.edu.kw; Tel.: +965-246-33278

**Abstract:** Legume intake has been associated with lower risk for a number of chronic disorders of high financial burden, and is advocated by dietary guidelines as an important part of healthy dietary patterns. Still, the intake of legumes generally falls short of the recommended levels in most countries around the world despite their role as an alternative protein source. The aim of this study was to assess the potential savings in costs of health care services that would follow the reduction in incidences of coronary heart disease (CHD) when adult consumers achieve a targeted level of 50 g/day of legumes intake in Australia. A cost-of-illness analysis was developed using estimates of current and targeted legumes intake in adults (age 25+ y), the estimated percent reduction in relative risk (95% CI) of CHD following legumes intake, and recent data on health care costs related to CHD in Australia. A sensitivity analysis of 'very pessimistic' through to 'universal' scenarios suggested savings in CHD-related health care costs equal to AUD 4.3 (95% CI 1.2–7.4) to AUD 85.5 (95% CI 23.3–147.7) million annually. Findings of the study suggest an economic value of incorporating attainable levels of legumes within the dietary behaviors of Australians. Greater prominence of legumes in dietary guidelines could assist with achieving broader sustainability measures in relation to diet, helping to bring together the environment and health as an important pillar in relation to sustainability.

**Keywords:** legumes; coronary heart disease; health care cost; cost-of-illness analysis; nutrition economics

**Citation:** Abdullah, M.M.H.; Hughes, J.; Grafenauer, S. Legume Intake Is Associated with Potential Savings in Coronary Heart Disease-Related Health Care Costs in Australia. *Nutrients* **2022**, *14*, 2912. https://doi.org/10.3390/nu14142912

Academic Editors: Stuart Johnson and Rewati Bhattarai

Received: 16 June 2022
Accepted: 14 July 2022
Published: 15 July 2022



## 1. Introduction

Legumes are increasingly the focus of discussions related to the future food supply due to their sustainability credentials and unique nutritional profile. Research suggests that although there is widespread promotion through food-based dietary guidelines, global intake patterns vary greatly. Whereas the term 'legumes' is the most inclusive word for the group of foods from the Fabacae (or Leguminosae) botanical family and is mentioned in 94 guidelines, fewer (n = 87) choose to depict legumes, beans, peas, or pulses only in the accompanying visual guide (unpublished data). Dietary guidelines act as a guide aimed at desirable intake for helping achieve and maintain health and although the extent to which dietary guidelines represent the range of foods normally consumed could be argued, however, the inclusion of the legume food group is well supported in the scientific literature. Relevant health outcome studies have focused on glycaemic control, blood pressure, and chronic disease [1–3], and although a unified daily target is lacking [4], studies support the inclusion of legumes at a dose of at least 50 g/day [5]. Furthermore, the risk of all-cause mortality has been shown to decrease by ~16% with increasing intake of legumes up to 150 g/day [6], with one 50 g serving possibly resulting in a 10% risk reduction in all-cause mortality. Intake globally is less than half that target at 21 g/day [7]. The lowest intake was recorded in Uzbekistan, with less than 1 g/day for adults over 25 years, and the highest in

Rwanda at 115.8 g/day according to the 2019 Global Burden of Disease Study [8]. Greater emphasis and perhaps repositioning of legumes in dietary guidelines may be required to encourage intake for health, environmental and economic benefits [9,10].

The economic benefits of pulses intake have been examined in Canada, but rather than 50 g/day, 100 g was modeled to suggest possible annual savings equal to Can$38–370 million in health care and related costs of type 2 diabetes- and cardiovascular disease combined [11]. Our previous nutrition economics research has assessed whole grains [12,13], but the combination with legumes is an important consideration, providing a complement of amino acids, which is important in plant-based and flexitarian dietary patterns. However, limiting the potential of legumes as a replacement for meat is also debated. In an examination of 94 food based dietary guidelines (FBDG), yet to be published, we found the classification of legumes highly variable, and while 38% of countries categorized legumes in the protein rich food group, 20% were in a group on their own and 15% were in the starchy staples group. Regardless of the categorization, the inclusion of legumes in FBDG is essential, and the specific phrasing even more so. The addition of legumes to the vegetable and the meat group is utilized in the Australian guidelines, however, the addition of 'legumes and/or beans' at the conclusion of each guideline statement is thought to do little in terms of inspiring intake [9].

As the body of evidence is relatively small for legumes, providing evidence of the health care cost savings based on the regular inclusion in diets could help guide decisions about the positioning and emphasis in dietary guidelines. Fifty grams per day, or 350 g over a week may be reasonably added to diets, considering that an international collaboration suggested a universal 100 g serve (or $\frac{1}{2}$ cup) [4]. The aim of this study was to estimate the annual health care cost savings related to the inclusion of 50 g/day of legumes relevant to reductions in coronary heart disease (CHD) in adult Australians.

## 2. Materials and Methods

### 2.1. Study Design

A three-step cost-of-illness analysis was developed on the basis of (1) estimates of current per capita [8] and a targeted level of legumes intake [6] among Australian adults (age 25+ y), (2) estimates of percent reductions in relative risk (95% CI) of CHD following legumes intake [14], and (3) recent data on annual costs related to CHD management within the Australian health care system [15]. To assess the uncertainty factor, a sensitivity analysis of four scenarios (very pessimistic, pessimistic, optimistic, and universal) was conducted, as modelled previously [12,13]. Input parameters are summarized in Table 1.

**Table 1.** Summary of the cost-of-illness analysis input parameters and corresponding references.

| Parameter | Men and Women | Reference |
|---|---|---|
| Current per capita legumes intake, g/day | 19.3 | Global Burden of Disease Study [8] |
| Target legumes intake, g/day | 50 | Schwingshackl et al. [6] |
| Gap amount, g/day | 30.7 | |
| Proportions of prospective consumers [1] | 5%, 15%, 50%, 100% | Estimates |
| CHD relative risk (95% CI) per 100 g/day legumes intake, no. of studies | 0.89 (0.81–0.97), n = 10 | Bechthold et al. [14] |
| CHD% risk reduction (95% CI) per 30.7 g/day legumes intake [2] | −3.4% (0.9–5.8) | |

Abbreviations: CHD, coronary heart disease. [1] Estimates of proportions of Australian adults (age 25+ y) who would increase their current estimated per capita legumes intake (19.3 g/day) to the targeted level of 50 g/day over the short term (very pessimistic), short-to-medium term (pessimistic), medium-to-long term (optimistic), and long term (universal) scenarios. [2] Percent risk reduction (95% CI) per 30.7 g/day was calculated based on the summary relative risk (95% CI) values per 100 g/day by Bechthold et al. [14] assuming a linear relationship.

### 2.2. Step 1: Employing Estimates of Current per Capita and Targeted Level of Legumes Intake

Any public health model that attempts to assess a potential benefit of healthy dietary patterns should consider the consumer's perception and behavior in the marketplace and at the dining table. Based on the 2019 Global Burden of Disease data [8], in the first step of this analysis, the current estimated per capita legumes intake of 19.3 g/day for adults (age 25+ y) was compared to a targeted level of 50 g/day [6] and a calculation was built on estimates of proportions of Australian adults (age 25+ y) who are likely to increase their legumes intake by the gap amount of 30.7 g daily. As previously [12,13], here, the sensitivity analysis assumed very pessimistic, pessimistic, optimistic, and universal scenarios to represent 5%, 15%, 50%, and 100% (all) of prospective consumers who would reach the targeted daily level of legumes intake in the short term, short-to-medium-term, medium-to-long-term, and long-term, respectively.

### 2.3. Step 2: Establishing Percent Reductions in Relative Risk of Coronary Heart Disease with Legumes Intake

Only a few meta-analyses of prospective cohort studies have recently assessed the relationship between legumes intake and CHD as a hard endpoint. The majority of these studies only report data on highest vs. lowest intakes [16–18], with no specific serving sizes (i.e., g/day amounts) provided thereof, and some only examine specific types of legumes (e.g., soy) or legumes intake within the context of certain dietary patterns such as the Mediterranean diet [19]. It appears that at this stage the evidence on such a relationship, and related health benefits thereof, is moderate overall. Upon a keyword search for the relevant English-language literature on PubMed, in the second step of the analysis, the dose-response figures by Bechthold et al. [14] of ~10% reduction in CHD risk with 1 serving intake were utilized. The systematic review and dose-response meta-analysis of prospective studies included 123 reports (8 studies on non-linear dose-response association between legumes intake and CHD risk) and suggested a summary relative risk (RR) per 1 serving (~100 g/day) equal to 0.89 (95% CI 0.81–0.97) [14]. Similar rates of risk reduction were reported by other meta-analyses that only provided evidence on highest vs. lowest intakes [6,20,21]. Building on this while assuming a linear relationship, a 3.4% (95% CI 0.9–5.8) reduction in CHD risk per 30.7 g/day of legumes intake was established and utilized in the final step of the analysis.

### 2.4. Step 3: Calculalting Annual Savings in Direct Health Care Costs Related to Coronary Heart Disease in Australia

The third and final step of the analysis calculated the annual savings in CHD-related costs within the Australian health care system that would potentially follow the targeted level of legumes intake (Step 1) and the estimated reduction in CHD risk (Step 2). As previously described [12,13], the most recent estimates of direct health expenditure (the year 2018–2019) reported by the Australian Institute of Health and Welfare (AIHW) [15] were first inflated to the year 2022 equivalent levels, based on adjustment of rates according to the Australian Bureau of Statistics (ABS) Consumer Price Index (Health group) [22] (Table 2), and then utilized within arithmetic calculations where components of the cost categories were examined individually for assessment of savings, with 1% reduction in costs assumed to correspond to each 1% decrease in CHD risk. Additionally, as previously outlined [12,13], using the net present value equation, a 7% real discount rate was applied to the sum of savings in present day costs related to CHD management to assess the discounted value of different scenarios of legumes intake over a 20-year time frame at five-year increments after the year 2022 (year 0).

**Table 2.** Summary of coronary heart disease direct health expenditures in Australia (AUD million), age 25+ y population [1].

| | Coronary Heart Disease | |
|---|---|---|
| | **2018–19** | **2022** [2] |
| Direct health expenditure | | |
| Allied health and other services | 1.8 | 1.9 |
| General practitioner services | 71.6 | 77.6 |
| Medical imaging | 23.3 | 25.3 |
| Pathology | 21.2 | 23.0 |
| Pharmaceutical benefits scheme | 155.1 | 168.2 |
| Private hospital services | 892.2 | 967.2 |
| Public hospital admitted patient | 823.0 | 892.2 |
| Public hospital emergency department | 103.8 | 112.5 |
| Public hospital outpatient | 142.6 | 154.6 |
| Specialist services | 101.2 | 109.7 |
| **All areas** | **2335.8** | **2532.1** |

Abbreviations: AUD, Australian dollar. [1] From the Australian Institute of Health and Welfare (AIHW) disease expenditure database (2018–2019) [15]. [2] Current dollars based on adjustment of inflation rates according to the Australian Bureau of Statistics (ABS) Consumer Price Index (Health group) [22].

## 3. Results

Savings that could be predicted in CHD-related health care costs when legumes intake is increased from the current per capita level of 19.3 g/day to the 50 g/day target level across proportions of the Australian adult population are summarized in Table 3. Under the very pessimistic scenario, assuming a 5% uptake rate over the short term, our analysis predicted total health care savings equal to AUD 4.3 (95% CI 1.2–7.4) million in CHD cost annually. With a 15% uptake rate over the short-to-medium-term and a 50% uptake rate over the medium-to-long-term, the pessimistic and optimistic scenarios suggested cost savings equal to AUD 12.8 (95% CI 3.5–22.2) million and AUD 42.8 (95% CI 11.7–73.8) million, respectively. And, under the universal scenario, assuming a 100% uptake rate and long-term estimate of potential savings with the targeted increase in legumes intake, some total annual health care savings of AUD 85.5 (95% CI 23.3–147.7) million may be realized in avoided CHD costs.

**Table 3.** Potential annual savings in direct health expenditures of coronary heart disease in Australian adults (age 25+ y) with 50 g/day legumes intake (AUD million) [1].

| | Scenario | | | |
|---|---|---|---|---|
| | **Very Pessimistic** | **Pessimistic** | **Optimistic** | **Universal** |
| Direct health expenditure savings | | | | |
| Allied health and other services | <0.1 (<0.1–<0.1) | <0.1 (<0.1–<0.1) | <0.1 (<0.1–0.1) | 0.1 (<0.1–0.1) |
| General practitioner services | 0.1 (<0.1–0.2) | 0.4 (0.1–0.7) | 1.3 (0.4–2.3) | 2.6 (0.7–4.5) |
| Medical imaging | <0.1 (<0.1–0.1) | 0.1 (<0.1–0.2) | 0.4 (0.1–0.7) | 0.9 (0.2–1.5) |
| Pathology | <0.1 (<0.1–0.1) | 0.1 (<0.1–0.2) | 0.4 (0.1–0.7) | 0.8 (0.2–1.3) |
| Pharmaceutical benefits scheme | 0.3 (0.1–0.5) | 0.9 (0.2–1.5) | 2.8 (0.8–4.9) | 5.7 (1.5–9.8) |
| Private hospital services | 1.6 (0.4–2.8) | 4.9 (1.3–8.5) | 16.3 (4.5–28.2) | 32.7 (8.9–56.4) |
| Public hospital admitted patient | 1.5 (0.4–2.6) | 4.5 (1.2–7.8) | 15.1 (4.1–26.0) | 30.1 (8.2–52.0) |
| Public hospital emergency department | 0.2 (0.1–0.3) | 0.6 (0.2–1.0) | 1.9 (0.5–3.3) | 3.8 (1.0–6.6) |
| Public hospital outpatient | 0.3 (0.1–0.5) | 0.8 (0.2–1.4) | 2.6 (0.7–4.5) | 5.2 (1.4–9.0) |
| Specialist services | 0.2 (0.1–0.3) | 0.6 (0.2–1.0) | 1.9 (0.5–3.2) | 3.7 (1.0–6.4) |
| **All areas** | **4.3 (1.2–7.4)** | **12.8 (3.5–22.2)** | **42.8 (11.7–73.8)** | **85.5 (23.3–147.7)** |

Abbreviations: AUD, Australian dollar. [1] Data (95% CI) are potential monetary savings following coronary heart disease risk reduction with 50 g/day intake of legumes (Table 1). The very pessimistic, pessimistic, optimistic, and universal scenarios are modeled to represent short term, short-to-medium-term, medium-to-long-term, and long-term of estimates of potential savings in CHD-related health care costs that could follow when, respectively, 5%, 15%, 50%, and 100% of Australian adults (age 25+ y) consume the targeted daily level of legumes.

As shown in Table 4, with a 7% discount rate, as per the Australian Governments recommendations [23], the sensitivity analysis of 'very pessimistic' through to 'universal' scenarios suggested total discounted savings in CHD-related health care costs of AUD 48.5 (95% CI 13.2–83.7) million to AUD 969.3 (95% CI 264.4–1674.3) million following the 50 g/day intake of legumes over a 20 year period (from 2022 through to 2041). Additionally, assuming adoption of each of the four scenarios every five years, i.e., the very pessimistic during years 0–4, pessimistic during years 5–9, optimistic during years 10–14, and universal scenario during years 15–19, the sum of total incremental discounted savings was estimated at AUD 290.2 (95% CI 79.1–501.3) million over the 20-year time frame.

**Table 4.** Sum of potential total discounted savings on direct health care expenditures of coronary heart disease in Australian adults (age 25+ y) with 50 g/day legumes intake over short- and long-term periods (AUD million) [1].

| | Scenario | | | |
|---|---|---|---|---|
| | **Very Pessimistic** | **Pessimistic** | **Optimistic** | **Universal** |
| Years 0 to 4 | 18.8 (5.1–32.4) | 56.3 (15.3–97.2) | 187.6 (51.2–324.0) | 375.2 (102.3–648.0) |
| Years 5 to 9 | 13.4 (3.6–23.1) | 40.1 (10.9–69.3) | 133.7 (36.5–231.0) | 267.5 (72.9–462.0) |
| Years 10 to 14 | 9.5 (2.6–16.5) | 28.6 (7.8–49.4) | 95.4 (26.0–164.7) | 190.7 (52.0–329.4) |
| Years 15 to 19 | 6.8 (1.9–11.7) | 20.4 (5.6–35.2) | 68.0 (18.5–117.4) | 136.0 (37.1–234.9) |
| **Total discounted savings** | **48.5 (13.2–83.7)** | **145.4 (39.7–251.1)** | **484.7 (132.2–837.1)** | **969.3 (264.4–1674.3)** |

Abbreviations: AUD, Australian dollar. [1] Data (95% CI) are potential total discounted monetary savings following coronary heart disease risk reduction with 50 g/day intake of legumes. The very pessimistic, pessimistic, optimistic, and universal scenarios are modeled to represent short term, short-to-medium-term, medium-to-long-term, and long-term of estimates of potential savings in CHD-related health care costs that could follow when, respectively, 5%, 15%, 50%, and 100% of Australian adults (age 25+ y) consume the targeted daily level of legumes.

## 4. Discussion

This cost-of-illness analysis supports a greater focus on legumes as a regular inclusion in the dietary pattern and demonstrates a predicted substantial savings for the Australian health care system in relation to costs for CHD from AUD 4.3 (95% CI 1.2–7.4) to AUD 85.5 (95% CI 23.3–147.7) million per year. Naturally, these figures are lower than the approximately $370 million in the Canadian study from 2017, which was based on an analysis of both type 2 diabetes and cardiovascular disease, at a level where 50% of the Canadian population would consume 100 g of legumes per day, in combination with a low glycaemic index or high fiber diet [11]. In consideration of the current Australian dietary pattern, and known lower current levels of consumption compared with Canada, the present economic analysis was based on the dose response data of 50 g (1/4 cup) of legumes per day by Schwingshackl et al. 2017 [6] who utilized the smallest serving with significant results for risk-decreasing foods in a systematic review and meta-analysis of prospective studies. The authors found that there was an absence of a linear association between legumes and all-cause mortality risk, with the nonlinear analysis showing ~16% reduced risk of all-cause mortality when consuming up to 150 g/day. However, this amount was considered unrealistic as a daily target for Australia, where the latest estimates from 2019 suggest 19.3 g/day [8]. Although 100 g/day was considered an unrealistically high daily target in light of current consumption and typical dietary patterns, we did calculate the potential yield savings in annual health care costs based on a target of 100 g of legumes per day, and thus a gap of 80.7 g/day compared to consumption estimates, as AUD 11.2 (95% CI 3.1–19.4) to AUD 224.8 (95% CI 61.3–388.3) million. This was more than double the savings predicted from reaching 50 g per day over time (Appendix A).

International serve size guidance has suggested 100 g or half a cup of cooked legumes [4], rather than 75 g (as a serving of vegetables) or 150 g (as a serving of lean meat) in the Australian Guide to Healthy Eating. However, this amount may still not be relevant as a daily target for Australians, although the amount (1/2 cup) could easily be consumed at a meal. The more reasonable target of 50 g/day (1/4 cup) indicates that our results may have further attributable savings, as consumers would not need to consume the food daily, yet could easily consume 350 g over one week. Increased consumption of legumes was supported through the dietary modeling performed for the purpose of informing the 2013 Australian Dietary Guidelines (ADGs), which recommends a 470% increase to meet the levels proposed [24].

Although determining the frequency of intake for legumes over a week presents more of a challenge than serving size, this analysis supports an achievable food volume for daily or weekly consumption. The main concern is setting a target being that amounts need to feasibly align with the cultural acceptability of a food, routine dietary patterns, and the infrastructure of the food systems [4,25]. Australian research points to consumers being receptive to increasing their intake, and "consumer attitudes towards legumes were positive, particularly in relation to their perceived health attributes" [10]. Others have found that knowledge of health benefits assisted consumers with planned dietary changes particularly "after reading the informational messages, 25–42% of all the participants said they planned to eat more legumes in the future" [26]. Favorable research has also been documented by Röös et al. (2022) [27] where "legumes were generally considered healthy and suitable in diets and many respondents stated an intention to increase consumption" and although they do not use them regularly, many consumers in New Zealand were also open to considering legumes as meat substitutes [28].

Legume consumption in the ADGs is promoted via the statements "consume plenty of vegetables, including different types and color, and legumes/beans" and "consume lean meat and poultry, fish, eggs, tofu, nuts and seeds, and legumes/beans" [29]. The inclusion of legumes as part of both the vegetable and meat alternatives groups aims to encourage visibility and consumption, however, there are equally valid concerns that legumes, in being included in both groups, do not give them the prominence they deserve. There have been suggestions that individualizing advice for the vegetable group, in particular for legumes, may improve consumer understanding and consumption [4,30,31]. A recent publication examined consumer preferences regarding the categorization and wording of both whole grain and legume statements in dietary guidelines (n = 314) [9]. When asked what would be helpful in achieving an increase in legume intake, the majority of participants preferred legumes to feature in their own food group (45%), and fewer suggested as part of the protein group (22%). When asked about wording, there was a significant preference for the statement "each day, consume at least one serve of legumes either as a serve of vegetables or as an alternative to meat" ($p < 0.05$). This statement provides both a specific frequency and quantification for legume consumption. Throughout the study, participants emphasized a preference for quantifiable recommendations expressed in cup measures, stating that "grams were less relevant and poorly visualized" [9].

In the planned revision of current ADGs, there has been a call to reflect environmental sustainability objectives as an opportunity to bring together environmental and population health goals [32,33]. A global shift towards plant-based foods, in their most natural form, would be considered beneficial for the health of humans and the planet as a dietary strategy [33,34]. Reflective of this, Canada's dietary guidelines have recently shifted towards more of a plant-based diet, emphasizing and clearly depicting vegetables and fruits, whole grain food choices, and protein foods, including legumes together with meat, eggs, lower fat dairy, nuts, and seeds [35].

The changing food supply, increasing the number and type of plant-protein products, and legumes in other forms such as flour, pasta, snack food, bread products, convenient lunch portions, alternative packaging to traditional canned, and dried forms [36] may do more to stimulate intake than government guidelines per se. When used as an ingredient though, legumes within the dietary pattern become far more difficult to identify and distinguish in intake studies, even though complex methods of data collection are used at the national level. Plant-based meats are another obvious opportunity for legumes, however, the value of extracted protein from legumes, would need to be evaluated in comparison to the research supporting whole legume intake, as the nutrient profiles in these more processed foods are significantly altered. Research to date has found that of 137 plant-based meat alternatives available on Australian supermarket shelves only 4% were low in sodium (58–1200 mg/100 g), and less than a quarter of products (24%) were fortified with vitamin $B_{12}$, 20% with iron, and 18% with zinc [37].

The analysis presented is based on a data-driven approach utilized previously [11–13], where nationally representative data was utilized, increasing the validity of the results. However, we were aware that the cost data was from 2018–2019, so this was inflated to accommodate the age of the data. We also utilized a discounted method [38] to accommodate time, as reported by the Australian Government. A meta-analysis of prospective data pertaining to CHD was used in preference to a similar study of randomized controlled trials in order to obtain overall disease outcomes. The risk reduction ranged from our chosen reference utilized (RR: 0.81–0.97), which approximates the RR from a meta-analysis of RCTs based on the Mediterranean diet and Cardiovascular Disease (CVD), where legumes were one of four dietary components with the most positive effects along with olive oil, vegetables and fruit (RR: 0.91; 95% CI: 0.83, 0.98; $I^2 = 33\%$) [19]. It is important to note that our analysis assumed a linear relationship that has not been shown to occur in nature [6]. Although changes in diet would not be immediate in terms of disease reduction, periods of one to three years lag time could be expected before individual or population benefits would be realized [39]. Adjustment for this time lag was not considered in our model, however, policy makers should not be dissuaded and view dietary change positively in light of the potential cost savings that could be realized through such very small changes. Finally, our sensitivity analysis was based on population-based adoption over time and utilized the reported risk reduction range from Bechtold et al. [14], however, effect sizes from observational studies have been shown to be more generous than randomized controlled trials [40], although these tend to report biomarkers rather than disease outcomes. This should be kept in mind when interpreting the analysis presented.

## 5. Conclusions

Nutrition economic analyses provide key support to directing efforts of population-based initiatives in relation to nutrition, dietary patterns, and health outcomes. The analyses combine metrics relevant to the specific population, incorporating known measures of consumption, and known health care costs, together with published risk reduction evidence, providing logical support and guidance for policy. This analysis is limited to capturing the impact of just one food type on one disease with substantial predicted savings annually and supports earlier work by our group on other key food groups. The potential economic benefits of incorporating legumes within the dietary patterns of Australians should provide greater impetus to promote this food group, consumed as part of a diet low in saturated fat and high in dietary fiber, in a more overt manner as part of revised dietary guidelines. But for Australians, this food could also assist with achieving broader sustainability measures in relation to diet, helping to bring together the environment and health as an important pillar in relation to sustainability. Although the plant-based protein movement seeks to view legumes in terms of nutrient components, protein, and perhaps dietary fiber, it is our assertion that food synergy, and the complex relationships within the food matrix, mean that the legume would be ideally consumed whole. Whole legume products are already plentiful within the Australian food supply in dried, frozen, canned, oven roasted, and as the main ingredient in dips.

**Author Contributions:** Conceptualization, M.M.H.A. and S.G.; methodology, M.M.H.A.; formal analysis, M.M.H.A.; writing-original draft preparation, M.M.H.A. and S.G.; writing—review and editing, J.H. All authors have read and agreed to the published version of the manuscript.

**Funding:** This research was funded by the Grains & Legumes Nutrition Council, a not-for-profit, registered charity.

**Institutional Review Board Statement:** Not applicable.

**Informed Consent Statement:** Not applicable.

**Data Availability Statement:** All data for this study is contained within the article.

**Conflicts of Interest:** S.G. and J.H. are employed by the Grains & Legumes Nutrition Council, a not-for-profit charity. M.A declares no conflict of interest.

## Appendix A

**Table A1.** Potential annual savings in direct health expenditures of coronary heart disease in Australian adults (age 25+ y) with 100 g/day legumes intake (AUD million) [1].

| | Scenario | | | |
|---|---|---|---|---|
| | **Very Pessimistic** | **Pessimistic** | **Optimistic** | **Universal** |
| Direct health expenditure savings | | | | |
| Allied health and other services | <0.1 (<0.1–<0.1) | <0.1 (<0.1–<0.1) | 0.1 (<0.1–0.1) | 0.2 (<0.1–0.3) |
| General practitioner services | 0.3 (0.1–0.6) | 1.0 (0.3–1.8) | 3.4 (0.9–5.9) | 6.9 (1.9–11.9) |
| Medical imaging | 0.1 (<0.1–0.2) | 0.3 (0.1–0.6) | 1.1 (0.3–1.9) | 2.2 (0.6–3.9) |
| Pathology | 0.1 (<0.1–0.2) | 0.3 (0.1–0.5) | 1.0 (0.3–1.8) | 2.0 (0.6–3.5) |
| Pharmaceutical benefits scheme | 0.7 (0.2–1.3) | 2.2 (0.6–3.9) | 7.5 (2.0–12.9) | 14.9 (4.1–25.8) |
| Private hospital services | 4.3 (1.2–7.4) | 12.9 (3.5–22.2) | 42.9 (11.7–74.1) | 85.9 (23.4–148.3) |
| Public hospital admitted patient | 4.0 (1.1–6.8) | 11.9 (3.2–20.5) | 39.6 (10.8–68.4) | 79.2 (21.6–136.8) |
| Public hospital emergency department | 0.5 (0.1–0.9) | 1.5 (0.4–2.6) | 5.0 (1.4–8.6) | 10.0 (2.7–17.3) |
| Public hospital outpatient | 0.7 (0.2–1.2) | 2.1 (0.6–3.6) | 6.9 (1.9–11.8) | 13.7 (3.7–23.7) |
| Specialist services | 0.5 (0.1–0.8) | 1.5 (0.4–2.5) | 4.9 (1.3–8.4) | 9.7 (2.7–16.8) |
| **All areas** | **11.2 (3.1–19.4)** | **33.7 (9.2–58.2)** | **112.4 (30.7–194.1)** | **224.8 (61.3–388.3)** |

Abbreviations: AUD, Australian dollar. [1] Data (95% CI) are potential monetary savings following coronary heart disease risk reduction with 100 g/day intake of legumes. The very pessimistic, pessimistic, optimistic, and universal scenarios are modeled to represent short term, short-to-medium-term, medium-to-long-term, and long-term of estimates of potential savings in CHD-related health care costs that could follow when, respectively, 5%, 15%, 50%, and 100% of Australian adults (age 25+ y) consume the targeted daily level of legumes.

**Table A2.** Sum of potential total discounted savings in direct health care expenditures of coronary heart disease in Australian adults (age 25+ y) with 100 g/day legumes intake over short- and long-term periods (AUD million) [1].

| | Scenario | | | |
|---|---|---|---|---|
| | **Very Pessimistic** | **Pessimistic** | **Optimistic** | **Universal** |
| Years 0 to 4 | **49.3 (13.4–85.2)** | 147.9 (40.3–255.5) | 493.1 (134.5–851.7) | 986.2 (269.0–1703.4) |
| Years 5 to 9 | 35.2 (9.6–60.7) | **105.5 (28.8–182.2)** | 351.6 (95.9–607.2) | 703.1 (191.8–1214.5) |
| Years 10 to 14 | 25.1 (6.8–43.3) | 75.2 (20.5–129.9) | **250.7 (68.4–432.9)** | 501.3 (136.7–865.9) |
| Years 15 to 19 | 17.9 (4.9–30.9) | 53.6 (14.6–92.6) | 178.7 (48.7–308.7) | **357.4 (97.5–617.4)** |
| **Total discounted savings** | **127.4 (34.7–220.1)** | **382.2 (104.2–660.2)** | **1274.0 (347.5–2200.5)** | **2548.0 (694.9–4401.1)** |
| **Total incremental discounted savings with adoption of each scenario every 5 years (years 2022–2041)** | | | | **762.9 (208.1–1317.7)** |

Abbreviations: AUD, Australian dollar. [1] Data (95% CI) are potential total discounted monetary savings following coronary heart disease risk reduction with 100 g/day intake of legumes. The very pessimistic, pessimistic, optimistic, and universal scenarios are modeled to represent short term, short-to-medium-term, medium-to-long-term, and long-term of estimates of potential savings in CHD-related health care costs that could follow when, respectively, 5%, 15%, 50%, and 100% of Australian adults (age 25+ y) consume the targeted daily level of legumes.

## References

1. Schwingshackl, L.; Hoffmann, G.; Iqbal, K.; Schwedhelm, C.; Oeing, H. Food groups and intermediate disease markers: A systematic review and network meta-analysis of randomized trials. *Am. J. Clin. Nutr.* **2018**, *108*, 576–586. [CrossRef] [PubMed]
2. Bielefeld, D.; Grafenauer, S.; Rangan, A. The Effects of Legume Consumption on Markers of Glycaemic Control in Individuals with and without Diabetes Mellitus: A Systematic Literature Review of Randomised Controlled Trials. *Nutrients* **2020**, *12*, 2123. [CrossRef] [PubMed]
3. Jayalath, V.H.; de Souza, R.J.; Sievenpiper, J.L.; Ha, V.; Chiavaroli, L.; Mirrahimi, A.; Di Buono, M.; Bernstein, A.M.; Leiter, L.A.; Kris-Etherton, P.M.; et al. Effect of Dietary Pulses on Blood Pressure: A Systematic Review and Meta-analysis of Controlled Feeding Trials. *Am. J. Hypertens.* **2014**, *27*, 56–64. [CrossRef] [PubMed]

4. Marinangeli, C.P.F.; Curran, J.; Barr, S.I.; Slavin, J.; Puri, S.; Swaminathan, S.; Tapsell, L.; Patterson, C.A. Enhancing nutrition with pulses: Defining a recommended serving size for adults. *Nutr. Rev.* **2017**, *75*, 17. [CrossRef]
5. Willett, W.; Rockström, J.; Loken, B.; Springmann, M.; Lang, T.; Vermeulen, S.; Garnett, T.; Tilman, D.; DeClerck, F.; Wood, A.; et al. Food in the Anthropocene: The EAT-Lancet Commission on healthy diets from sustainable food systems. *Lancet* **2019**, *393*, 447–492. [CrossRef]
6. Schwingshackl, L.; Schwedhelm, C.; Hoffmann, G.; Lampousi, A.-M.; Knüppel, S.; Iqbal, K.; Bechthold, A.; Schlesinger, S.; Boeing, H. Food groups and risk of all-cause mortality: A systematic review and meta-analysis of prospective studies. *Am. J. Clin. Nutr.* **2017**, *105*, 1462–1473. [CrossRef]
7. Rawal, V.; Navarro, D.K. *The Global Economy of Pulses*; FAO: Rome, Italy, 2019.
8. Global Burden of Disease Collaborative Network. *Global Burden of Disease Study 2019 (GBD 2019) Dietary Risk Exposure Estimates 1990–2019*; Global Burden of Disease Collaborative Network: Seattle, DC, USA, 2021. [CrossRef]
9. Reyneke, G.; Hughes, J.; Grafenauer, S. Consumer Understanding of the Australian Dietary Guidelines: Recommendations for Legumes and Whole Grains. *Nutrients* **2022**, *14*, 1753. [CrossRef]
10. Figueira, N.; Curtain, F.; Beck, E.; Grafenauer, S. Consumer Understanding and Culinary Use of Legumes in Australia. *Nutrients* **2019**, *11*, 1575. [CrossRef]
11. Abdullah, M.; Marinangeli, C.P.F.; Jones, P.J.H.; Carlberg, J.G. Canadian Potential Healthcare and Societal Cost Savings from Consumption of Pulses: A Cost-Of-Illness Analysis. *Nutrients* **2017**, *9*, 793. [CrossRef]
12. Abdullah, M.M.H.; Hughes, J.; Grafenauer, S. Healthcare Cost Savings Associated with Increased Whole Grain Consumption among Australian Adults. *Nutrients* **2021**, *13*, 1855. [CrossRef]
13. Abdullah, M.M.H.; Hughes, J.; Grafenauer, S. Whole Grain Intakes Are Associated with Healthcare Cost Savings Following Reductions in Risk of Colorectal Cancer and Total Cancer Mortality in Australia: A Cost-of-Illness Model. *Nutrients* **2021**, *13*, 2982. [CrossRef] [PubMed]
14. Bechthold, A.; Boeing, H.; Schwedhelm, C.; Hoffmann, G.; Knüppel, S.; Iqbal, K.; De Henauw, S.; Michels, N.; Devleesschauwer, B.; Schlesinger, S.; et al. Food groups and risk of coronary heart disease, stroke and heart failure: A systematic review and dose-response meta-analysis of prospective studies. *Crit. Rev. Food Sci. Nutr.* **2019**, *59*, 1071–1090. [CrossRef] [PubMed]
15. Australian Institute of Health Welfare. Disease Expenditure in Australia 2018–19. Available online: https://www.aihw.gov.au/reports/health-welfare-expenditure/disease-expenditure-australia/data-1 (accessed on 7 June 2022).
16. Viguiliouk, E.; Glenn, A.J.; Nishi, S.K.; Chiavaroli, L.; Seider, M.; Khan, T.; Bonaccio, M.; Iacoviello, L.; Mejia, S.B.; Jenkins, D.J.A.; et al. Associations between Dietary Pulses Alone or with Other Legumes and Cardiometabolic Disease Outcomes: An Umbrella Review and Updated Systematic Review and Meta-analysis of Prospective Cohort Studies. *Adv. Nutr.* **2019**, *10*, S308–S319. [CrossRef] [PubMed]
17. Viguiliouk, E.; Blanco Mejia, S.; Kendall, C.W.C.; Sievenpiper, J.L. Can pulses play a role in improving cardiometabolic health? Evidence from systematic reviews and meta-analyses. *Ann. New York Acad. Sci.* **2017**, *1392*, 43–57. [CrossRef] [PubMed]
18. Becerra-Tomás, N.; Papandreou, C.; Salas-Salvadó, J. Legume Consumption and Cardiometabolic Health. *Adv. Nutr.* **2019**, *10*, S437–S450. [CrossRef]
19. Grosso, G.; Marventano, S.; Yang, J.; Micek, A.; Pajak, A.; Scalfi, L.; Galvano, F.; Kales, S.N. A comprehensive meta-analysis on evidence of Mediterranean diet and cardiovascular disease: Are individual components equal? *Crit. Rev. Food Sci. Nutr.* **2017**, *57*, 3218–3232. [CrossRef]
20. Afshin, A.; Micha, R.; Khatibzadeh, S.; Mozaffarian, D. Consumption of nuts and legumes and risk of incident ischemic heart disease, stroke, and diabetes: A systematic review and meta-analysis. *Am. J. Clin. Nutr.* **2014**, *100*, 278–288. [CrossRef]
21. Marventano, S.; Izquierdo Pulido, M.; Sanchez-Gonzalez, C.; Godos, J.; Speciani, A.; Galvano, F.; Grosso, G. Legume consumption and CVD risk: A systematic review and meta-analysis. *Public Health Nutr.* **2017**, *20*, 245–254. [CrossRef]
22. Australian Bureau of Statistics. Consumer Price Index, Australia. Available online: https://www.abs.gov.au/statistics/economy/price-indexes-and-inflation/consumer-price-index-australia/latest-release (accessed on 7 June 2022).
23. Australian Government 2007. Best Practice Regulation Handbook. Available online: http://regulationbodyofknowledge.org/wp-content/uploads/2013/03/AustralianGovernment_Best_Practice_Regulation.pdf (accessed on 7 June 2022).
24. National Health and Medical Research Council. *A Modelling System to Inform the Revision of the Australian Guide to Healthy Eating*; Department of Health and Ageing: Canberra, Australia, 2011.
25. Herforth, A.; Arimond, M.; Álvarez-Sánchez, C.; Coates, J.; Christianson, K.; Muehlhoff, E. A Global Review of Food-Based Dietary Guidelines. *Adv. Nutr.* **2019**, *10*, 590–605. [CrossRef]
26. Marette, S.; Roosen, J. Just a little bit more legumes! Results of an online survey in Europe. *Int. Food Agribus. Manag. Rev.* **2022**, *25*, 329–345. [CrossRef]
27. Röös, E.; de Groote, A.; Stephan, A. Meat tastes good, legumes are healthy and meat substitutes are still strange–The practice of protein consumption among Swedish consumers. *Appetite* **2022**, *174*, 106002. [CrossRef] [PubMed]
28. Lemken, D.; Spiller, A.; Schulze-Ehlers, B. More room for legume–Consumer acceptance of meat substitution with classic, processed and meat-resembling legume products. *Appetite* **2019**, *143*, 104412. [CrossRef] [PubMed]
29. National Health and Medical Research Council. Australian Dietary Guidelines. Available online: https://www.nhmrc.gov.au/adg (accessed on 24 May 2022).

30. Venter, C.S.; Vorster, H.H.; Ochse, R.; Swart, R. "Eat dry beans, split peas, lentils and soya regularly": A food-based dietary guideline. *South Afr. J. Clin. Nutr.* **2013**, *26*, S36–S45.
31. Havemeier, S.; Erickson, J.; Slavin, J. Dietary guidance for pulses: The challenge and opportunity to be part of both the vegetable and protein food groups. *Ann. NY Acad. Sci.* **2017**, *1392*, 58–66. [CrossRef]
32. Williams, P. Broadening influence on the food supply and environmental sustainability. *Nutr. Diet. J. Dietit. Assoc. Aust.* **2019**, *76*, 247–249. [CrossRef]
33. Foley, J. Diets for a Better Future-Scientific Report-EAT Knowledge. Available online: https://eatforum.org/knowledge/diets-for-a-better-future/ (accessed on 7 June 2022).
34. Fanzo, J. Healthy and Sustainable Diets and Food Systems: The Key to Achieving Sustainable Development Goal 2? *Food Ethics.* **2019**, *4*, 159–174. [CrossRef]
35. Health Canada. Canada's Dietary Guidelines. Available online: https://food-guide.canada.ca/en/ (accessed on 20 June 2022).
36. Bielefeld, D.; Hughes, J.; Grafenauer, S. The Changing Landscape of Legume Products Available in Australian Supermarkets. *Nutrients* **2021**, *13*, 3226. [CrossRef]
37. Curtain, F.; Grafenauer, S. Plant-Based Meat Substitutes in the Flexitarian Age: An Audit of Products on Supermarket Shelves. *Nutrients* **2019**, *11*, 2603. [CrossRef]
38. O'Mahony, J.F.; Newall, A.T.; van Rosmalen, J. Dealing with Time in Health Economic Evaluation: Methodological Issues and Recommendations for Practice. *Pharm. Econ.* **2015**, *33*, 1255–1268. [CrossRef]
39. Mozaffarian, D.; Micha, R.; Wallace, S. Effects on coronary heart disease of increasing polyunsaturated fat in place of saturated fat: A systematic review and meta-analysis of randomized controlled trials. *PLoS Med.* **2010**, *7*, e1000252. [CrossRef]
40. Tzoulaki, I.; Siontis, K.C.; Ioannidis, J.P. Prognostic effect size of cardiovascular biomarkers in datasets from observational studies versus randomised trials: Meta-epidemiology study. *Bmj* **2011**, *343*, d6829. [CrossRef] [PubMed]

*nutrients*

Review

# Legumes—A Comprehensive Exploration of Global Food-Based Dietary Guidelines and Consumption

**Jaimee Hughes [1,*], Emma Pearson [2] and Sara Grafenauer [3]**

1 Grains & Legumes Nutrition Council, 1 Rivett Rd., North Ryde 2113, Australia
2 School of Medical, Indigenous and Health Sciences, University of Wollongong, Northfields Avenue, Wollongong 2522, Australia; emma.pearson01@gmail.com
3 School of Health Science, Faculty of Medicine and Health, University of New South Wales, Randwick 2052, Australia; s.grafenauer@unsw.edu.au
* Correspondence: jaimee.hughes2@gmail.com

**Abstract:** Despite the well-known human and planetary health benefits of legumes, consumption is often low. This scoping review aimed to evaluate the inclusion of legumes in global food-based dietary guidelines (FBDG), and to review consumption data against global food group classifications for legumes. The review of FBDG from 94 countries identified legume-based key messaging, the key terms used to define legumes, recommended serving size and frequency of consumption and the classification of legumes into food groups as depicted by food guides. The 2018 Global Dietary Database isolated consumption data of legumes and beans using individual-level, nationally representative dietary survey data for matched countries. Food-based dietary guidelines from 40/94 countries most often identified legumes utilising the term legumes, followed by beans ($n$ = 13), pulses ($n$ = 10), or as beans, peas and lentils ($n$ = 5). The serving size recommendations for legume consumption varied widely, and there was no consistency in the suggested frequency of consumption. Median bean and legume consumption for countries with FBDG ranged from 1.2 g/d (Norway) to 122.7 g/d (Afghanistan). Classification of legumes into food groups varied, with 38% of countries categorising legumes in the protein-rich food group, 20% were in a group on their own and 15% were in the starchy staples group. In countries where legumes were together with either nuts or seeds had the greatest range in intake (11.6–122.7 g/day), followed by those that grouped legumes together with protein-rich foods (4.0–104.7 g/day), while countries that grouped legumes into two food groups, in an attempt to promote consumption, tended to have a lower consumption. Greater emphasis and perhaps repositioning of legumes in dietary guidelines may be required to encourage consumption for health, environmental and economic benefits.

**Keywords:** legumes; dietary guidelines; consumption; sustainability; chronic disease

**Citation:** Hughes, J.; Pearson, E.; Grafenauer, S. Legumes—A Comprehensive Exploration of Global Food-Based Dietary Guidelines and Consumption. *Nutrients* **2022**, *14*, 3080. https://doi.org/10.3390/nu14153080

Academic Editors: Stuart Johnson and Rewati Bhattarai

Received: 6 July 2022
Accepted: 25 July 2022
Published: 27 July 2022

**Publisher's Note:** MDPI stays neutral with regard to jurisdictional claims in published maps and institutional affiliations.



## 1. Introduction

Legume, from the Fabacae (or Leguminosae) botanical family, is the more inclusive classification for non-oil seed (pulse) and oil-seed crops (peanuts and soybeans) and includes both fresh and dried forms [1]. Common types of legumes include dried beans, broad beans, peas, chickpeas, cowpeas, lentils, lupins, peanuts and soybeans. Abundant in various vitamins and minerals, including B vitamins, iron, magnesium, potassium and zinc, legumes are an economical source of dietary fibre, phytonutrients and protein important for plant-based or flexitarian dietary patterns [2,3]. According to the Food and Agriculture Organization of the United Nations ("FAO"), legumes are an important inclusion in the diet with known benefits for human and planetary health [3]. Regular consumption has been shown to improve the nutrient density of the diet in a US population [4], and has been linked to reductions in the risk of disease, particularly coronary heart disease [5] and all-cause mortality at a serving of 50 g per day [6]. Soy intake specifically is linked to a reduced risk of certain cancers including prostate and breast cancer [7], where population studies indicate soy consumption has a role in both preventing breast cancer and reducing risk of reoccurrence in breast cancer survivors due to the isoflavone content [8,9]. Legumes also play a valuable role in sustainable food production, and are well placed to form part of nutritious and environmentally sustainable dietary patterns [10,11], as highlighted in the recent Eat Lancet Commission Planetary Health Diet, where it is recommended to consume 100 g/day (50 g of dried beans, lentils and peas, 25 g of soybeans and 25 g of peanuts) [12]. Increased intakes have also been shown to have a significant impact on socioeconomic burden [13], with a recent Canadian study reporting potential combined annual health care and lost productivity cost savings of CAD377.9 million in the prevention of type 2 diabetes and cardiovascular disease with the daily consumption of 100 g of legumes [13]. Lower targets, of 50 g per day based only on Coronary Heart Disease (CHD) in an Australian population have also been shown to produce cost savings of AUD 4.3 (95% CI 1.2–7.4) to AUD 85.5 (95% CI 23.3–147.7) million annually [14].

National food-based dietary guidelines ("FBDG") provide advice on foods, food groups, and dietary patterns for populations to promote overall health [15]. Although included in earlier dietary guidance as part of the Five Food Groups, legumes were first mentioned in the 1992 Australian Dietary Guidelines ("ADG") as part of the vegetable food group ("Eat plenty of breads and cereals (preferably whole grain), vegetables (including legumes), and fruits") [16]. In the most recent revision of the ADG in 2013, legumes were placed into two food groups; as both a vegetable (75 g serving size) and as an alternative to meat within the protein-rich food group (150 g serving size), with the aim of encouraging consumption [17]. Despite this, modelling as part of the development of the ADGs suggests that Australian's would need to increase consumption by 470% in order to meet the proposed recommendations [18]. Although there is no universally accepted daily target, a recent review conducted by Marinangeli et al. (2016) concluded that a reasonable serve size to be promoted internationally could be based on 100 g (1/2 metric cup or 125 mL) of cooked beans, lentils, chickpeas, or peas per day [19]. This serve size guide was deemed appropriate based on the nutrient content, particularly the delivery of iron, folate and zinc, and is sufficient to facilitate nutrient content claims across multiple jurisdictions [19].

While the classification of legumes as both a vegetable and a protein source in the ADGs was intended to encourage consumption, such placement may be confusing for consumers and nutrition professionals due to their distinct nutritional profile [11]. It has been suggested that future recommendations to encourage legume intake into dietary patterns may require more explicit emphasis in FBDG [20]. Therefore, this scoping review aimed to evaluate the inclusion of legumes in FBDG globally, and to review country- and regional-level consumption data against global food group classifications for legumes.

## 2. Materials and Methods

### *2.1. Legume Messaging in Food-Based Dietary Guidelines*

To review legume-based messaging within National Dietary Guidelines, the FAO online repository of FBDG was utilised to source data for countries with dietary recommendations [15]. Only those related to the general healthy population were included for review. In July and August 2021, country-specific webpages from the FAO online repository were individually reviewed, and relevant data were extracted and transcribed into a Microsoft® Excel® spreadsheet (Microsoft 365 MSO Version 16.0.13426.20306, Redmond, WA, USA) for analysis. The following data were extracted for each country: Region, as defined by the FAO (Asia and the Pacific "APAC", Latin America and the Caribbean "LAC", Europe, Africa, Near East and North America), the official name of the dietary guidelines, date of most recent revision, verbatim legume-based key messages, the specific terminology used to define legumes, the recommended dietary intake of legumes, serving size suggestions and the classification of legumes into food group/s using the provided food guides. Food guides are a visual representation of the dietary guidelines and depict the recommended food groups, and often frequency of consumption for a nutritious diet [15]. Verbatim key messages related to legumes were further classified as either qualitative or quantitative (e.g., recommended frequency of consumption) based messaging. Where available, full FBDG documents in English language were consulted to obtain additional information such as recommended servings sizes and/or frequency of consumption.

The pictorial food guides for each country were visually examined to determine the classification of legumes into one of six food group/s: legumes (own group), fruits/vegetables, vegetables only, starchy staples (together with grain, cereals, tubers and/or cassava), protein-rich foods (together with lean meats, poultry, fish, eggs, tofu and/or nuts and seeds), legumes, nuts and seeds (own group), or a combination of the above mentioned. The precise terminology used to define legumes in the guidelines was also noted.

Using Microsoft® Excel® (Microsoft 365 MSO Version 16.0.13426.20306, Redmond, WA, USA), description statistics were applied to explore how legumes were classified and encouraged within dietary guidelines.

### *2.2. Legume Consumption and Comparison to Legume Food Group Classifications*

Global bean and legume consumption data were extracted from the 2018 Global Dietary Database ("GDD 2018") in June 2022 [21]. The GDD 2018 prediction model provides estimate consumption data for 54 dietary factors, including bean and legume data for 185 countries stratified by age, sex, country, region and time period [22]. Details on data collection for the GDD have been described in detail elsewhere [23–26]; however briefly, estimates are derived from a combination of sources including the use of ~1500 national and subnational surveys of individual-level dietary intakes from public and private sources and from over 800 *covariates* to supplement individual level dietary intake data including food and nutrient availability data and food product sales [27]. The first iteration, GDD 2010, was used to establish the calculations for the 2010 and 2013 Global Burden of Disease Study [28]. According to the GDD 2018, bean and legume consumption is defined as the total intake of beans and legumes (beans, lentils), including fresh, frozen, cooked, canned or dried beans/legumes and includes soybeans but excludes soy milk and soy protein. This definition also excludes peanuts and peanut butter [29].

Median legume intake data (in grams/day) and 95% confidence interval was extracted and analysed in Microsoft® Excel® (Microsoft 365 MSO Version 16.0.13426.20306, Redmond, WA, USA). Intake data for 185 countries were available; however, only those with FBDG were used in the present analysis ($n$ = 94). Following extraction of data, countries were recategorised according to the FAO regional categorisation scheme as previously defined (APAC, LAC, Europe, Africa, Near East and North America) to determine regional-level findings. Intake data were analysed for individuals aged 20 years and over (both sexes), from all education levels (low (0–6 years formal), medium (6.01–12 years) and high (12.01+ years) and all residences (rural and urban).

Country and regional-level consumption data were compared to a daily target of 50 g of cooked legumes, as this target was most commonly utilised in the literature [5,6,12].

## 3. Results

A total of 94 countries with FBDG were included for review to explore how legumes were classified and encouraged within dietary guidelines (excludes Cambodia as the FBDG only related to school children aged 6 to 17 years). Of the 94 included countries, 35% were in Europe ($n$ = 33), 31% were in LAC ($n$ = 29), 18% were from the APAC region ($n$ = 17), 7% were in Africa ($n$ = 7), 6% were in the Near East ($n$ = 6) and 2% in North America ($n$ = 2).

### *3.1. Legume-Based Messaging in Dietary Guidelines*

Most countries had at least one key message related to legumes (96.8%, $n$ = 91); however, there was great variability in the language used and the detail provided (Table 1). In 33/91 countries, legumes were not specifically mentioned in the set of key messages; however, consumption was implied via a general message pertaining to the intake of all food groups, e.g., "Focus on meeting food group needs with nutrient-dense foods and beverages ... " (United States). Legumes were mentioned as a unique dietary component in 20 dietary guidelines, e.g., "Eat dry beans, split peas, lentils and soya regularly" (South Africa), "Eat legumes like beans, lentils and green beans, daily" (Mexico) and "Eat less meat—choose legumes and fish" (Denmark), while 15 countries grouped legumes together with other protein-rich foods, e.g., "Eat some beans, pulses, fish, eggs, meat and other proteins"(The United Kingdom), "Lean meats and poultry, fish, eggs, tofu, nuts and seeds, and legumes/beans" (Australia). Countries in Africa more often highlighted legumes as a separate key message ($n$ = 4/7), while countries in the APAC region more often group legumes together with other protein sources ($n$ = 10/17).

Of the 91 countries with at least one legume-based key message, 42 utilised qualitative-based messaging and often used descriptors such as regularly, frequently or often to encourage consumption, e.g., "Consume legumes frequently" (Greece). Less than a quarter of countries outlined quantitative-based messages ($n$ = 24) and highlighted specific serving frequencies, such as daily consumption, e.g., "Eat legumes daily" (Qatar) or encouraged intake at every meal, e.g., "Integrate grains and legumes, other grain and potatoes—in every main meal" (Albania).

### 3.2. Key Terms Used to Define Legumes

'Legume' was the most commonly used term to define the category in 42.6% of dietary guidelines ($n$ = 40/94), followed by 'Beans' (13.8%, $n$ = 13), 'Pulses' (10.6%, $n$ = 10), 'Peas and Beans' (6.4%, $n$ = 6), 'Beans, Peas and Lentils (5.3%, $n$ = 5) and 'Pulses and Beans' (2.1%, $n$ = 2) (Table 1). Other terms used included 'Beans, Lentils, Soybeans, Chickpeas' (Slovenia), 'Legume Seeds' (Poland), 'Legumes and Pulses' (Thailand), 'Soybeans' (China), 'Legumes/Beans' (Australia) and a combination of 'Legumes' and 'Peas, Kidney Beans, Lentils', all of which were only used by one country each. Twelve countries did not have specific terminology, with no mention of legumes.

### 3.3. Recommended Serving Sizes and Frequency of Consumption

Overall, 39 countries provided a recommended serving frequency and 25 provided a recommended serving size for legumes (Table 1). Frequency of consumption ranged from two servings per day (Chile, Colombia and Paraguay) to two servings per week (Albania, France). In the United States, individuals should aim for 1–3 cups/week depending on energy needs and in Denmark the recommendation is to simply consume 100 g per day. The most common serving size was one-half a cup of cooked legumes (Kenya, Benin, Sierra Leone, Saudi Arabia, Oman, Australia (as a vegetable), Bangladesh and Austria), and ranged from 30 g fresh or cooked/10 g dried legumes (Estonia) to one cup of cooked legumes (Greece, Australia and New Zealand). Australia was the only country with two serving sizes for legumes: 75 g as a vegetable and 150 g as a meat alternative.

### 3.4. Food Group Classification

Overall, 87/94 countries with FBDG had visual food guides. Legumes were mostly commonly included in the protein-rich food group (38%, $n$ = 33), together with lean meats, poultry, fish, eggs, tofu and/or nuts and seeds. Legumes were a separate food group in 20% of countries ($n$ = 17) and 13 countries grouped legumes together with other starchy staples (variously defined, however mostly included grains, cereals, tubers and/or cassava) (Figure 1). Seven FBDG grouped legumes with nuts and seeds (8%) and a smaller proportion included legumes in the fruits/ vegetable food group (5%, $n$ = 4) and vegetable food group only (4%, $n$ = 3). Overall, five FBDG grouped legumes into two foods groups. In Australia and the United States, legumes were considered both a vegetable and a meat alternative in the protein-rich food group. In Albania, legumes were classified in the protein-rich food group and the starchy staples group, while in Poland and Sweden, legumes featured in the "eat more" food group and the "replace/switch" food group (Table 1).

**Table 1.** Legume-based messaging, food group classification, key legume terminology, recommended serving frequency, recommended serving size and median bean and legume consumption (in grams/day) for countries with FBDG ($n$ = 94).

| Country | Food Group Classification in Food Guide | Legume-Based Message/s | Key Legume Term | Recommended Serving Frequency [a] | Recommended Serving Size | GDD Consumption (Median; 95% CI) |
|---|---|---|---|---|---|---|
| | | | Africa | | | |
| Benin (2015) | Protein-rich foods | "When there is no meat, fish or eggs in a given day, you can replace them with pulses, peanuts, soybeans, soya, cheese or peas." | Pulses | 2–3 servings of food group/day * | ½ cup (140 g) * | 26.7 (22.7–31.5) |
| Kenya (2017) | Legumes, nuts and seeds (own group) | "Eat beans, peas, lentils, cowpeas, pigeon peas, soya, nuts and edible seeds regularly (at least four times a week)." | Beans, peas, lentils | 4 serving/week | ½ cup (125 mL) cooked dried beans, peas or lentils | 29.2 (25.3–34) |
| Namibia (2000) | Protein-rich foods | "Eat beans or meat regularly" | Beans | n/s | n/s | 13.4 (11.2–16.2) |
| Nigeria (2001) | Starchy staples | The diet should contain as wide a variety of foods as possible, e.g., legumes | Legumes | n/s | n/s | 21.6 (18.7–24.8) |
| Seychelles (2006) | Fruits and/or vegetable | "Eat pulses (peas, beans and lentils) at least 4 times a week" | Pulses | 4 servings/week | n/s | 20.4 (17.1–24.4) |
| Sierra Leone (2016) | Legumes (own group) | "Eat beans, peas and lentils everyday" | Beans, peas and lentils | n/s | ½ cup cooked beans/lentils | 23.4 (20.0–27.7) |
| South Africa (2012) | Legumes (own group) | "Eat dry beans, split peas, lentils and soya regularly" | Beans, split peas, lentils and soya | n/s | n/s | 33.6 (25.2–44.3) |
| | | | Asia and the Pacific | | | |
| Afghanistan (2015) | Legumes, nuts and seeds (own group) | "Eat different types of food daily" | Pulses and beans | 0.5–2 servings/day depending on energy needs. | ½ cup (100 g) boiled lentil/peas, ¼ cup raw dry lentils/peas | 122.7 (59.5–237.3) |
| Australia (2013) | Vegetable | "Plenty of vegetables, including different types and colours, and legumes/beans" | Legumes/beans | 5 servings of the food group/day | ½ cup (75 g) cooked dried or canned beans, peas or lentils | 26.1 (18.7–36.5) |
| | Protein-rich foods | "Lean meats and poultry, fish, eggs, tofu, nuts and seeds, and legumes/beans" | | 2.5 servings of the food group/day | 1 cup (150 g) cooked or canned legumes/beans | |
| Bangladesh (2013) | Protein-rich foods | "Consume required amounts of fish, meat, poultry, egg and legumes daily" | Legumes | 1 serving/day. Combine cereals with legumes in a 3:1 ratio | ⅓–½ cup pulses | 31.7 (28.6–35.4) |
| China (2016) | Starchy staples | "Consume plenty of vegetables, milk, and soybeans" | Soybeans | 250–400 g of the food group/day | n/s | 15.8 (12.8–19.6) |
| Fiji (2018) | Protein-rich foods | "Eat body building foods such as dhal, dried peas and beans … " | Dhal, dried peas and beans | n/s | n/s | 30.7 (17.9–53.9) |
| India (2011) | Starchy staples | "Eat variety of foods to ensure a balanced diet" | Pulses | 2 servings/day for vegetarians, 1 serving/day for non-vegetarian | 30 g | 27.2 (24.6–30.2) |
| Indonesia (2014) | Protein-rich foods | "Eat high-protein foods (animal or vegetable source)" | n/s | 2–4 servings of the food group/day | n/s | 29.0 (25.5–33) |

**Table 1.** *Cont.*

| Country | Food Group Classification in Food Guide | Legume-Based Message/s | Key Legume Term | Recommended Serving Frequency [a] | Recommended Serving Size | GDD Consumption (Median; 95% CI) |
|---|---|---|---|---|---|---|
| Asia and the Pacific | | | | | | |
| Japan (2010) | Cannot be determined ** | "Combine vegetables, fruits, milk products, beans and fish in your diet." | Beans | n/s | n/s | 61.1 (53.5–70.5) |
| Malaysia (2010) | Protein-rich foods | "Consume moderate amounts of fish, meat, poultry, eggs, legumes and nuts." | Legumes | ½–1 serving/day | n/s | 26.6 (22.6–31.1) |
| Mongolia (2010) | Cannot be determined ** | "Consume a variety of nutrient-dense foods and beverages." | n/s | n/s | n/s | 32.0 (18.8–56) |
| Nepal (2012) | No food guide | "Eat pulses, fish, poultry, eggs and a little meat regularly." | Pulses | n/s | n/s | 64.7 (56.5–74.4) |
| New Zealand (2020) | Protein-rich foods | "Enjoy a variety of nutritious foods every day including: some legumes . . . " | Legumes | 2 servings of the food group/day | 1 cup (150 g) cooked or canned | 68.0 (53.8–85.6) |
| Philippines (2012) | Protein-rich foods | "Consume fish, lean meat, poultry, eggs, dried beans or nuts daily for growth and repair of body tissues." | Beans | 3–4 servings of the food group/day | n/s | 9.7 (8.6–11) |
| Republic of Korea (2016) | Protein-rich foods | "Eat a variety of foods including rice & other grains, vegetables, fruits, milk & dairy products, meat, fish, eggs, and beans." | Beans | 3–4 servings of the food group/day | n/s | 28.5 (26.4–30.9) |
| Sri Lanka (2011) | Protein-rich foods | "Eat pulses, fish, dried fish, eggs, poultry and lean meat." | Pulses | 3–4 servings of the food group/day | 3 Tbsp cooked pulses | 89.6 (76.6–104.4) |
| Thailand (1998) | Protein-rich foods | "Eat fish, lean meat, eggs, legumes and pulses regularly." | Legumes and pulses | n/s | n/s | 43.4 (17.8–104.9) |
| Vietnam (2013) | Protein-rich foods | "Eat protein-rich foods from a good balance of vegetable and animal sources. Increase the intake of . . . beans/peas" | Beans/peas | n/s | n/s | 104.7 (69.8–157.2) |
| Near East | | | | | | |
| Iran (2015) | Legumes, nuts and seeds | "Eat legumes and dishes made with legumes once a day" | Legumes | n/s | n/s | 16.4 (14.6–18.2) |
| Lebanon (2013) | Protein-rich foods | "Consume legume-based dishes regularly . . . " | Legumes | 5–6.5 servings of the food group/day | ¼ cup cooked legumes, 2 Tbsp hummus, 1 baked falafel | 28.9 (24–34.7) |
| Oman (2009) | Legumes (own group) | "Consume one serving of legumes daily." | Legumes | 1 serving/day | ½ cup cooked lentils, beans or peas, ¼ cup dried beans or tofu | 45.5 (23.4–92.1) |
| Qatar (2015) | Legumes (own group) | "Eat legumes daily" "Choose legumes, nuts and seeds as alternative protein sources" | Legumes | n/s | n/s | 39.0 (19.5–85.9) |

**Table 1.** *Cont.*

| Country | Food Group Classification in Food Guide | Legume-Based Message/s | Key Legume Term | Recommended Serving Frequency [a] | Recommended Serving Size | GDD Consumption (Median; 95% CI) |
|---|---|---|---|---|---|---|
| | | Near East | | | | |
| Saudi Arabia (2013) | Protein-rich foods | "Enjoy a variety of food items from major food groups daily—Include legumes, nuts, seeds, poultry, and lean meats in your eating pattern" | Legumes | 2–3 servings of the food group/day | ½ cup cooked legumes | 58.9 (31.1–110.7) |
| United Arab Emirates (2019) | Cannot be determined ** | "Consume diversified nutrient-rich foods and beverages." | n/s | n/s | n/s | 41.7 (21.5–84.2) |
| | | Europe | | | | |
| Albania (2008) | Starchy staples | "Integral grains and legumes, other grain and potatoes—in every main meal" | Legumes | 2 servings/ day | 60–100 g legumes | 42.4 (33.3–54) |
| | Protein-rich foods | "Substitute greasy meat and meat by-products with peas, kidney beans, lentils . . . " | Peas, kidney beans, lentils | | | |
| Austria (2015) | Vegetables, legumes and fruits | "Eat five servings of vegetables, legumes and fruits every day. The ideal would be to eat three servings of vegetables and/or legumes . . . " | Legumes | 3 servings of vegetables and/or legumes/day * | 70–100 g dry pulses (150–200 g cooked) * | 7.0 (4.6–10.8) |
| Belgium (French region) (2020) | Protein-rich foods | n/s | Legumes | n/s | n/s | 5.0 (3.4–7.4) |
| Bosnia and Herzegovina (2004) | Cannot be determined ** | "Eat meat, poultry, eggs and legumes several times a week" | Legumes | n/s | n/s | 14.6 (8.7–24.8) |
| Bulgaria (2006) | Protein-rich foods | "Replace meat and meat products often with fish, poultry or pulses" | Pulses | 2 servings per week | 200–300 g | 18.9 (13.7–25.9) |
| Croatia (2002) | Starchy staples | n/s | n/s | n/s | n/s | 8.6 (3.4–21.8) |
| Cyprus (2007) | Legumes (own group) | "Consume a traditional Mediterranean diet with lots of legumes . . . " "Increase your intake of fibre and complex carbohydrates by using whole grains, legumes, vegetables and fruit with the peel on" | Legumes | n/s | n/s | 34.3 (13.3–88.3) |
| Denmark (2021) | Legumes (own group) | "Eat less meat—choose legumes and fish" | Legumes | 100 g/day | n/s | 25.5 (15.6–42.2) |
| Estonia (2017) | Vegetable | "Increase the consumption of vegetables, including legumes." | Legumes | 3–4 servings/ week * | 30 g fresh or cooked, 10 g dried legumes * | 6.3 (4.9–8.2) |
| Finland (2014) | Vegetable, fruits and berries | "Eat vegetables, fruits and berries frequently . . . " | n/s | 500 g of the food group/day | n/s | 6.7 (5.4–8.5) |
| France (2019) | No guide | "The consumption of pulses (beans, lentils, chickpeas, etc), at least twice a week" | Pulses | 2 servings/week | n/s | 14.9 (12.7–17.5) |

**Table 1.** *Cont.*

| Country | Food Group Classification in Food Guide | Legume-Based Message/s | Key Legume Term | Recommended Serving Frequency [a] | Recommended Serving Size | GDD Consumption (Median; 95% CI) |
|---|---|---|---|---|---|---|
| | | Europe | | | | |
| Georgia (2005) | Protein-rich foods | "Replace fatty meat and meat products with legumes . . . " | Legumes | 1–3 servings/d (150–200 g) | ¼ cup of beans | 36.9 (21.5–65.3) |
| Germany (2017) | Vegetable | "Enjoy a variety of foods". | n/s | n/s | n/s | 5.7 (5.1–6.5) |
| Greece (2014) | Legumes, nuts and seeds | "Consume legumes frequently" | Legumes | 3 servings/week * | 1 cup of cooked drained legumes (150–200 g) * | 15.2 (11.5–19.9) |
| Hungary (2004) | Fruits and vegetables | "Eat dark green vegetables, citrus fruits, tomato and legumes often.." | Legumes | n/s | n/s | 16.9 (5.4–51) |
| Iceland (2014) | Protein-rich foods | "Variety of foods in reasonable quantity" | Beans | 2 servings of food group/day * | $\frac{3}{4}$ cup beans or lentils * | 6.0 (4.7–7.7) |
| Ireland (2015–16) | Protein-rich foods | "Choose eggs, beans and nuts" | Beans | n/s | n/s | 25.5 (11–56.4) |
| Israel (2008) | Protein-rich foods | "Choose fibre-containing foods such as . . . legumes . . . " | Legumes | n/s | n/s | 75.3 (64–88.5) |
| Italy (2019) | No guide | "Eat whole grain and legumes" | Legumes | n/s | n/s | 14.9 (13.2–16.9) |
| Latvia (2008) | Protein-rich foods | "Eat legumes, fish or lean meat. The recommended daily amount of those products is 2–3 servings" | Legumes | 2–3 servings of food group/day | n/s | 10.5 (4.9–23.5) |
| Malta (2016) | Protein-rich foods | "Include legume-based dishes throughout the week. These could take the form of home-made dips (bigilla, red kidney dip and hummus). Salads (bean and chickpea salad), stews, vegetable soups (minestra) and home-made torta talful." | Legumes | 2+ servings/week | 70 g (raw), 140 g (cooked/canned) | 4.0 (2.1–7.5) |
| The Netherlands (2016) | Protein-rich foods | "Eat less meat and more plant-based foods, and vary with fish, pulses, nuts, eggs and vegetarian products" | Pulses | n/s | n/s | 5.9 (5.0–6.9) |
| The Republic of North Macedonia (2014) | No guide | "Substitute meat and meat products with fish, poultry, beans and bean-based products" | Beans | n/s | n/s | 22.4 (13.6–38.2) |
| Norway (2014) | No guide | n/s | n/s | n/s | n/s | 1.2 (0.5–3.1) |
| Poland (2020) | Eat more food group | "Eat more legume seeds (e.g., beans, peas, chickpeas, lentils, broad beans)" | Legume seeds | n/s | n/s | 3.0 (2.6–3.4) |
| | Replace food group | "Replace red meat and processed meat with fish, poultry, eggs, legume seeds and nuts" | | n/s | n/s | |
| Portugal (2003) | Legumes (own group) | "Eat foods from each food group every day to have a complete diet." | Legumes | 1–2 servings/day | 1 Tbsp raw dried (25 g), 3 Tbsp raw fresh (80 g), 3 Tbsp dried/cooked fresh (80 g) | 12.8 (11.3–14.4) |

**Table 1.** *Cont.*

| Country | Food Group Classification in Food Guide | Legume-Based Message/s | Key Legume Term | Recommended Serving Frequency [a] | Recommended Serving Size | GDD Consumption (Median; 95% CI) |
|---|---|---|---|---|---|---|
| Europe | | | | | | |
| Romania (2006) | Cannot be determined ** | "Eat a variety of foods" | n/s | n/s | n/s | 34.6 (27.8–43.5) |
| Slovenia (2011) | Protein-rich foods | "Eat a variety of foods originating mainly from plants, rather than animals". | Beans, lentils, soybeans, chickpeas. | 3–5 servings of the food groups/day * | 4 tablespoons of beans, lentils, soybeans, chickpeas * | 20.1 (8.9–44.4) |
| Spain (2008) | Protein-rich foods | "Enjoy a variety of foods" | Legumes | n/s | n/s | 22.9 (10.7–49.5) |
| Sweden (2015) | "Eat More" | "Eat more vegetables and fruit—Ideally, choose high fibre vegs such as . . . beans." | Beans | n/s | n/s | 17.0 (14.7–19.6) |
| | Switch To | "Soups, pies and stir fries can easily be made without meat" | | n/s | n/s | |
| Switzerland (2011) | Starchy staples | "Consume three portions of grains, potatoes and pulses per day" | Pulses | 3 servings of the food group/day | n/s | 2.9 (2.1–4.1) |
| Turkey (2014) | Protein-rich foods | "Increase consumption of wholegrain cereals and leguminous seeds" | Legumes | 2 servings of food group/day | 90 g | 32.6 (21.9–49.3) |
| United Kingdom (2016) | Protein-rich foods | "Eat some beans, pulses, fish, eggs, meat and other proteins.." | Beans and pulses | n/s | 80 g | 44.8 (36.8–55.2) |
| Latin America and the Caribbean | | | | | | |
| Antigua and Barbuda (2013) | Legumes, nuts and seeds | "Choose to eat a variety of foods every day" | Peas and beans | n/s | n/s | 44.2 (26.2–75.7) |
| Argentina (2014) | Starchy staples | "Eat legumes; cereals, preferably wholemeal; potato; sweet potato; corn or cassava" | Legumes | n/s | n/s | 3.7 (2.7–4.9) |
| Bahamas (2002) | Legumes (own group) | "Make starchy vegetables, peas and beans a part of your diet" | Peas and beans | n/s | n/s | 18 (10.1–32.4) |
| Barbados (2017) | Legumes (own group) | "Enjoy a wide variety of foods every day" | Legumes | n/s | n/s | 6.9 (4.4–11) |
| Belize (2012) | Legumes (own group) | "Choose different types of foods from all the food groups daily" | Legumes | 1–2 servings/day depending on energy needs | ¼ cup red beans or lentils | 28.2 (16.4–49.7) |
| Bolivia (2013) | Starchy staples | "Consume a varied diet daily, including foods from all groups" | n/s | n/s | n/s | 11.7 (10–13.6) |
| Brazil (2014) | No guide | "Make natural or minimally processed foods part of your diet" | Beans | n/s | n/s | 83.0 (71.6–96.2) |
| Chile (2013) | Protein-rich foods | "To keep your heart healthy . . . eat legumes at least twice a week, without mixing them with cold or cured meats" | Legumes | 2 servings/ week | n/s | 13.1 (7.4–23.2) |
| Colombia (2015) | Protein-rich foods | "To complement your diet, eat pulses like beans, lentils, peas and chickpeas at least two times per week" | Pulses | 2 servings/ week | n/s | 26.1 (22.8–29.9) |

**Table 1.** *Cont.*

| Country | Food Group Classification in Food Guide | Legume-Based Message/s | Key Legume Term | Recommended Serving Frequency [a] | Recommended Serving Size | GDD Consumption (Median; 95% CI) |
|---|---|---|---|---|---|---|
| | | Latin America and the Caribbean | | | | |
| Costa Rica (2010) | Starchy staples | "Eat rice and beans; they are the basis of the everyday diet" | Beans | n/s | n/s | 28.4 (16.9–49.7) |
| Cuba (2009) | Protein-rich foods | "A variety of foods during the day is pleasant and necessary for good health" | Legumes | n/s | n/s | 26.4 (16–44.8) |
| Dominica (2007) | Legumes (own group) | "Always try to eat a variety of foods every day. Use the basket to help you make the right choices" | Peas and Beans | n/s | n/s | 30.9 (18.6–53.5) |
| Dominican Republic (2009) | Legumes (own group) | "Increase consumption of beans, grains, fish, eggs and dairy to keep your bones and organs healthy" | Beans | n/s | n/s | 46.8 (40.9–54.1) |
| Ecuador (2018) | Protein-rich foods | "Let's include animal source foods or legumes in our daily dishes to develop and strengthen our bodies" "Let's eat better by combining legumes with cereals like rice, maize or quinoa" | Legumes | n/s | n/s | 15.8 (8.9–28) |
| El Salvador (2012) | No guide | "Prepare varied meals using natural foods every day" | Beans | n/s | n/s | 36.8 (21.8–63.8) |
| Grenada (2006) | Legumes (own group) | "Eat a variety of foods" | Legumes | n/s | n/s | 29.1 (17.7–50.2) |
| Guatemala (2012) | Starchy Staples | "Eat beans and tortillas every day: eat two tablespoons of beans per tortilla, because they provide more nutrients and fill you up more." | Beans | n/s | n/s | 48.1 (41.7–55.5) |
| Guyana (2018) | Legumes (own group) | "Eat different types of foods from all the food groups daily" | Legumes | n/s | n/s | 21.9 (17.6–27) |
| Honduras (2013) | Starchy staples | "Eat foods from all food groups to enjoy good health" | n/s | n/s | n/s | 42.4 (36.7–49.1) |
| Jamaica (2015) | Legumes, nuts and seeds | "Include peas, beans and nuts in your daily meals" | Legumes | 3 servings of food group/day | ¼ cup cooked peas, beans, baked beans, stewed peas and chickpeas (1 Tbspn), ½ cup canned green peas and lentils | 11.6 (7.9–16.9) |
| Mexico (2015) | Legumes (own group) | "Include the three food groups: fruits and vegetables, legumes and animal source foods in your breakfast, lunch and dinner" "Eat legumes like beans, lentils and green beans, daily" | Legumes | n/s | n/s | 41.4 (37.3–45.9) |
| Panama (2013) | Starchy staples | "Eat a variety of foods every day" | n/s | n/s | n/s | 66.1 (40.8–109.8) |

**Table 1.** *Cont.*

| Country | Food Group Classification in Food Guide | Legume-Based Message/s | Key Legume Term | Recommended Serving Frequency [a] | Recommended Serving Size | GDD Consumption (Median; 95% CI) |
|---|---|---|---|---|---|---|
| Latin America and the Caribbean | | | | | | |
| Paraguay (2015) | Protein-rich foods | "To have a healthy diet, eat from all 7 food groups ( … meats, legumes and eggs … ) every day" "Eat cereals and legumes 2 times a week because together they are more nutritious" | Legumes | 2 servings/week consumed with cereals | n/s | 16.7 (10.2–28.9) |
| Peru (2019) | Starchy staples | "Don't miss legumes, they are tasty, healthy and can be prepared in many ways" | Legumes | n/s | n/s | 34.4 (30.3–39.5) |
| Saint Kitts and Nevis (2010) | Legumes, nuts and seeds (own group) | "Every day, choose foods from each of the groups (shown on the mill)" | Peas and beans | n/s | n/s | No data |
| Saint Lucia (2007) | Legumes (own group) | "Always try to eat ground provisions, peas and beans in your meals every day" | Peas and beans | n/s | n/s | 57.6 (34.2–98.3) |
| Saint Vincent and the Grenadines (2006) | Legumes (own group) | "Eat a variety of foods from the foods groups in the breadfruit" | Legumes | n/s | n/s | 46.6 (28.2–79.2) |
| Uruguay (2016) | Vegetable | "Incorporate vegetables and fruits in all your meals. This will help you to feel good and to maintain a healthy weight" | Legumes | n/s | n/s | 45.6 (26.4–77.6) |
| Venezuela (1991) | Starchy staples | "Eat a varied diet. Get the fibre that your body needs from plant foods on a daily basis" | n/s | n/s | n/s | 40.4 (23.9–71.7) |
| North America | | | | | | |
| Canada (2019) | Protein-rich foods | "Eat plenty of vegetables and fruits, whole grain foods and protein foods. Choose protein foods that come from plants more often" | Legumes | n/s | n/s | 20.8 (18–24.2) |
| United States (2020) | Vegetable<br>Protein-rich foods | "Focus on meeting food group needs with nutrient-dense foods and beverages … " | Beans, Peas, Lentils | 1–3 cups/ week depending on energy needs | n/s<br>n/s | 21.8 (20.2–23.6) |

* Translated; ** Food guide image unclear; [a] Recommended serving frequency relates to legumes specifically unless stated otherwise. n/s = not specific.

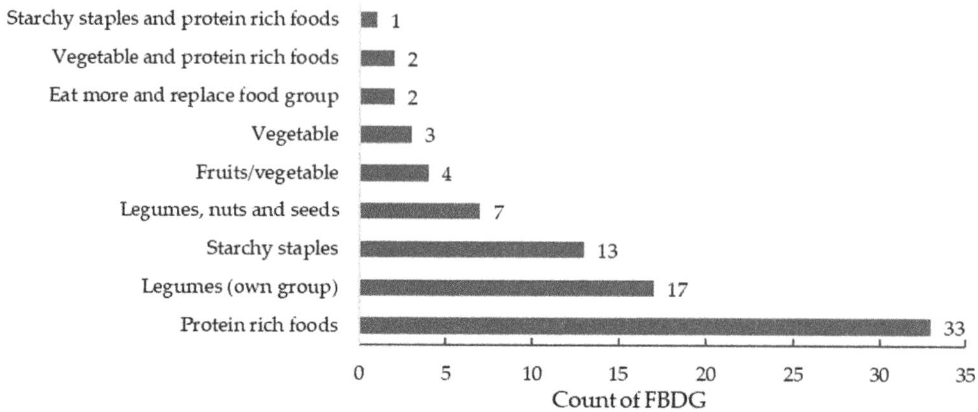


**Figure 1.** Classification of legumes by food group ($n$ = 82). Excludes seven countries that did not have a food guide and five countries where the food group classification could not be determined (image of food guide was unclear).

### *3.5. Legume Intake Data and Comparison to Legume Food Group Classification*

Consumption data were available for 98.9% of countries with FBDG ($n$ = 93/94) (Table 1) and of which, median intake ranged from 1.2 g/day (Norway) to 122.7 g/day (Afghanistan). At the regional level, the greatest range in intake for those with FBDG was seen in the APAC region (9.7–122.7 g/day) (Appendix A Figure A1), followed by LAC (3.7–83.0 g/day) (Appendix B Figure A2), Europe (1.2–75.3 g/day) (Appendix C Figure A3), Near East (16.4–58.9 g/day) (Appendix D Figure A4), Africa (13.4–33.6 g/day) (Appendix E Figure A5) and North America (20.8 to 21.8 g/day) (Appendix F Figure A6). Only 11 countries had a median intake >50 g/day, most of which were in the APAC region ($n$ = 6/11).

All countries in the European region ($n$ = 33) fell below 50 g of legumes per day, except Israel, and 36% (12/33) fell below 10 g/day indicating that Europe had the lowest level of reported consumption overall. In APAC, 65% of countries fell below 50 g/day ($n$ = 11/17) but only two countries fell below 25 g/day. No country in Africa met the 50 g daily target, and 4/7 consumed below 25 g/day. More than two-thirds of countries in LAC (89%) fell below the consumption target of 50 g/day ($n$ = 25/28), although most countries ($n$ = 19/28) consumed above 25 g/day. Most countries in the Near East consumed above 25 g/day ($n$ = 5/6), with just one country, Saudi Arabia, exceeding the 50 g/day target. Neither of the two countries representing North America met the 50 g target.

At the country-specific level, highest intakes were reported in Afghanistan (122.7 g/day), Vietnam (104.7 g/day), Sri Lanka (89.6 g/day), Brazil (83.0 g/day) and Israel (75.3 g/day), while Norway (1.2 g/day), Switzerland (2.9 g/day), Poland (3.0 g/day), Argentina (3.7 g/day) and Malta (4.0 g/day) reported the lowest consumption overall. In Australia and New Zealand, median legume intake was 26.1 and 68.0 g/day, respectively, and lower intakes were reported in Canada (20.8 g/day) and the United States (21.8 g/day). The UK reported a median intake of 44.8 g/day.

Median bean and legume consumption was compared to the legume food group classification for 81 countries with matched data. This excludes seven countries that did not have a food guide, five countries where the food group classification could not be determined and one country where consumption data was not available. In Albania, where legumes were classified as both a starchy staple and as a meat alternative in the protein-rich food group, median consumption was 42.4 g/day, higher than any other food group classification, although it should be noted that only one country grouped legumes as both a starchy staple and alternative to meat. Countries that placed legumes together with either nuts or seeds had the greatest range in intake (11.6–122.7 g/day), followed by those that grouped legumes together with protein-rich foods (4.0–104.7 g/day), which also had the greatest number of countries exceeding the 50 g/day target (5/33). More than two-thirds of those that classified legumes in their own food group (71%) consumed more than 25 g/day.

## 4. Discussion

Food-based dietary guidelines reflect a country's particular pattern of eating while trying to optimise the health of the general population [30]. This global examination of dietary guidelines, focused on legumes, found that there was no consistency in the terminology used to define the category, with twelve different variations in the language used. While 'legumes', the more inclusive term, was utilised more frequently to define the category than the terms 'beans' or 'pulses' (referring to the dried, mature seeds), the discrepancy in the language used to categorise legumes may be a barrier to the dissemination of clear dietary guideline messaging, as previously mentioned by Didinger and Thompson (2021) [31]. Findings from this review, and previously published research [31,32], may provide impetus to revise the language and the classification system used to define legumes in dietary guidelines globally to move towards a more consistent, and accurate definition.

This review points to a particular opportunity for Australia to refer more specifically to 'Beans, peas and lentils', as opposed to 'legumes/beans' with the suggestion that the general terms, legumes and pulses, may not be familiar to consumers. This was reflected in a recent study involving 505 Australian respondents, in which legumes were most commonly associated with foods such as "beans" (72%), "lentils" (55%), "chickpeas" (42%), and "peas" (24%) [20]. Interestingly, a change in the terminology used to classify and define legumes has been adopted in the most recent revision of The United States 2020–2025 Dietary Guidelines, where the vegetable subgroup changed from 'legumes (beans and peas)' to 'beans, peas and lentils' [31,33]. A shift in the language used to define legumes in the ADG may assist in the translation of the guidelines and help consumers to understand what is expected in terms of consumption. However, it should be noted that consumer understanding and interpretation of dietary guidelines was not assessed in this review and further research is warranted.

Global bean and legume consumption drawn from the GDD 2018 was lower than suggested targets, with only 11 of those with FBDG reporting a median intake greater than 50 g/day. Although countries in the APAC region reported the highest level of consumption, this region also had the greatest variability (9.7–122.7 g/day), whereas surveys from North America had very little variation (20.8–21.8 g/day). Variability in consumption may be due to differences in cultural dietary patterns, where there may be a larger number of pulse consumers. For example, the highest level of consumption was reported in Afghanistan (122.7 g/day), where legumes are commonly incorporated into vegetarian and meat dishes as a thickener; however, this country was also one of few that presented the most explicit recommendations for the inclusion of legumes in the dietary pattern (Table 1). As a region, Europe, made up of 33 countries, had by far the lowest consumption of legumes, with more than one-third of countries (36%) reporting intakes less than 10 g/day.

In Australia, low consumption may be due to unfamiliarity with legumes and the tendency to use them as an ingredient rather than as a traditional pattern of eating or being central to the dish [34]. It has been suggested that increasing pulse consumption requires a multifaceted approach, with a focus on addressing the barriers to consumption, such as changing purchasing habits, or the provision of culinary education to improve knowledge and understanding on how to prepare and cook legumes [20]. The impact of COVID-19 on consumption of legumes is yet to be accounted for, with some indication that consumers may have pantry-stocked canned and dried legumes due to the uncertainty related to food supply issues. Future national nutrition surveys will need to consider methodological issues in detecting changes in consumption over the 2020–2021 time period, particularly considering food groups such as legumes. In any case, the history of poor intake of legumes in Australia may also be attributable, in part, to their lack of adequate representation within dietary guidelines. Findings from the present analysis provide some support for the need to address this in order to influence change and shift consumption.

Over time, traditional legume consumption may have been displaced through changes in dietary patterns and availability of alternate protein sources. A historical reflection from Germany in 1850 to 1970 reported a decrease in legume consumption and a rise in meat and fat intake, a trend that may be continuing today in some countries [35]. In Spain, over a shorter timeframe, legume consumption had decreased from 20.2 g/day (1991) to 11.9 g/day (2010) [36], and since then, intake has increased to 22.0 g/day following the recategorisation of legumes into the protein-rich food group. Changes in economic circumstances may also encourage a change from legumes to more animal proteins and the reverse may also be true. More recent attention on dietary sustainability issues, plant protein and flexitarian eating patterns may help draw attention to the relative proportion of animal and plant protein within dietary patterns. A historical review of legume recommendations in dietary guidelines with associated changes in consumption would be a suggested area for future research to determine the impact of dietary guidelines on intake.

Almost all food guides visually represented legumes, which were most often categorised as a meat alternative within the protein-rich food group (38%) or within their own group (20%). In contrast, when comparing placement of legumes in written guidelines with the relevant consumption data, higher consumption appeared aligned with placement of legumes in their own category or together with nuts and seeds (23 countries in total). This is closely followed by countries where legumes were recognised as a protein food (33 countries) and together with starchy staples. Australia and the US tended to have lower reported consumption, although legumes were classified in two groups which is based on their nutritional profile [11,19]; grouped together with vegetables despite being far higher in protein, and with meat and protein foods despite the high dietary fibre content. Including legumes in two separate food groups is purportedly to encourage consumption, although this approach was only utilised by these two countries. Furthermore, the phrase utilised in the Australian guidelines, "and/or legumes/beans" at the end of each food group statement could be considered tokenistic and may understate the importance of this food group in future dietary patterns.

While FBDG are built on the available scientific evidence supporting foods and dietary patterns, the statements within guidelines are designed to specifically suit the population. Similar to the terminology used to define the category, it was clear there was variation in the way guidelines were expressed based on culturally based dietary patterns. Countries such as Mexico, Qatar, Dominican Republic, Albania, Guatemala and Malta encouraged legume intake by categorising legumes as their own food group or provide detailed information on the culinary use of legumes to encourage consumption [30]. Other influences such as food security, historical effects, and the different structure of food chains in each country may also influence legume classification within FBDG [19].

This analysis represents the most comprehensive review of global FBDG and consumption data for legumes, however, should be viewed in light of the limitations. Full FBDG documents could not be reviewed for 49 countries, as they were not in English language which limits the scope the present study. The GDD 2018 dietary intake data are based on estimates using a combination of various sources, particularly self-reported dietary intake data. As such, there needs to be some consideration for the self-reported nature of dietary data collection, although this is true of all large population dietary data sets. In comparison to the GDD 2018, FAO estimates have shown to underestimate legume consumption by 50% [37], providing support for the use of the GDD. Furthermore, we acknowledge that legume content may be hidden as an ingredient in foods [38], particularly in countries where there is a more complex food supply system, this may perpetuate issues with underestimation of intake. We were unable to draw direct links between consumption data and the categorisation of legumes in dietary guidelines due to the variety of classification systems and low levels of consumption across numerous countries. However, there were some apparent patterns; particularly when legumes were grouped together with protein-rich foods or where legumes were classified in their own food group, consumption more often exceeded 25 g/day.

## 5. Conclusions

Dietary guidelines aim to facilitate easy translation to dietary patterns specific to a population. Legumes are potentially difficult to position due to their nutrient composition, being high in dietary fibre and plant-based protein. With greater emphasis on sustainable eating patterns, novel placement of legumes or at least repositioning of legumes may be a consideration within FBDG. Future research should test messaging in dietary guidelines, particularly in countries where increased consumption is desired and where legumes have been added to multiple food groups such as in Australia.

**Author Contributions:** Conceptualization, S.G.; methodology, S.G. and J.H.; formal analysis, J.H.; writing—original draft preparation, J.H. and E.P.; writing—review and editing, S.G., J.H. and E.P. All authors have read and agreed to the published version of the manuscript.

**Funding:** This research received no external funding but was supported by the Grains & Legumes Nutrition Council, a not-for-profit charity.

**Institutional Review Board Statement:** Not applicable.

**Informed Consent Statement:** Not applicable.

**Data Availability Statement:** All data on food-based dietary guidelines are contained within the article. Publicly available datasets on bean and legume consumption were analysed and presented as Figures A1–A6. Consumption data can be found via the Global Dietary Database: https://www.globaldietarydatabase.org/ (accessed on 15 June 2022).

**Acknowledgments:** The authors acknowledge the Honours research supervision of EP by Eleanor Beck, including original data analysis and review of this work as part of an Honours thesis.

**Conflicts of Interest:** J.H. and S.G. were employed by the Grains & Legumes Nutrition Council during the period of time when this research was planned and conducted.

## Appendix A

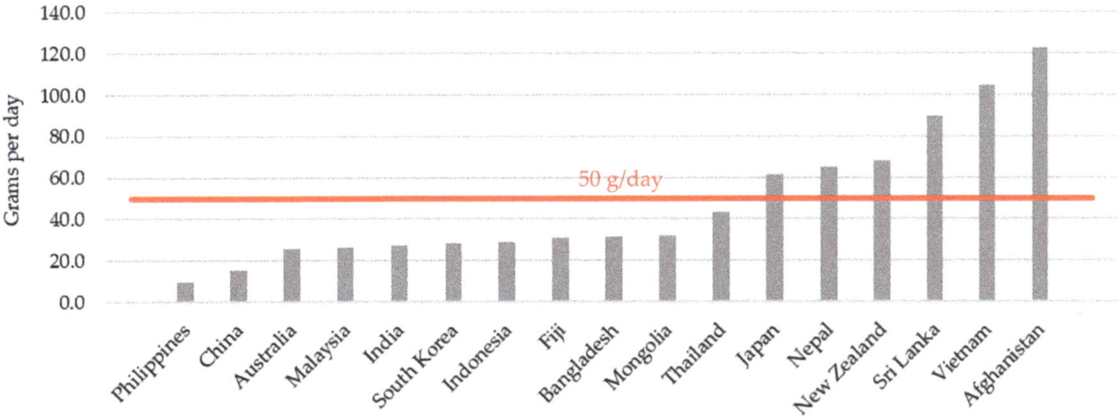


**Figure A1.** Median bean and legume intake for countries in the Asia and Pacific Region with food-based dietary guidelines ($n$ = 17).

## Appendix B

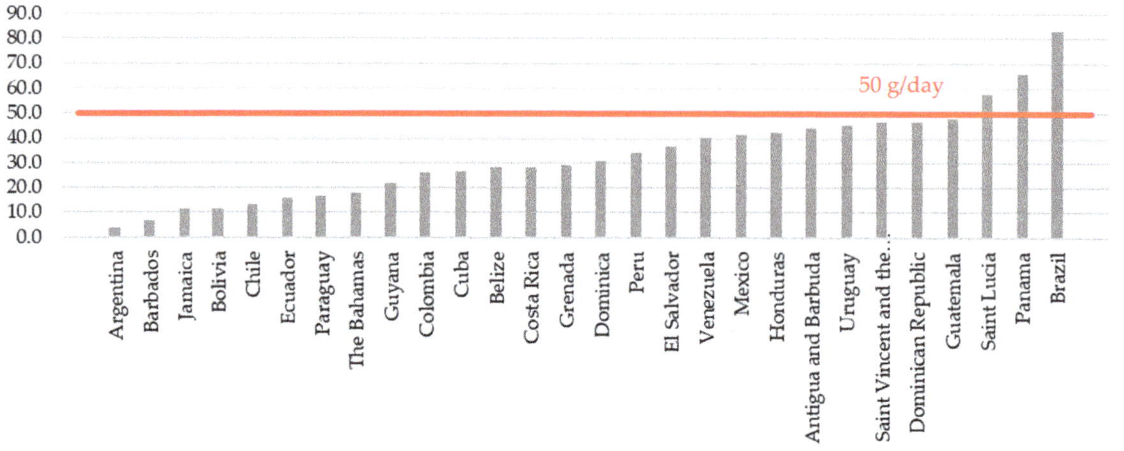


**Figure A2.** Median bean and legume intake for countries in Latin America and the Caribbean with food-based dietary guidelines ($n$ = 28).

## Appendix C

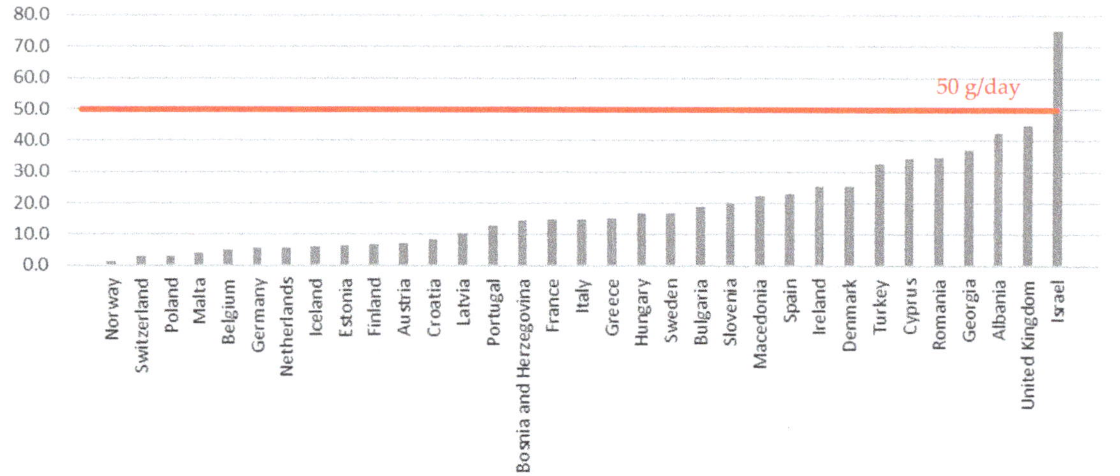


**Figure A3.** Median bean and legume intake for countries in Europe with food-based dietary guidelines ($n$ = 33).

## Appendix D

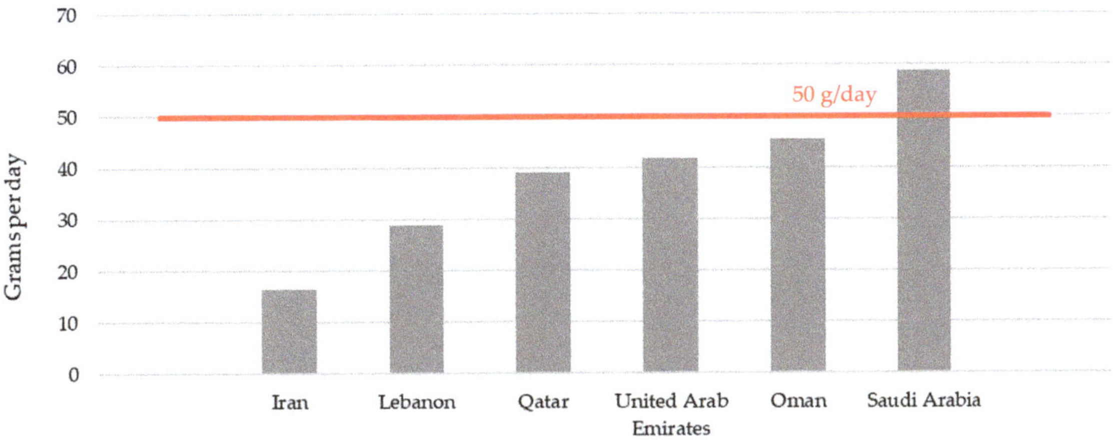


**Figure A4.** Median bean and legume intake for countries in the Near East with food-based dietary guidelines ($n = 6$).

## Appendix E

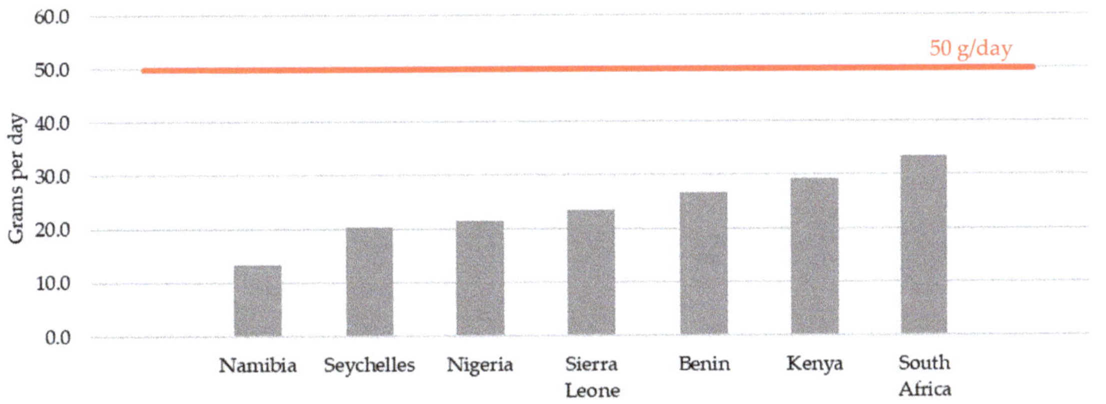


**Figure A5.** Median bean and legume intake for countries in Africa with food-based dietary guidelines ($n = 7$).

## Appendix F

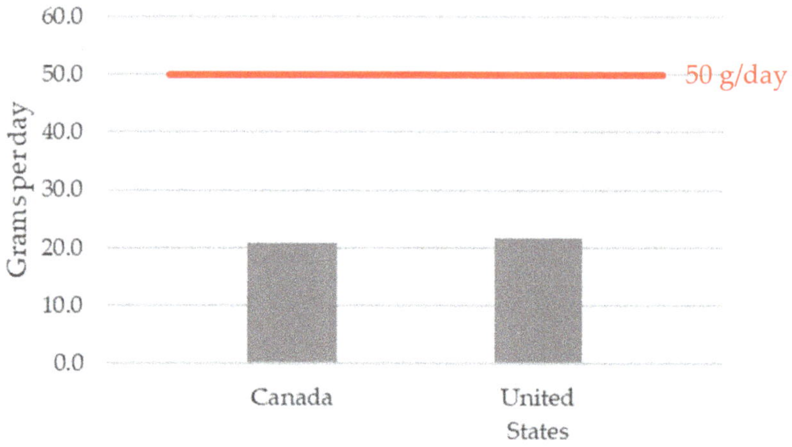


**Figure A6.** Median bean and legume intake for countries in North America with food-based dietary guidelines ($n = 2$).

## References

1. Food and Agriculture Organization of the United Nations. Definition and Classification of Commodities, Pulses and Derived Products. Available online: http://www.fao.org/es/faodef/fdef04e.htm (accessed on 20 September 2021).
2. Margier, M.; Georgé, S.; Hafnaoui, N.; Remond, D.; Nowicki, M.; Du Chaffaut, L.; Amiot, M.-J.; Reboul, E. Nutritional Composition and Bioactive Content of Legumes: Characterization of Pulses Frequently Consumed in France and Effect of the Cooking Method. *Nutrients* **2018**, *10*, 1668. [CrossRef] [PubMed]
3. Calles, T.; del Castello, R.; Baratelli, M.; Xipsiti, M.; Navarro, D.K. *The International Year of Pulses-Final Report*; FAO: Rome, Italy, 2019.
4. Mitchell, D.C.; Marinangeli, C.P.F.; Pigat, S.; Bompola, F.; Campbell, J.; Pan, Y.; Curran, J.M.; Cai, D.J.; Jaconis, S.Y.; Rumney, J. Pulse Intake Improves Nutrient Density among US Adult Consumers. *Nutrients* **2021**, *13*, 2668. [CrossRef]
5. Bechthold, A.; Boeing, H.; Schwedhelm, C.; Hoffmann, G.; Knüppel, S.; Iqbal, K.; De Henauw, S.; Michels, N.; Devleesschauwer, B.; Schlesinger, S.; et al. Food groups and risk of coronary heart disease, stroke and heart failure: A systematic review and dose-response meta-analysis of prospective studies. *Crit. Rev. Food Sci. Nutr.* **2019**, *59*, 1071–1090. [CrossRef]
6. Schwingshackl, L.; Schwedhelm, C.; Hoffmann, G.; Lampousi, A.-M.; Knüppel, S.; Iqbal, K.; Bechthold, A.; Schlesinger, S.; Boeing, H. Food groups and risk of all-cause mortality: A systematic review and meta-analysis of prospective studies. *Am. J. Clin. Nutr.* **2017**, *105*, 1462–1473. [CrossRef] [PubMed]
7. Applegate, C.C.; Rowles, J.L.; Ranard, K.M.; Jeon, S.; Erdman, J.W. Soy Consumption and the Risk of Prostate Cancer: An Updated Systematic Review and Meta-Analysis. *Nutrients* **2018**, *10*, 40. [CrossRef] [PubMed]
8. Qin, L.-Q.; Xu, J.-Y.; Wang, P.-Y.; Hoshi, K. Soyfood Intake in the Prevention of Breast Cancer Risk in Women: A Meta-Analysis of Observational Epidemiological Studies. *J. Nutr. Sci. Vitaminol.* **2006**, *52*, 428–436. [CrossRef] [PubMed]
9. Trock, B.J.; Hilakivi-Clarke, L.; Clarke, R. Meta-Analysis of Soy Intake and Breast Cancer Risk. *J. Natl. Cancer Inst.* **2006**, *98*, 459–471. [CrossRef]
10. Strid, A.; Hallström, E.; Sonesson, U.; Sjons, J.; Winkvist, A.; Bianchi, M. Sustainability Indicators for Foods Benefiting Climate and Health. *Sustainability* **2021**, *13*, 3621. [CrossRef]
11. Havemeier, S.; Erickson, J.; Slavin, J. Dietary guidance for pulses: The challenge and opportunity to be part of both the vegetable and protein food groups. *Ann. N. Y. Acad. Sci.* **2017**, *1392*, 58–66. [CrossRef]
12. Willett, W. Food in the Anthropocene: The EAT-Lancet Commission on healthy diets from sustainable food systems. *Lancet Glob. Health* **2019**, *393*, 447–492. [CrossRef]
13. Abdullah, M.; Marinangeli, C.P.F.; Jones, P.J.H.; Carlberg, J.G. Canadian Potential Healthcare and Societal Cost Savings from Consumption of Pulses: A Cost-Of-Illness Analysis. *Nutrients* **2017**, *9*, 793. [CrossRef] [PubMed]
14. Abdullah, M.M.H.; Hughes, J.; Grafenauer, S. Legume Intake Is Associated with Potential Savings in Coronary Heart Disease-Related Health Care Costs in Australia. *Nutrients* **2022**, *14*, 2912. [CrossRef]
15. Food and Agriculture Organisation of the United Nations. Food-Based Dietary Guidelines. 2020. Available online: http://www.fao.org/nutrition/nutrition-education/food-dietary-guidelines/en/ (accessed on 20 September 2021).

16. Curtain, F.; Grafenauer, S. Historical and Global Perspectives on Grains and Whole Grains within Dietary Guidelines. *Cereal Foods World* **2020**, *65*, 29. [CrossRef]
17. National Health and Medical Research Council. *Australian Dietary Guidelines Summary*; National Health and Medical Research Council: Canberra, Australia, 2013. Available online: https://www.eatforhealth.gov.au/guidelines (accessed on 15 June 2022).
18. National Health and Medical Research Council. *A Modelling System to Inform the Revision of the Australian Guide to Healthy Eating*; National Health and Medical Research Council: Canberra, Australia, 2011; p. 68. Available online: https://www.eatforhealth.gov.au/file/modelling-system-inform-revision-australian-guide-healthy-eating-fms (accessed on 15 June 2022).
19. Marinangeli, C.; Curran, J.; Barr, S.; Slavin, J.; Puri, S.; Swaminathan, S.; Tapsell, L.; Patterson, C. Enhancing nutrition with pulses: Defining a recommended serving size for adults. *Nutr. Rev.* **2017**, *75*, 990–1006. [CrossRef]
20. Figueira, N.; Curtain, F.; Beck, E.; Grafenauer, S. Consumer Understanding and Culinary Use of Legume in Australia. *Nutrients* **2019**, *11*, 1575. [CrossRef] [PubMed]
21. Global Dietary Database. GDD 2018 Final Estimates-Beans and Legumes. Available online: https://www.globaldietarydatabase.org/data-download (accessed on 7 June 2022).
22. Global Dietary Database. The GDD Prediction Model. Available online: https://www.globaldietarydatabase.org/gdd-prediction-model (accessed on 7 June 2022).
23. Khatibzadeh, S.; Saheb Kashaf, M.; Micha, R.; Fahimi, S.; Shi, P.; Elmadfa, I.; Kalantarian, S.; Wirojratana, P.; Ezzati, M.; Powles, J.; et al. A global database of food and nutrient consumption. *Bull. World Health Organ.* **2016**, *94*, 931–934. [CrossRef]
24. Micha, R.; Khatibzadeh, S.; Shi, P.; Andrews, K.G.; Engell, R.E.; Mozaffarian, D. Global, regional and national consumption of major food groups in 1990 and 2010: A systematic analysis including 266 country-specific nutrition surveys worldwide. *BMJ Open* **2015**, *5*, e008705. [CrossRef]
25. Miller, V.; Singh, G.M.; Onopa, J.; Reedy, J.; Shi, P.; Zhang, J.; Tahira, A.; Shulkin Morris, M.L.; Marsden, D.P.; Kranz, S.; et al. Global Dietary Database 2017: Data availability and gaps on 54 major foods, beverages and nutrients among 5.6 million children and adults from 1220 surveys worldwide. *BMJ Glob. Health* **2021**, *6*, e003585. [CrossRef]
26. Miller, V.; Reedy, J.; Cudhea, F.; Zhang, J.; Shi, P.; Erndt-Marino, J.; Coates, J.; Micha, R.; Webb, P.; Mozaffarian, D.; et al. Global, regional, and national consumption of animal-source foods between 1990 and 2018: Findings from the Global Dietary Database. *Lancet Planet. Health* **2022**, *6*, e243–e256. [CrossRef]
27. Global Dietary Database. Model Covariates. Available online: https://www.globaldietarydatabase.org/model-covariates (accessed on 7 June 2022).
28. Forouzanfar, M.H.; Alexander, L.; Anderson, H.R.; Bachman, V.F.; Biryukov, S.; Brauer, M.; Burnett, R.; Casey, D.; Coates, M.M.; Cohen, A.; et al. Global, regional, and national comparative risk assessment of 79 behavioural, environmental and occupational, and metabolic risks or clusters of risks in 188 countries, 1990–2013: A systematic analysis for the Global Burden of Disease Study 2013. *Lancet* **2015**, *386*, 2287–2323. [CrossRef]
29. Global Dietary Database. GDD Variable Definitions. Available online: https://www.globaldietarydatabase.org/GDD/VariableDefinitions (accessed on 7 June 2022).
30. Painter, J.; Rah, J.; Lee, Y. Comparision of international food guide pictorial representations. *J. Am. Diet. Assoc.* **2002**, *102*, 483–489. [CrossRef]
31. Didinger, C.; Thompson, H.J. Defining Nutritional and Functional Niches of Legumes: A Call for Clarity to Distinguish a Future Role for Pulses in the Dietary Guidelines for Americans. *Nutrients* **2021**, *13*, 1100. [CrossRef] [PubMed]
32. Reyneke, G.; Hughes, J.; Grafenauer, S. Consumer Understanding of the Australian Dietary Guidelines: Recommendations for Legumes and Whole Grains. *Nutrients* **2022**, *14*, 1753. [CrossRef]
33. U.S. Department of Agriculture; U.S. Department of Health and Human Services. Dietary Guidelines for Americans, 2020–2025, 9th ed. Available online: https://www.dietaryguidelines.gov/ (accessed on 28 September 2021).
34. Gilham, B.; Hall, R.; Woods, J.L. Vegetables and legumes in new Australasian food launches: How are they being used and are they a healthy choice? *Nutr. J.* **2018**, *17*, 104. [CrossRef] [PubMed]
35. Huebbe, P.; Rimbach, G. Historical Reflection of Food Processing and the Role of Legumes as Part of a Healthy Balanced Diet. *Foods* **2020**, *9*, 1056. [CrossRef]
36. Varela-Moreiras, G.; Ruiz, E.; Valero, T.; Avila, J.M.; del Pozo, S. The Spanish diet: An update. *Nutr. Hosp.* **2013**, *28*, 13–20.
37. Del Gobbo, L.C.; Khatibzadeh, S.; Imamura, F.; Micha, R.; Shi, P.; Smith, M.; Myers, S.S.; Mozaffarian, D. Assessing global dietary habits: A comparison of national estimates from the FAO and the Global Dietary Database. *Am. J. Clin. Nutr.* **2015**, *101*, 1038–1046. [CrossRef]
38. Bielefeld, D.; Hughes, J.; Grafenauer, S. The Changing Landscape of Legume Products Available in Australian Supermarkets. *Nutrients* **2021**, *13*, 3226. [CrossRef]

*nutrients*

MDPI

*Article*

# Food Fortification of Instant Pulse Porridge Powder with Improved Iron and Zinc Bioaccessibility Using Roselle Calyx

**Sandhya Subedi, Uthaiwan Suttisansanee, Aikkarach Kettawan, Chaowanee Chupeerach, Chanakan Khemthong, Sirinapa Thangsiri and Nattira On-nom ***

Food and Nutrition Academic and Research Cluster, Institute of Nutrition, Mahidol University, Salaya, Phuttamonthon, Nakhon Pathom 73170, Thailand

* Correspondence: nattira.onn@mahidol.ac.th; Tel.: +66-(0)-2800-2380 (ext. 426)

**Abstract:** Undernutrition and mineral deficiencies negatively impact both the health and academic performance of school children, while diets high in phytic acid and some phenolics inhibit the absorption of minerals such as iron and zinc. This study developed instant porridge powders rich in iron and zinc using pregelatinized chickpea flour (PCPF) and pregelatinized foxtail millet flour (PFMF) and assessed the potential of utilizing roselle calyx powder (RCP) as a source of organic acids to enhance its iron and zinc bioaccessibility. Physical properties, nutrients, mineral inhibitors and in vitro iron and zinc bioaccessibility of different proportions of PCPF, PFMF and RCP in instant porridge powders were evaluated. Three instant porridge powder formulations including instant chickpea powder (ICP) using PCPF, instant composite flour (ICF) using PCPF and PFMF and instant pulse porridge powder (IPP) using PCPF, PFMF and RCP were developed. Results show that all instant porridge powders were accepted by sensory evaluation, while different ingredients impacted color, consistency and the viscosity index. Addition of RCP improved the bioaccessibility of iron (1.3–1.6-fold) and zinc (1.3–1.9-fold). A 70 g serving of these instant porridge powders substantially contributed to daily protein, iron and zinc requirement for children aged 7–9 years. These porridge powders hold potential to serve as school meals for young children in low-to-middle income countries.

**Keywords:** *Setaria italica* L.; *Cicer arietinum* L.; *Hibiscus sabdariffa* L.; chickpea flour; foxtail millet flour; minerals; organic acids; pregelatinization; sensory evaluation; nutritional values; in vitro bioaccessibility

**Citation:** Subedi, S.; Suttisansanee, U.; Kettawan, A.; Chupeerach, C.; Khemthong, C.; Thangsiri, S.; On-nom, N. Food Fortification of Instant Pulse Porridge Powder with Improved Iron and Zinc Bioaccessibility Using Roselle Calyx. *Nutrients* **2022**, *14*, 4070. https://doi.org/10.3390/nu14194070

Academic Editors: Stuart Johnson and Rewati Bhattarai

Received: 7 September 2022
Accepted: 28 September 2022
Published: 30 September 2022

**Publisher's Note:** MDPI stays neutral with regard to jurisdictional claims in published maps and institutional affiliations.



## 1. Introduction

Iron deficiency anemia is one of the five leading causes of individuals living with disability burden worldwide [1]. Iron deficiency anemia has been widely researched, with the global prevalence of iron deficiency without anemia expected to be at least double that of iron deficiency anemia [2]. Zinc deficiency is also a major public health problem impacting almost all low-to-middle income countries [3]. These deficiencies lead to impaired metal and physical and cognitive development, reduced work productivity, an increased risk of perinatal complications, morbidity from infectious disease and mortality [2,3]. At both individual and population levels, deficiencies of iron and zinc coexist in low-to-middle income countries such as Nepal, Ethiopia and Nigeria. This is because they share similar food sources and dietary elements, hindering or facilitating the absorption of these two minerals [3,4]. The problem is more severe in southern Asia, primarily among women and children, even though this region is witnessing improved agricultural output, poverty alleviation and access to education [4]. Iron and zinc deficiencies arise from the consumption of monotonous, predominantly unrefined plant-based diets. They are usually due to the presence of high amounts of mineral binders such as phytic acid and condensed tannin [5,6]. Wholegrain cereals and legumes may contain sufficient quantities of these minerals to meet daily intake requirements, but their bioavailability is generally low [5].

Food-based strategies include the integration of dietary diversity, with the principles of bioavailability translated to efficient food synergies, and due emphasis on food accessibility and affordability [7]. Such strategies are recognized as cost-effective, viable and sustainable solutions to combat the double burden of malnutrition of at-risk populations, especially school children. They also promote self-reliance of the community [3,7]. The nutritional status of school children has a direct impact on their health, cognitive abilities and educational performance. School children are often far from mainstream beneficiaries in underdeveloped countries such as Nepal, and the available resources mostly go to younger children who are more vulnerable [8]. However, the importance of investment in school children has been recognized to improve nutrition and also derive long-term societal benefits through poverty reduction and diminishing the economic gap [8,9]. A copious amount of evidence suggests that integration of school meal programs to provide complete or supplemental meals to students promotes nutrition, school participation and enrollment and also enhances educational achievement [10–13]. It is recommended that school meals be prepared using locally available foods high in protein and micronutrients, such as cereal- and legume-based porridge powder [10–13], which are widely consumed in Nepal by all age groups, particularly children [14].

Foxtail millet (*Setaria italica* L.) is one of the oldest varieties of millet, and a neglected minor crop with great potential to help in coping with food insecurity and climate change in remote areas of Nepal [15]. This gluten-free grain is rich in calcium, dietary fiber, polyphenols and protein [16]. Chickpea (*Cicer arietinum* L.) is one of the major pulses consumed as a reliable source of plant-based protein, vitamins and minerals in Nepal [17]. Utilization of this pulse as a sustainable and ecofriendly source of protein is attracting increased attention, with countless possibilities to improve food insecurity worldwide [18]. To improve released iron and/or zinc contents from the food matrix through digestion and solubility in the gut lumen [19], addition of organic acids, especially from plant sources, is required. Roselle (*Hibiscus sabdariffa* L.) is a locally available phenolics-rich plant containing dietary components such as ascorbic acid, citric acid and other organic acids [20,21]. Combining such locally available plants with staple grains (food-to-food fortification) has been advocated as a sustainable method to improve iron and zinc contents and bioavailability [22]. However, a study of food-to-food fortification of instant porridge powder with improved iron and zinc bioaccessibility using roselle calyx for Nepalese school children is challenging due to the specific characterizations of each ingredient.

Hence, this study developed nutrient-dense instant porridge powder from pregelatinized chickpea flour (PCPF) and pregelatinized foxtail millet flour (PFMF) using roselle calyx powder (RCP) as a potential food-to-food fortification to improve the bioaccessibility of iron and zinc while also maintaining the desired sensory properties. Different proportions of pregelatinized chickpea flour, pregelatinized foxtail millet flour and roselle calyx powder were assessed for their physical properties, nutritive values and mineral inhibitor contents. Results provide crucial information concerning iron and zinc bioaccessibility of roselle calyx-enriched instant porridge powder products, using locally available and inexpensive food sources as possible future school meals.

## 2. Materials and Methods

### *2.1. Raw Materials*

Chickpea and foxtail millet were purchased from Young Smart Farmer, Songkhla, Thailand. The pregelatinized forms of these two pulses, PCPF and PFMF, were prepared according to Subedi et al. [23]. PCPF consisted of 4.85% moisture, 22.34% protein, 7.27% total fat, 63.68% total carbohydrates (comprising 17.07% dietary fiber), 4.90% iron and 2.62% zinc, while PFMP consisted of 5.88% moisture, 12.84% protein, 3.58% total fat, 76.60% total carbohydrates (comprising 3.41% dietary fiber), 3.29% iron and 2.52% zinc. The chemical composition of RCP was 9.11% moisture, 7.26% protein, 1.69% total fat, 73.99% total carbohydrates (comprising 41.89% dietary fiber) and 38.60% total organic acids (comprising 18.84% succinic acid, 13.61% formic acid, 2.21% acetic acid, 2.15% oxalic acid, 1.55% citric acid, 0.23% ascorbic acid and 0.01% malic acid). RCP was purchased from Huglamool Farm, Amnat Charoen, Thailand. Dried powders of seasoning herbs (onion powder, garlic powder, cumin seed powder, coriander seed powder, black pepper and turmeric) were bought from Khunsiri brand, Pathum Thani, Thailand. Chili powder was obtained from Prakriti Indian Trading Co., Ltd., Bangkok, Thailand. Soybean oil was purchased from Angoon brand, Thai vegetable oil PLC, Bangkok, Thailand. All reagents used in this study were obtained from Sigma-Aldrich (St. Louis, MO, USA).

### *2.2. Preparation of Three Different Instant Porridge Powder Formulations*

Three different formulations of instant porridge powders, including instant chickpea powder (ICP) using PCPF, instant composite flour (ICF) using PCPF and PFMF and instant pulse porridge power (IPP) using PCPF, PFMF and RCP, were prepared (Table 1) by mixing the ingredients with a laboratory blender (model HR1393/00, Philips (Thailand) Ltd., Bangkok, Thailand) for 30 s. The optimal ratio of RCP in the IPP formulation was selected based on sensory evaluation and in vitro bioaccessibility of iron and zinc, as shown in Supplementary Tables S2 and S3. The mixtures were then sealed in aluminum foil bags and kept at 4 °C until use. Seasoning spices mixed powder (SSMP) consisting of 52.83% onion powder, 3.33% garlic powder, 0.67% turmeric powder, 1.67% coriander powder, 0.66% cumin powder, 1.67% chili powder and 0.33% black pepper powder, was prepared using a laboratory blender, packed in aluminum foil bags and kept at 4 °C until use.

**Table 1.** Ingredients of three different instant porridge powder formulations.

| Ingredients (%) | Instant Porridge Powders | | |
|---|---|---|---|
| | ICP | ICF | IPP |
| Pregelatinized chickpea flour (PCPF) | 82.66 | 49.74 | 46.76 |
| Pregelatinized foxtail millet flour (PFMF) | – | 33.45 | 31.44 |
| Oil | 8.51 | 7.23 | 7.23 |
| Roselle calyx powder (RCP) | – | – | 4.99 |
| Seasoning spices mix powder (SSMP) | 6.70 | 7.32 | 7.32 |
| Salt | 2.13 | 2.26 | 2.26 |

ICP: instant chickpea powder using PCPF; ICF: instant composite flour using PCPF and PFMF; IPP: instance pulse porridge power using PCPF, PFMF and RCP.

### *2.3. Analysis of Nutritive Values*

Proximate analysis (moisture content, fat, protein, ash, carbohydrates, energy and total dietary fiber) was determined according to the method of AOAC [24]. Moisture content was evaluated by drying the samples in a hot-air oven (Memmert model UNE 500, Eagle, WI, USA) at 100 °C until constant weight was achieved (AOAC 930.04). Total fat content was determined by acidic digestion and solvent extraction with petroleum ether using a Tecator Soxtec System (model 1043, Hoganas, Sweden) (AOAC 922.06). Protein content was analyzed using the Kjeldahl method utilizing digestion and distillation units (BÜCHI Corporation, New Castle, DE, USA), and then calculated using a conversion factor of 6.25 (AOAC 991.20). Ash content was analyzed by incineration of the organic matter in a muffle furnace (Carbolite model CWF 1100, Hope, UK) at 550 °C (AOAC 930.30). Total carbohydrates were calculated by the subtraction of moisture, fat, protein and ash contents from 100. Energy value was attained from the integration of total energy from carbohydrates, protein and fat as 4, 4 and 9 kcal/g samples, respectively. Total dietary fiber was evaluated using the enzyme gravimetric method (AOAC 985.29). Iron and zinc were analyzed using an Inductively Coupled Plasma–Optical Emission Spectrophotometer (ICP–OES) (AOAC 984.27) [24]. All samples were determined in the Accredited Laboratory, complying with ISO/IEC 17025:2017, at the Institute of Nutrition, Mahidol University (Nakhon Pathom, Thailand). Results as per 100 g fresh weight (FW) are shown in Supplementary Table S1. To accurately compare the nutritive values among the three instant porridge powders, compositions per 100 g dry weight (DW) were calculated.

## 2.4. Sensory Evaluation

Sensory evaluation was conducted by 40 untrained panelists comprising Nepali people living in Thailand. The test was performed in an individual testing booth of the sensory science laboratory at the Institute of Nutrition, Mahidol University, and lasted 10–15 min. All samples were prepared on the same day of the test and served fresh. The porridge was prepared by reconstituting porridge powder in warm water (60 $\pm$ 2 °C) at a ratio of 1:3 (*w/w*). The panelists evaluated the products for satisfaction based on appearance, color, odor, taste, texture and overall liking using a nine-point hedonic scale (1 = dislike extremely to 9 = like extremely) [25].

## 2.5. Physical Analysis

Physical properties were analyzed in terms of color (L* (lightness), a* (red–green color) and b* (yellow–blue color)) using a Colorflex EZ Spectrophotometer (HunterLab, Reston, VA, USA), water activity ($A_w$) by a water activity measurement instrument (model ms1–1 M, Novasina, Lachen, Switzerland) and pH using a pH meter (Ohaus Corporation, Morris Country, NJ, USA). Bulk density (BD) was determined by weighing 50 g of sample into a 100 mL measuring cylinder before tapping continuously until a constant volume was obtained. BD was calculated using Equation (1) [26].

$$\text{Bulk density } (\text{g/cm}^3) = \frac{\text{Weight of the sample}}{\text{Volume of the sample after tapping}} \tag{1}$$

The water solubility index (WSI) and water absorption index (WAI) were determined following Anderson, 1999 [27]. The sample (1.25 g) was weighed into a preweighed 50 mL centrifuge tube, followed by the addition of 20 mL of deionized water. The centrifuge tube with the sample mixture was then placed in a SWB20 water bath shaker (Australian Scientific Instruments PTY. Ltd., Fyshwick, Canberra, Australia) at 30 °C for 30 min. The sample was then centrifuged using a Hettich®ROTINA 16R centrifuge (Andreas Hettich GmbH, Tuttlingen, Germany) at 1190$\times$ *g* for 10 min at 25 °C. The supernatant was decanted for determination of its solid content, and the sediment was weighed using a weighing balance (Mettler Toledo, Toronto, Canada). The supernatant was dried at 105 °C until reaching a stable weight (approximately 3 h). WSI and WAI were calculated using Equations (2) and (3), respectively.

$$\text{WSI} = \left(\frac{\text{Weight of dissolved solid in supernatant}}{\text{Weight of dry solids}}\right) \times 100 \tag{2}$$

$$\text{WAI} = \frac{\text{Weight of sediment}}{\text{Volume of dry solid}} \tag{3}$$

The texture of porridge samples as determined by the consistency and viscosity index was analyzed by a back extrusion test using a TA–XT plus Texture Analyzer (Stable Micro Systems, Godalming, Surrey, UK) followed by the method of Syahariza and Yong (2017) [28]. The texture analysis was carried out at 25 °C and replicated ten times for each sample.

## 2.6. Anti-Nutrient Analysis

Phytic acid content (total phosphorus) was measured using a phytic acid assay kit (Megazyme, Wicklow, Ireland) as described by McKie and McCleary (2016) [29], with results expressed as g/100 g DW. To determine total phenol and tannin contents, the Folin–Ciocalteu method of The Food and Agriculture Organization of the United Nations (FAO) and the International Atomic Energy Agency (IAEA) (2000) was utilized [30]. Results are expressed as mg tannic acid equivalent (TAE)/100 g DW.

## 2.7. Determination of Iron and Zinc Bioaccessibility

The samples were analyzed for bioaccessibility by the in vitro dialyzability method of Ting and Loh (2016) [31]. Peptic and pancreatic digestion were simulated by the enzyme pepsin (P7000, from porcine stomach mucosa), pancreatin (P1750, from porcine pancreas) and bile extract (B8631, porcine). Mineral contents of the dialysates were analyzed by ICP-OES as described in the iron and zinc contents analysis.

### 2.8. Microbiological Analysis

Total plate count (TPC) and yeast and mold counts (YMCs) were performed following the standard protocols of the Bacteriological Analytical Manual (BAM) [32]. The numbers of colonies appearing on the dilution plates were counted, averaged and reported as colony forming units (CFUs)/g.

### 2.9. Statistical Analysis

All measurements were performed in triplicate or as stated otherwise, with results expressed as mean ± standard deviation (SD). Mean differences of $p < 0.05$ were determined by one-way analysis of variance (ANOVA), followed by Duncan's multiple comparison test for more than two data sets or Student's unpaired $t$-test for two data sets using IBM SPSS Statistics for Windows version 26.0 (IBM Corp., Armonk, NY, USA).

## 3. Results

### 3.1. Nutritional Analysis

The proximate analysis results as well as iron and zinc contents (per 100 g DW) of ICP, ICF and IPP are shown in Table 2, while contents per 100 g FW are shown in Supplementary Table S1. The energy, fat, carbohydrates and total dietary fiber of all instant porridge powders varied significantly. ICP exhibited slightly but significantly higher energy (1-fold higher than the others), protein (1.2-fold higher than the others), fat (1.2-fold higher than the others) and total dietary fiber (1.2–1.5-fold higher than the others), while containing the lowest carbohydrate content (1.1-fold lower than the others). The ash content of IPP was also slightly but significantly higher than ICP and ICF (1.1-fold higher), possibly due to the supplementation of RCP. The same trends with ash content were observed in iron contents of all instant porridge powders, while no significant difference in zinc content (2.72–3.95 mg/100 g DW) was observed among the instant porridge powders.

**Table 2.** Proximate, iron and zinc contents of instant porridge powders (per 100 g dry weight).

| Formulations | Energy (kcal) | Protein (g) | Fat (g) | Carbohydrates (g) | Total DF (g) | Ash (g) | Iron (mg) | Zinc $^{ns}$ (mg) |
|---|---|---|---|---|---|---|---|---|
| ICP | 458.76 ± 0.44 $^{a}$ | 20.15 ± 0.07 $^{a}$ | 15.03 ± 0.13 $^{a}$ | 60.72 ± 0.25 $^{c}$ | 17.15 ± 0.53 $^{a}$ | 4.10 ± 0.05 $^{b}$ | 4.15 ± 0.01 $^{b}$ | 3.58 ± 1.09 |
| ICF | 447.07 ± 0.89 $^{c}$ | 16.60 ± 0.11 $^{b}$ | 12.67 ± 0.23 $^{c}$ | 66.66 ± 0.20 $^{a}$ | 11.54 ± 0.24 $^{c}$ | 4.07 ± 0.08 $^{b}$ | 4.23 ± 0.16 $^{b}$ | 2.72 ± 0.13 |
| IPP | 450.70 ± 0.79 $^{b}$ | 16.50 ± 0.00 $^{b}$ | 13.73 ± 0.00 $^{b}$ | 65.40 ± 0.07 $^{b}$ | 13.80 ± 0.00 $^{b}$ | 4.44 ± 0.01 $^{a}$ | 4.89 ± 0.25 $^{a}$ | 3.95 ± 0.21 |

All data are shown as the mean ± standard deviation (SD) of triplicate determination ($n = 3$). Different lowercase letters denote significantly different contents of the same proximate composition or iron content at $p < 0.05$, while $^{ns}$ denotes no significant differences in zinc content at $p \geq 0.05$ in different instant porridge powder formulations using one-way ANOVA, followed by Duncan's multiple comparison test. DF: dietary fiber; ICP: instant chickpea powder using pregelatinized chickpea flour (PCPF); ICF: instant composite flour using PCPF and pregelatinized foxtail millet flour (PFMF); IPP: instant pulse porridge powder using PCPF, PFMF and roselle calyx powder (RCP).

### 3.2. Sensory Evaluation of Instant Porridge Powders

Sensory evaluation results of ICP, ICF and IPP are shown in Table 3, indicating no significant differences ($p \geq 0.05$) in sensory attribute scores among the three instant porridge powders, including in appearance (6.45–6.73), color (6.68–6.85), odor (6.68–7.10), taste (6.28–6.75), texture (6.20–6.73) and overall liking (6.43–6.85). The sensory attribute scores of all instant porridge powders were higher than 6 (like slightly).

**Table 3.** Sensory evaluation of instant porridge powders.

| Instant Porridge Powders | Sensory Attributes | | | | | |
|---|---|---|---|---|---|---|
| | Appearance $^{ns}$ | Color $^{ns}$ | Odor $^{ns}$ | Taste $^{ns}$ | Texture $^{ns}$ | Overall Liking $^{ns}$ |
| ICP | 6.45 ± 1.21 | 6.85 ± 1.09 | 7.10 ± 1.66 | 6.28 ± 1.81 | 6.20 ± 1.41 | 6.43 ± 1.69 |
| ICF | 6.65 ± 1.37 | 6.68 ± 1.21 | 6.68 ± 1.44 | 6.75 ± 1.11 | 6.55 ± 1.11 | 6.68 ± 1.27 |
| IPP | 6.73 ± 1.38 | 6.70 ± 1.36 | 7.10 ± 1.21 | 6.60 ± 1.35 | 6.73 ± 0.93 | 6.85 ± 1.10 |

All data are shown as the mean ± standard deviation (SD) of 40 untrained panelists ($n$ = 40). The $^{ns}$ denotes no significantly different values at $p \geq 0.05$ for the same sensory attributes in different instant porridge powder formulations using one-way ANOVA, followed by Duncan's multiple comparison test. ICP: instant chickpea powder using pregelatinized chickpea flour (PCPF); ICF: instant composite flour using PCPF and pregelatinized foxtail millet flour (PFMF); IPP: instant pulse porridge powder using PCPF, PFMF and roselle calyx powder (RCP).

### 3.3. Physical Analysis of Instant Porridge Powders

Physical property results including color, water activity ($A_w$), water solubility index (WSI), water absorption index (WAI), bulk density (BD) and texture (consistency and viscosity index) are shown in Table 4. Color analysis indicated that IPP was significantly darker and redder than ICP and ICF, possibly due to the addition of RCP. All instant porridge powders exhibited $A_w$ ranging from 0.13 to 0.18, with IPP exhibiting the highest value. No significant differences in WSI (6.13–6.40%), WAI (3.77–4.04 g/g) and BD (0.72 g/cm$^3$) were observed among the instant porridge powders. For texture analysis, the highest consistency (1.3–4.3-fold higher than the others) and viscosity index (1.6–25.1-fold higher than others) were recorded in ICF, while ICP exhibited the lowest consistency and viscosity index.

**Table 4.** Physical properties of instant porridge powders.

| Formulations | Color | | | $A_w$ | WSI (%) $^{ns}$ | WAI (g/g) $^{ns}$ | BD (g/cm$^3$) $^{ns}$ | Consistency (N s) | Index of Viscosity (N s) |
|---|---|---|---|---|---|---|---|---|---|
| | L* | a* | b* | | | | | | |
| ICP | 35.40 ± 0.27 $^a$ | 3.69 ± 0.09 $^b$ | 35.88 ± 0.09 $^a$ | 0.17 ± 0.00 $^a$ | 6.40 ± 3.67 | 4.04 ± 0.12 | 0.72 ± 0.01 | 96.65 ± 4.37 $^c$ | 2.47 ± 2.10 $^c$ |
| ICF | 34.06 ± 0.57 $^b$ | 3.42 ± 0.17 $^c$ | 32.47 ± 0.26 $^b$ | 0.13 ± 0.01 $^b$ | 6.40 ± 1.39 | 4.00 ± 0.09 | 0.72 ± 0.00 | 413.94 ± 35.38 $^a$ | 62.07 ± 8.93 $^a$ |
| IPP | 27.80 ± 0.40 $^c$ | 4.13 ± 0.02 $^a$ | 19.64 ± 0.31 $^c$ | 0.18 ± 0.01 $^a$ | 6.13 ± 1.22 | 3.77 ± 0.17 | 0.72 ± 0.01 | 306.64 ± 22.65 $^b$ | 37.65 ± 5.18 $^b$ |

All data are shown as the mean ± standard deviation (SD) of triplicate determination ($n$ = 3). Different lowercase letters denote significant differences in color, $A_w$ or texture at $p < 0.05$, while $^{ns}$ denotes no significant differences in WSI, WAI or BD at $p \geq 0.05$ in different instant porridge powder formulations using one-way ANOVA, followed by Duncan's multiple comparison test. ICP: instant chickpea powder using pregelatinized chickpea flour (PCPF); ICF: instant composite flour using PCPF and pregelatinized foxtail millet flour (PFMF); IPP: instant pulse porridge powder using PCPF, PFMF and roselle calyx powder (RCP); $A_w$: water activity; WSI: water solubility index; WAI: water absorption index; BD: bulk density; N s: Newton second; L*: lightness; a*: red/green value; b*: blue/yellow value.

### 3.4. Mineral Inhibitor Contents in Instant Porridge Powders

A plant-based diet constitutes an important source of all nutrients, including minerals. However, plants are also sources of mineral inhibitors such as phytate, the main inhibitor of mineral absorption (iron, zinc and calcium) [33]. Thus, the mineral inhibitor contents of ICP, ICF and IPP, including phytic acid, tannin and phenols, were evaluated (Table 5). Results indicate no significant difference in phytic acid (1.45–1.58 g/100 g DW) among the instant porridge powders. However, the tannin and total phenol contents of all instant porridge powders varied significantly. IPP presented the highest tannin (1.8–3.6-fold higher than the others) and total phenols (1.7–2.5-fold higher than the others), while the lowest tannin and total phenol content were recorded in ICF.

### 3.5. In Vitro Iron and Zinc Bioaccessibility of Instant Porridge Powders

Iron and zinc bioaccessibility and pH values of the three instant porridge powders were determined using the in vitro digestion assay (Table 6). IPP presented the highest bioaccessibility and percentage of bioaccessibility of iron (1.5–1.6-fold and 1.3–1.4-fold, respectively, higher than the others) with zinc (1.9-fold and 1.3–1.5-fold, respectively, higher than the others). However, no significant differences in the bioaccessibility or percentage of iron and zinc bioaccessibility were observed between ICP and ICF. For pH, IPP exhibited the lowest values (1.3-fold lower than the others), while no significant pH difference was recorded between ICP (5.78) and ICF (5.83).

**Table 5.** Mineral inhibitor content of instant porridge powders.

| Formulation | Mineral Inhibitors | | |
|---|---|---|---|
| | Phytic Acid (g/100 g DW) $^{ns}$ | Tannin (mg TAE/100 mg DW) | Total Phenols (mg TAE/100 mg DW) |
| ICP | 1.48 ± 0.05 | 1.39 ± 0.01 $^{b}$ | 1.67 ± 0.01 $^{b}$ |
| ICF | 1.45 ± 0.13 | 0.71 ± 0.01 $^{c}$ | 1.15 ± 0.02 $^{c}$ |
| IPP | 1.58 ± 0.16 | 2.58 ± 0.13 $^{a}$ | 2.83 ± 0.13 $^{a}$ |

All data are shown as the mean ± standard deviation (SD) of triplicate determination ($n$ = 3). Different lowercase letters denote significantly different values of tannin or total phenol at $p < 0.05$, while $^{ns}$ denotes no significant differences in phytic acid content at $p \geq 0.05$ in different instant porridge powder formulations using one-way ANOVA, followed by Duncan's multiple comparison test. ICP: instant chickpea powder using pregelatinized chickpea flour (PCPF); ICF: instant composite flour using PCPF and pregelatinized foxtail millet flour (PFMF); IPP: instant pulse porridge powder with PCPF, PFMF and roselle calyx powder (RCP); TAE: tannin acid equivalent; DW: dry weight.

**Table 6.** In vitro iron and zinc bioaccessibility and pH of instant porridge powders.

| Formulations | Iron | | Zinc | | pH |
|---|---|---|---|---|---|
| | Bioaccessible (mg/100 g) | % Bioaccessibility | Bioaccessible (mg/100 g) | % Bioaccessibility | |
| ICP | 0.32 ± 0.01 $^{b}$ | 7.83 ± 0.08 $^{b}$ | 0.87 ± 0.01 $^{b}$ | 30.97 ± 2.38 $^{b}$ | 5.78 ± 0.02 $^{a}$ |
| ICF | 0.33 ± 0.02 $^{b}$ | 8.20 ± 0.07 $^{b}$ | 0.90 ± 0.02 $^{b}$ | 34.65 ± 2.66 $^{b}$ | 5.83 ± 0.02 $^{a}$ |
| IPP | 0.51 ± 0.01 $^{a}$ | 11.08 ± 0.86 $^{a}$ | 1.70 ± 0.02 $^{a}$ | 45.83 ± 2.90 $^{a}$ | 4.55 ± 0.03 $^{b}$ |

All data are shown as the mean ± standard deviation (SD) of triplicate determination ($n$ = 3). Different lowercase letters denote significantly different values of in vitro iron and zinc bioaccessibility and pH at $p < 0.05$ in different instant porridge powder formulations using one-way ANOVA, followed by Duncan's multiple comparison test. ICP: instant chickpea powder using pregelatinized chickpea flour (PCPF); ICF: instant composite flour using PCPF and pregelatinized foxtail millet flour (PFMF); IPP: instant pulse porridge powder using PCPF, PFMF and roselle calyx powder (RCP).

### 3.6. Microbiological Analysis

In this study, the total plate count (TPC) and yeast and mold counts (YMCs) of the instant porridge powders were assessed according to the Thai Community Product Standard based on instant rice porridge powder. This states that the TPC of the product must be less than $1 \times 10^6$ CFU/g, while the YMC must not exceed 100 CFU/g [34]. Results show that the TPC and YMC values of the three instant porridge powders were in line with the standard (Table 7).

**Table 7.** Microbiological quality assessment of instant porridge powders.

| Formulations | Microbiological Quality | |
|---|---|---|
| | Total Plate Count (CFU/g) | Yeast and Mold (CFU/g) |
| ICP | $3.0 \times 10^4 \pm 0.06$ | Less than 10 |
| ICF | $7.2 \times 10^4 \pm 0.03$ | Less than 10 |
| IPP | $6.8 \times 10^3 \pm 0.01$ | Less than 10 |

All data are shown as the mean ± standard deviation (SD) of triplicate determination ($n$ = 3). ICP: instant chickpea powder using pregelatinized chickpea flour (PCPF); ICF: instant composite flour using PCPF and pregelatinized foxtail millet flour (PFMF); IPP: instant pulse porridge powder using PCPF, PFMF and roselle calyx powder (RCP).

## 4. Discussion

To combat climate change, promotions of sustainable plant-based food sources are gaining emphasis worldwide. Plant-based food also contains calories and macronutrients.

To prevent malnutrition and improve health and cognitive outcomes, the macronutrient balance (ratios between protein, carbohydrates and fat) is a key factor [35,36]. Minerals such as iron and zinc are found abundantly in plant-based foods such as legumes, while mineral inhibitors such as phytate have lower bioavailability compared to animal sources. However, household processing methods such as soaking and addition of enhancers help to ensure that the body uses the minerals we consume. Growing children require higher amounts of these nutrients. Our bodies should be able to easily utilize nutrients to gain the benefits of consuming bioactive compounds from plant-based food sources. This study developed instant porridge powder using pregelatinized foxtail millet and chickpea flours as sources of protein, carbohydrates and fat according to the World Health Organization (WHO) recommendation and fortified the product using roselle calyx powder as a source of organic acids to improve the bioaccessibility of the contained iron and zinc.

The proximate analysis results show that ICP had the highest protein and fat contents owing to having the highest PCPF content. Chickpea is a good source of protein and contains significant amounts of all the essential amino acids. All the instant porridge powders met the recommended values of protein (≥15%) according to FAO and WHO (1991) [37]. The WHO (2009) also suggested that values higher than 15% protein may increase renal solute load and interfere with appetite [38]. The recommended dietary allowances (RDAs) and absolute requirement (AR) of protein in the Indian Council of Medical Research (ICMR) (2020) [39] states that the RDA of protein for children aged 7–9 years is 23 g protein/day. One serving size (70 g) of ICP provided a 1.2-fold higher protein contribution to RDA than ICF and IPP (Table 8). All three instant porridge powders demonstrated a higher content of protein than the RDA, while for fat content, all instant porridge powders met the recommended FAO and WHO (1991) values of fat (10–25%). [37]. The fat content in chickpea varies from 3.80 to 10.20% as a relatively good source of nutritionally important polyunsaturated fatty acids, linoleic acid and monounsaturated oleic acid [40]. The carbohydrate contents of ICF and IPP were also within the recommended levels of FAO/WHO (1991), while ICP did not meet the recommended level (64 ± 4 g/100 g) [37]. This observation suggests that legume and cereal combinations provide more balanced nutrition than a single plant-based food material. The dietary fiber contents of the three instant porridge powders also showed significant differences, with the highest value recorded in ICP. According to the ICMR (2020) recommendation, dietary fiber intake should be 25 g/day [39]. IPP presented the highest dietary fiber (1.2–1.5-fold higher than the others), possibly due to the high proportion of PCPF. All instant porridge powders were high in fiber products, with dietary fiber contents of more than 6 g/100 g [41]. Energy is provided by proteins, fats and carbohydrates. The highest energy was recorded in ICP, due to the relatively high protein and fat contents in PCPF. All instant porridges were energy-dense foods that provide many calories in a small serving amount.

Ash content represents the total mineral content in foods. The three instant porridge powders varied significantly in ash content due to the differences in amounts of individual ingredients. Ash content was highest in IPP, possibly due to the addition of RCP during formulation, as RCP is a food source of minerals [42]. This result was confirmed by the observation that the highest amounts of iron (1.2-fold higher than the others) and zinc (1.1–1.4-fold higher than the others) occurred in IPP. According to the RDA and AR values for iron and zinc in the ICMR (2020) [39], children aged 7–9 years require 15 mg iron/day and 5.90 mg zinc/day. Our results show that IPP had the highest contribution of iron (1.1-fold higher than the others) and zinc (1.3–1.4-fold higher than the other) to the RDA for children aged 7–9 years (Table 8), possibly due to the RCP supplemented in the formulation. The AR of iron (requirement for growth + basal losses + menstrual losses) was based on the WHO/UNICEF/UNU 2004 (95th percentile) for children aged 7–10 years [43]. They reported that the AR of iron for children aged 7–10 years is 0.89 mg/day. The highest contribution to the AR of iron was recorded in IPP (40.10%), followed by ICF (25.96%) and ICP (25.17%) (Table 8). The International Zinc Nutrition Consultative Group (IZiNCG) (2004) set the AR of zinc for children aged 4–8 years with a reference body weight of

21 kg as 0.83 mg/day and the AR of zinc for children aged 9–13 years with a reference body weight of 38 kg as 4.53 mg/day [44]. Total endogenous zinc losses were calculated as 0.034 mg/kg/day for children one year of age and older (i.e., urinary losses, surface losses and intestinal losses), based on their respective reference body weights and rates of weight gain [44]. IPP showed the highest AR of zinc for both children aged 4–8 years with a reference body weight of 21 kg and children aged 9–13 years with a reference body weight of 38 kg (1.9–2.0-fold higher than the others), while ICP had the lowest contribution to the AR of zinc in both age groups. A food is considered to be a good source of nutrients if it fulfills more than 20% of the RDA [39]. All instant porridges were excellent sources of protein, iron and zinc, and could benefit children who suffer from macronutrient (protein) and micronutrient (mineral) deficiency.

**Table 8.** Recommended dietary allowances (RDAs) and absolute requirement (AR) of protein, iron and zinc of instant porridge powders and their contribution to RDA and AR.

| Formulations | Minerals | RDA [1] | Contribution (%) per Serving (70 g) | AR (mg/d) | Contribution (%) per Serving (70 g) |
|---|---|---|---|---|---|
| ICP | Protein | 23 | 58.89 | NA | NA |
| | Iron | 15 | 19.04 | 0.89 [2] | 25.17 [2] |
| | Zinc | 5.90 | 33.46 | 0.83 [3] (1.53 [4]) | 73.38 [3] (39.87 [4]) |
| ICF | Protein | 23 | 48.51 | NA | NA |
| | Iron | 15 | 18.99 | 0.89 [2] | 25.96 [2] |
| | Zinc | 5.90 | 30.97 | 0.83 [3] (1.53 [4]) | 75.90 [3] (41.18 [4]) |
| IPP | Protein | 23 | 47.17 | NA | NA |
| | Iron | 15 | 21.40 | 0.89 [2] | 40.10 [2] |
| | Zinc | 5.90 | 44.02 | 0.83 [3] (1.53 [4]) | 143.40 [3] (77.80 [4]) |

[1] Recommended dietary allowances of protein, iron and zinc for children aged 7–9 years are according to Indian Council of Medical Research (ICMR) 2020; [2] Absolute requirement of iron (requirement for growth + basal losses + menstrual losses) was based on WHO/UNICEF/UNU 2004 (95th percentile) for children aged 7–10 years; [3] Absolute requirement of zinc for children aged 4–8 years with reference body weight of 21 kg as according to IZiNCG (2004); [4] Absolute requirement of zinc for children aged 9–13 years with reference body weight of 38 kg as according to IZiNCG (2004); ICP: instant chickpea powder using pregelatinized chickpea flour (PCPF); ICF: instant composite flour using PCPF and pregelatinized foxtail millet flour (PFMF); IPP: instant pulse porridge powder using PCPF, PFMF and roselle calyx powder (RCP); RDA: recommended dietary allowances; AR: absolute requirement; NA: not available.

The highest moisture content was also found in IPP, and attributed to the high moisture content of the RCP added in this formulation. Moisture content is important to maintain food quality. High moisture content of food results in microbial growth and ultimately destroys quality. El Wakeel (2007) stated that dried materials with lower than 10% moisture content, such as instant soup ingredients, had better storage qualities [45]. All instant porridge powders were deemed microbiologically safe to consume according to the criteria of the Thai community product standard based on instant rice porridge powder [34].

As final products, all instant porridge powders were accepted by the consumers and received sensory attribute scores above 6, or 'like slightly'. Previous research used an average value of 6 (like slightly) on a nine-point hedonic scale as the minimum acceptable limit for consumers liking a product [46–48]. Therefore, ICP, ICF and IPP showed promise to be developed with consumer acceptance. The hedonic rating obtained for IPP was also higher than that for ICP and ICF. Results suggest that the addition of RCP did not adversely affect product sensory perception.

To determine the physical properties, the color of the instant porridge powders was measured. IPP was darker and redder than ICP and ICF, probably due to the intense red pigment of anthocyanins in RCP [49]. Anthocyanins are water-soluble pigments belonging to the phenolic group. They are responsible for red, purple and blue colors in fruits and vegetables [50]. Similar results were reported by Beswa and Singo (2019) [51], who revealed that a high concentration of roselle extracts increased the redness of ice cream, while lightness decreased. Color changes in instant porridge powders occurred because of the unique colors of the ingredients.

Water activity of ICP, ICF and IPP was low (0.13–0.18), and less than the required water activity level for microbe growth [52]. The water solubility index (WSI) and water absorption index (WAI) are parameters used to explain the hydration properties of porridge. The WSI determines the amount of soluble degradation from starch upon the addition of excess of water [27,53] and also indicates the water penetration into starch granules [54]. A lower WSI indicates minor degradation of starch [55]. The WSI values of instant porridge powders in this study, at 6.13–6.40%, were lower than reported those by Mahgoub et al. (2020) [55]. They found that WSI values of instant porridge supplemented with different types of mung bean were 16.12–22.5% [55], while a similar result was found by Walle and Moges (2017) for legume cereal blended with complementary porridge powder [56]. WAI is an indirect measure of the degree of instantaneous reconstitutability of cereals or powders in excess water [57]. No significant differences for the WAI were found among the three instant porridge powders (3.77–4.04 g/g); however, the WAI increased when the protein content increased. Kinsella (1976) explained that the polar amino acid residues of protein bind to water molecules, resulting in a higher water absorption index [58].

Bulk density (BD) represents the structural change of a material. All the instant porridge powders exhibited low BD (0.72 $g/cm^3$). Similar results were observed for porridge powders supplemented with soybean and mung bean, with a BD ranging from 0.57 to 0.70 $g/cm^3$ [59], and yellow maize porridge, with a BD ranging from 0.69 to 0.81 $g/cm^3$ [60]. Low BD is advantageous in the formulation of food for our target demographic, namely school children, because a minimal quantity of powder is necessary to achieve the desired bulkiness of the food products. Higher-BD foods are suitable for use as thickeners, gels and other high-viscosity food products [60,61].

The texture of the instant porridge powders was determined by consistency and the viscosity index. Consistency relates to the firmness, thickness and viscosity of a liquid or fluid semi-solid product. The consistency analysis revealed significant differences among the three instant porridge powders. The higher the PFMF content, the more work or energy was needed to compress the instant porridge powders due to the higher starch content. The viscosity index is regarded as the extrusion energy or work of adhesion. Higher values mean more resistance is required when pulling out the sample. The viscosity index is highly related to resistance to flow and also to cohesiveness and consistency [62,63]. Results indicate that PCPF reduced porridge viscosity, supporting the observation that ICP had the lowest viscosity index.

Chickpea and foxtail millets contain high amounts of macronutrients and micronutrients; however, anti-nutritional factors were also detected. The anti-nutrients determined in this study include phytates, tannins and total phenols. They can combine with nutrients and reduce nutrient bioavailability. Phytate is a negatively charged structure that can bind with positively charged metal ions such as zinc, iron, magnesium and calcium in the digestive tract to form mineral complexes and reduce their bioavailability, eventually leading to mineral deficiency [64]. Thompson (1993) stated that consumption of a 10–60 mg/g phytate diet over a long period resulted in decreased bioavailability of minerals in monogastric animals [65]. Therefore, the phytate composition of all instant porridge powders might not pose a health hazard. Tannins are phenolic compounds that can precipitate protein and decrease protein digestibility and palatability. Tannins can form an insoluble complex with ferric ion ($Fe^{3+}$) and reduce the bioaccessibility of non-haem iron [66,67], while zinc absorption is not affected by tannin [68]. Our results reveal that the combination of chickpea and foxtail millet reduced tannin and total phenol content; however, supplementation of RCP increased these anti-nutrients due to the proportion of ingredients in the formulations. A high concentration of anti-nutrients has an adverse effect but may exert beneficial health effects at adequate amounts by reducing blood glucose levels and the insulin response to starchy foods and/or plasma cholesterol and triglyceride to reduce cancer risk [69,70].

Bioaccessibility is defined as the fraction of the total amount of a substance that is released from the food matrix during gastrointestinal digestion and is potentially available for being absorbed [71]. IPP showed a higher bioaccessibility and percentage of bioaccessibility of iron due to the higher concentration of organic acids in RCP. Organic acids can improve iron absorption and stabilize its soluble form in the small intestine by changing the state of iron oxidation from ferric ion ($Fe^{3+}$) to ferrous ion ($Fe^{2+}$) [72]. A Caco-2 cell study suggested mechanisms underlying the boosting effect of organic acids on two types of iron [73]. Organic acids can improve ferric iron uptake both by chelation and by decreasing the pH, but ferrous iron uptake is promoted only by lowering the pH. No significant difference in zinc content was found in the three instant porridge powders, but IPP showed the highest bioaccessibility and percentage of zinc bioaccessibility. This result suggests that the increase in the bioaccessibility and percentage of bioaccessibility occurred for iron as well as zinc. Organic acids prevented the production of insoluble zinc–phytate complexes [74], like in an in vitro (dialyzability) study by Van der Merwe et al. (2019), who suggested that adding 5.56% and 15% of RCP enhanced the percentage of bioaccessible iron (by 42% and 55%, respectively) and amount of bioaccessible iron (by 107% and 269%, respectively) [5]. Our study provides valuable information on how to improve the bioaccessibility of iron and zinc using RCP in plant-based foods. However, an intervention study needs to be conducted to confirm the improved status because bioavailability has a physiological endpoint [75].

## 5. Conclusions

To reduce the risk of malnutrition, iron deficiency anemia and zinc deficiency retardment, instant porridge powders were successfully developed using PCPF, PFMF and RCP. The ingredients used resulted in significant changes in the physical properties of the instant porridge powders. Addition of PFMF caused an increase in consistency and the viscosity index. Instant porridge powders (ICP, ICF and IPP) are high in protein, fiber and minerals and provide a reasonable percentage of children's daily requirements for protein, iron and zinc according to the RDA and AR. Addition of RCP in IPP augmented the bioaccessible content and percentage of bioaccessibility of iron and zinc. Roselle calyx powder could be used as a prominent source of iron and organic acids to improve iron content and bioaccessibility of iron and zinc from locally available but often mineral-inhibitor-rich plant-based food sources. The process of preparing instant porridge powders is suitable on a small-scale basis, and the product is safe for consumption. However, the stability studies on the instant pulse porridge powders under long-term and accelerated conditions were the limitations of this study, which should be further investigated. Additionally, even though in vitro dialyzability has been frequently used as a trustworthy predictor for evaluating mineral bioavailability or the contents of ingested iron or zinc that pass through intestinal cell, in vivo experiments should also be further performed to confirm the amount of iron and zinc being absorbed into our body. Our results demonstrate that the developed instant porridge powders had better nutritional profiles that could help to prevent and treat undernutrition and micronutrient deficiencies in young children living in low-to-middle income countries. Additionally, these instant porridge powders might be a prototype either for school meals or in the food industry for development and introduction of new food poducts on the market.

**Supplementary Materials:** The following supporting information can be downloaded at: https://www.mdpi.com/article/10.3390/nu14194070/s1. Table S1: Proximate iron and zinc contents of instant porridge powders (per 100 g fresh weight); Table S2: Sensory evaluation of instant porridge powders with different amount of roselle calyx powder; Table S3: In vitro iron and zinc bioaccessibility of instant porridge powders with different amount of roselle calyx powder.

**Author Contributions:** Conceptualization, S.S., U.S., C.C., A.K. and N.O.-n.; methodology, S.S., U.S, A.K., C.C., C.K., S.T. and N.O.-n.; validation, S.S. and N.O.-n.; formal analysis, S.S., U.S., A.K., C.C., C.K., S.T. and N.O.-n.; investigation, S.S, U.S. and N.O.-n.; resources, N.O.-n., A.K., U.S. and C.C.; data curation, S.S., U.S., A.K., C.C. and N.O.-n.; writing—original draft preparation, S.S. and N.O.-n.;

writing—review and editing, S.S., U.S., C.C. and N.O.-n.; supervision, N.O.-n.; project administration, N.O.-n.; funding acquisition, N.O.-n. All authors have read and agreed to the published version of the manuscript.

**Funding:** This research received no external funding.

**Institutional Review Board Statement:** The study was conducted in accordance with the Declaration of Helsinki, and approved by the Center Institutional Review Board (MU–CIRB) of Mahidol University (protocol code MU–CIRB 2020/376.1311; date of approval: 9 December 2020).

**Informed Consent Statement:** Informed consent was obtained from all subjects involved in the study.

**Data Availability Statement:** Data are contained within this article.

**Conflicts of Interest:** The authors declare no conflict of interest.

## References


1. GBD 2016 Disease and Injury Incidence and Prevalence Collaborators. Global, regional, and national incidence, prevalence, and years lived with disability for 328 diseases and injuries for 195 countries, 1990–2016: A systematic analysis for the Global Burden of Disease Study 2016. *Lancet* **2017**, *390*, 1211–1259. [CrossRef]
2. Camaschella, C. Iron deficiency. *Blood* **2019**, *133*, 30–39. [CrossRef] [PubMed]
3. Gupta, S.; Brazier, A.K.M.; Lowe, N.M. Zinc deficiency in low- and middle-income countries: Prevalence and approaches for mitigation. *J. Hum. Nutr. Diet* **2020**, *33*, 624–643. [CrossRef] [PubMed]
4. Harding, K.L.; Aguayo, V.M.; Webb, P. Hidden hunger in South Asia: A review of recent trends and persistent challenges. *Public Health Nutr.* **2018**, *21*, 785–795. [CrossRef]
5. van der Merwe, R.; Kruger, J.; Ferruzzi, M.G.; Duodu, K.G.; Taylor, J.R.N. Improving iron and zinc bioaccessibility through food-to-food fortification of pearl millet with tropical plant foodstuffs (moringa leaf powder, roselle calyces and baobab fruit pulp). *J. Food Sci. Technol.* **2019**, *56*, 2244–2256. [CrossRef]
6. Gibson, R.; Hotz, C.; Temple, L.; Yeudall, F.; Mtitimuni, B.; Ferguson, E. Dietary Strategies to Combat Deficiencies of Iron, Zinc, and Vitamin A in Developing Countries: Development, Implementation, Monitoring, and Evaluation. *Food Nutr. Bull.* **2000**, *21*, 219–231. [CrossRef]
7. Nair, M.K.; Augustine, L.F.; Konapur, A. Food-Based Interventions to Modify Diet Quality and Diversity to Address Multiple Micronutrient Deficiency. *Front. Public Health* **2015**, *3*, 277. [CrossRef]
8. Best, C.; Neufingerl, N.; van Geel, L.; van den Briel, T.; Osendarp, S. The nutritional status of school-aged children: Why should we care? *Food Nutr. Bull.* **2010**, *31*, 400–417. [CrossRef]
9. Human Capital Summit: Committing to Action to Drive Economic Growth. Available online: https://live.worldbank.org/human-capital-summit-2017 (accessed on 28 October 2021).
10. Dhungana, M. Mid-Day Meals Attract Children to Schools in Achham. Available online: https://kathmandupost.com/national/2019/02/17/mid-day-meals-attract-children-to-schools-in-achham (accessed on 13 October 2021).
11. Fiorentino, M. *Malnutrition in School-Aged Children and Adolescents in Senegal and Cambodia: Public Health Issues and Interventions*; Université Montpellier: Montpellier, France, 2015.
12. MoHP; UNICEF; USAID; EU; CDC; Era, N. *Nepal National Micronutrient Status Survey Report 2016*; Ministry of Health and Population: Kathmandu, Nepal, 2018.
13. Shrestha, R.M.; Schreinemachers, P.; Nyangmi, M.G.; Sah, M.; Phuong, J.; Manandhar, S.; Yang, R.-Y. Home-grown school feeding: Assessment of a pilot program in Nepal. *BMC Public Health* **2020**, *20*, 28. [CrossRef]
14. Shrestha, L.; Parmar, A.; Kulig, B.; Hensel, O.; Sturm, B. Feeding practices of pre-school children and associated factors in Kathmandu, Nepal. *J. Hum. Nutr. Diet* **2020**, *33*, 241–251. [CrossRef]
15. Ghimire, K.; Bhandari, B.; Gurung, S.B.; Dhami, N.; Baniya, B. Diversity and Utilization Status of Millets Genetic Resources in Nepal. In Proceedings of the 2nd National Workshop, Kathmandu, Nepal, 22–23 May 2017.
16. Yang, T.; Ma, S.; Liu, J.; Sun, B.; Wang, X. Influences of four processing methods on main nutritional components of foxtail millet: A review. *Grain Oil Sci. Technol.* **2022**, *5*, 156–165. [CrossRef]
17. Shrestha, R.; Neupane, R.K.; Adhikari, N.P. Status and Future Prospects of Pulses in Nepal. In Proceedings of the Regional Workshop on Pulse Production, Nepal Agricultural Research Council (NARC), Kathmandu, Nepal, 24–25 October 2011.
18. Bessada, S.M.F.; Barreira, J.C.M.; Oliveira, M.B.P.P. Pulses and food security: Dietary protein, digestibility, bioactive and functional properties. *Trends Food Sci. Technol.* **2019**, *93*, 53–68. [CrossRef]
19. Gibson, R.S. Zinc: The missing link in combating micronutrient malnutrition in developing countries. *Proc. Nutr. Soc.* **2006**, *65*, 51–60. [CrossRef]
20. Da-Costa-Rocha, I.; Bonnlaender, B.; Sievers, H.; Pischel, I.; Heinrich, M. *Hibiscus sabdariffa* L.—A phytochemical and pharmacological review. *Food Chem.* **2014**, *165*, 424–443. [CrossRef]
21. Jabeur, I.; Pereira, E.; Caleja, C.; Calhelha, R.C.; Soković, M.; Catarino, L.; Barros, L.; Ferreira, I. Exploring the chemical and bioactive properties of Hibiscus sabdariffa L. calyces from Guinea-Bissau (West Africa). *Food Funct.* **2019**, *10*, 2234–2243. [CrossRef]

22. Thompson, B.; Amoroso, L. *Combating Micronutrient Deficiencies: Food-Based Approaches*; CABI: Wallingford, UK, 2011.
23. Subedi, S.; Suttisansanee, U.; Chupeerach, C.; Chamchan, R.; Khemthong, C.; Kettawan, A.; On-nom, N. Changes in Nutritional and Physico-Chemical Properties of Chickpea Flour and Foxtail Millet Flour as a Result of Pregelatinizing. In Proceedings of the 23rd Food Innovation Asia Conference 2021 (FIAC 2021), Bangkok, Thailand, 17–18 June 2021.
24. AOAC. *Official Method of Analysis of AOAC Internation*, 21st ed.; AOAC International: Arlington, VA, USA, 2019.
25. Meilgaard, M.; Civille, G.V.; Carr, B.T. *Sensory Evaluation Techniques*, 4th ed.; CRC Press: Boca Raton, FL, USA, 2006.
26. Oladele, A.K.; Aina, J. Chemical composition and functional properties of flour produced from two varieties of tigernut (*Cyperus esculentus*). *Afr. J. Biotechnol.* **2007**, *6*, 2473–2476. [CrossRef]
27. Andersson, A.A.M.; Andersson, R.; Autio, K.; Åman, P. Chemical Composition and Microstructure of Two Naked Waxy Barleys. *J. Cereal Sci.* **1999**, *30*, 183–191. [CrossRef]
28. Syahariza, Z.A.; Yong, H.Y. Evaluation of rheological and textural properties of texture-modified rice porridge using tapioca and sago starch as thickener. *J. Food Meas. Charact.* **2017**, *11*, 1586–1591. [CrossRef]
29. McKie, V.A.; MccleAry, B.V. A Novel and Rapid Colorimetric Method for Measuring Total Phosphorus and Phytic Acid in Foods and Animal Feeds. *J. AOAC Int.* **2016**, *99*, 738–743. [CrossRef]
30. Joint FAO/IAEA Division of Nuclear Techniques in Food and Agriculture. *Quantification of Tannins in Tree Foliage A Laboratory Manual for the FAO/IAEA Co-Ordinated Research Project on 'Use of Nuclear and Related Techniques to Develop Simple Tannin Assays for Predicting and Improving the Safety and Efficiency of Feeding Ruminants on Tanniniferous Tree Foliage'*; International Atomic Energy Agency (IAEA): Vienna, Austria, 2000; p. 26.
31. Ting, S.R.; Loh, S.P. In vitro bioaccessibility of calcium, iron and zinc from breads and bread spreads. *Int. Food Res. J.* **2016**, *23*, 2175–2180.
32. FDA. *Bacteriological Analytical Manual (BAM)*; US Food and Drug Administration: Silver Spring, MD, USA, 2001.
33. Gibson, R.S.; Bailey, K.B.; Gibbs, M.; Ferguson, E.L. A review of phytate, iron, zinc, and calcium concentrations in plant-based complementary foods used in low-income countries and implications for bioavailability. *Food Nutr. Bull.* **2010**, *31*, S134–S146. [CrossRef] [PubMed]
34. TISI. *Thai Community Product Standard, Instant Rice Porridge*; Thai Industrial Standards Institute: Bangkok, Thailand, 2019.
35. Ejigui, J.; Savoie, L.; Marin, J.; Desrosiers, T. Improvement of the nutritional quality of a traditional complementary porridge made of fermented yellow maize (Zea mays): Effect of maize-legume combinations and traditional processing methods. *Food Nutr. Bull.* **2007**, *28*, 23–34. [CrossRef] [PubMed]
36. Suri, D.J.; Tano-Debrah, K.; Ghosh, S.A. Optimization of the nutrient content and protein quality of cereal-legume blends for use as complementary foods in Ghana. *Food Nutr. Bull.* **2014**, *35*, 372–381. [CrossRef] [PubMed]
37. FAO/WHO. Codex Alimentarius (CODEX CAC/GL 08): Guidelines on Formulated Supplementary Foods for Older Infants and Young Children. Available online: https://www.fao.org/fao-who-codexalimentarius/sh-proxy/zh/?lnk=1&url=https%253A%252F%252Fworkspace.fao.org%252Fsites%252Fcodex%252FStandards%252FCXG%2B8-1991%252FCXG_008e.pdf (accessed on 13 October 2021).
38. WHO. Proceedings of the WHO UNICEF WFP UNHCR informal consultation on the management of moderate malnutrition in under-5 children. *Food Nutr. Bull.* **2009**, *30*, S464–S474.
39. ICMR; NIN. *Nutrient Requirements for Indians*; The Indian Council of Medical Research (ICMR) and National Institute of Nutrition (NIN): Hyderabad, India, 2020.
40. Jukanti, A.K.; Gaur, P.M.; Gowda, C.L.; Chibbar, R.N. Nutritional quality and health benefits of chickpea (*Cicer arietinum* L.): A review. *Br. J. Nutr.* **2012**, *108* Suppl. 1, S11–S26. [CrossRef]
41. Thai Recommended Daily Intakes (Thai RDIs). Appendix No. 3. In *MOPH Notification No. 182 B.E. 2541 Re: Nutrition Labelling*; Royal Thai Government Gazette: Nonthaburi, Thailand, 1998.
42. Mgaya Kilima, B. Physiochemical, mineral composition and antioxidant properties of Roselle (*Hibiscus sabdariffa* L.) extract blended with tropical fruit juices. *Afr. J. Food Agric. Nutr. Dev.* **2014**, *14*, 8963–8978.
43. WHO; FAO. *Vitamin and Mineral Requirements in Human Nutrition*; Nutrition and Food Safety: Geneva, Switzerland, 2004.
44. Brown, K.; Rivera, J.; Bhutta, Z.; Gibson, R.; King, J.; Lonnerdal, B.; Ruel, M.; Sandtröm, B.; Wasantwisut, E.; Hotz, C. International Zinc Nutrition Consultative Group (IZiNCG) technical document #1. Assessment of the risk of zinc deficiency in populations and options for its control. *Food Nutr. Bull.* **2004**, *25*, S99–S203. [CrossRef]
45. El Wakeel, M.A. *Ultra Structure and Functional Properties of Some Dry Mixes of Food*; Ain Shams University: Cairo, Egypt, 2007.
46. Gámbaro, A.; Giménez, A.N.A.; Ares, G.; Gilardi, V. Influence of enzymes on the texture of brown pan bread. *J. Texture Stud.* **2006**, *37*, 300–314. [CrossRef]
47. Giménez, A.; Varela, P.; Salvador, A.; Ares, G.; Fiszman, S.; Garitta, L. Shelf life estimation of brown pan bread: A consumer approach. *Food Qual. Prefer.* **2007**, *18*, 196–204. [CrossRef]
48. Giménez, A.N.A.; Ares, G.; Gámbaro, A. Survival analysis to estimate sensory shelf life using acceptability scores. *J. Sens. Stud.* **2008**, *23*, 571–582. [CrossRef]
49. Abou-Arab, A.A.; Abu-Salem, F.M.; Abou-Arab, E.A. Physico- chemical properties of natural pigments (anthocyanin) extracted from Rosellecalyces (*Hibiscus subdariffa*). *J. Am. Sci.* **2011**, *7*, 445–456.
50. Khoo, H.E.; Azlan, A.; Tang, S.T.; Lim, S.M. Anthocyanidins and anthocyanins: Colored pigments as food, pharmaceutical ingredients, and the potential health benefits. *Food Nutr. Res.* **2017**, *61*, 1361779. [CrossRef]

51. Beswa, D.; Singo, T. Effect of roselle extracts on the selected quality characteristics of ice cream. *Int. J. Food Prop.* **2019**, *22*, 42–53. [CrossRef]
52. Bonazzi, C.; Dumoulin, E. *Quality Changes in Food Materials as Influenced by Drying Processes*; Wiley: Hoboken, NJ, USA, 2011.
53. Yousf, N.; Nazir, F.; Salim, R.; Ahsan, H.; Sirwal, A. Water solubility index and water absorption index of extruded product from rice and carrot blend. *J. Pharmacogn. Phytochem.* **2017**, *6*, 2165–2168.
54. Ikegwu, O.; Okechukwu, P.; Ekumankana, E. PhysicoChemical and Pasting Characteristics of Flour and Starch from Achi Brachystegia eurycoma Seed. *J. Food Technol.* **2010**, *8*, 58–66. [CrossRef]
55. Mahgoub, S.; Mohammed, A.; Mobarak, E.-A. Physiochemical, Nutritional and Technological Properties of Instant Porridge Supplemented with Mung Bean. *Food Nutr. Sci.* **2020**, *11*, 1078–1095. [CrossRef]
56. Walle, H.; Haile, D. Optimization of cereal-legume blend ratio to enhance the nutritional quality and functional property of complementary food. *Ethiop. J. Sci. Technol.* **2017**, *10*, 109. [CrossRef]
57. Sharma, C.; Singh, B.; Hussain, S.Z.; Sharma, S. Investigation of process and product parameters for physicochemical properties of rice and mung bean (Vigna radiata) flour based extruded snacks. *J. Food Sci. Technol.* **2017**, *54*, 1711–1720. [CrossRef]
58. Kinsella, J.E. Functional properties of proteins in foods: A survey. *Crit. Rev. Food Sci. Nutr.* **1976**, *1*, 219–280. [CrossRef]
59. Mayachiew, P.; Charunuch, C.; Devahastin, S. Physicochemical and Thermal Properties of Extruded Instant Functional Rice Porridge Powder as Affected by the Addition of Soybean or Mung Bean. *J. Food Sci.* **2015**, *80*, E2782–E2791. [CrossRef]
60. Ukom, A.N.; Adiegwu, E.C.; Ojimelukwe, P.C.; Okwunodulu, I.N. Quality and sensory acceptability of yellow maize ogi porridge enriched with orange-fleshed sweet potato and African yam bean seed flours for infants. *Sci. Afr.* **2019**, *6*, e00194. [CrossRef]
61. Nnam, N.M. Chemical, sensory and rheological properties of porridges from processed sorghum (*Sorghum bicolor*), bambara groundnut (*Vigna subterranea* L. Verdc) and sweet potato (*Ipomoea batatas*) flours. *Plant Foods Hum. Nutr.* **2001**, *56*, 251–264. [CrossRef]
62. Liu, H.; Xu, X.M.; Guo, S.D. Rheological, texture and sensory properties of low-fat mayonnaise with different fat mimetics. *LWT-Food Sci. Technol.* **2007**, *40*, 946–954. [CrossRef]
63. Reyes, V.G., Jr.; Jindal, V.K. A small sample back extrusion test for measuring texture of cooked-rice. *J. Food Qual.* **1990**, *13*, 109–118. [CrossRef]
64. Lopez, H.W.; Leenhardt, F.; Coudray, C.; Remesy, C. Minerals and phytic acid interactions: Is it a real problem for human nutrition? *Int. J. Food Sci. Technol.* **2002**, *37*, 727–739. [CrossRef]
65. Thompson, L.U. Potential Health Benefits and Problems Associated With Anti Nutrients in Foods. *Int. J. Food Resour.* **1993**, *26*, 131–149. [CrossRef]
66. Perron, N.R.; Brumaghim, J.L. A review of the antioxidant mechanisms of polyphenol compounds related to iron binding. *Cell Biochem. Biophys.* **2009**, *53*, 75–100. [CrossRef]
67. Shahidi, F.; Ambigaipalan, P. Phenolics and polyphenolics in foods, beverages and spices: Antioxidant activity and health effects—A review. *J. Funct. Foods* **2015**, *18*, 820–897. [CrossRef]
68. Afsana, K.; Shiga, K.; Ishizuka, S.; Hara, H. Reducing effect of ingesting tannic acid on the absorption of iron, but not of zinc, copper and manganese by rats. *Biosci. Biotechnol. Biochem.* **2004**, *68*, 584–592. [CrossRef]
69. Fekadu Gemede, H.; Retta, N. Antinutritional factors in plant foods: Potential health benefits and adverse effects. *Int. J. Nutr. Food Sci.* **2014**, *3*, 284–289. [CrossRef]
70. Nath, H.; Samtiya, M.; Dhewa, T. Beneficial attributes and adverse effects of major plant-based foods anti-nutrients on health: A review. *Hum. Nutr. Metab.* **2022**, *28*, 200147. [CrossRef]
71. Jacob, J.K.; Tiwari, K.; Correa-Betanzo, J.; Misran, A.; Chandrasekaran, R.; Paliyath, G. Biochemical basis for functional ingredient design from fruits. *Annu. Rev. Food Sci. Technol.* **2012**, *3*, 79–104. [CrossRef]
72. Torres, L.; Amaral, J.; Canniatti-Brazaca, S. Promoters effectiveness in the improvement in iron and zinc absorption from the rice and bean. *Food Sci. Technol.* **2020**, *40*, 363–368. [CrossRef]
73. Salovaara, S.; Sandberg, A.-S.; Andlid, T. Combined Impact of pH and Organic Acids on Iron Uptake by Caco-2 Cells. *J. Agric. Food Chem.* **2003**, *51*, 7820–7824. [CrossRef]
74. Ma, G.; Jin, Y.; Piao, J.; Kok, F.; Guusje, B.; Jacobsen, E. Phytate, calcium, iron, and zinc contents and their molar ratios in foods commonly consumed in China. *J. Agric. Food Chem.* **2005**, *53*, 10285–10290. [CrossRef]
75. Etcheverry, P.; Grusak, M.A.; Fleige, L.E. Application of in vitro bioaccessibility and bioavailability methods for calcium, carotenoids, folate, iron, magnesium, polyphenols, zinc, and vitamins B(6), B(12), D, and E. *Front. Physiol.* **2012**, *3*, 317. [CrossRef]

*nutrients*

MDPI

*Article*

# Inhibition of $\alpha_1$-Adrenergic, Non-Adrenergic and Neurogenic Human Prostate Smooth Muscle Contraction and of Stromal Cell Growth by the Isoflavones Genistein and Daidzein

**Ru Huang, Yuhan Liu, Sheng Hu, Alexander Tamalunas, Raphaela Waidelich, Frank Strittmatter, Christian G. Stief and Martin Hennenberg ***

Department of Urology, University Hospital, LMU Munich, 81377 Munich, Germany
* Correspondence: martin.hennenberg@med.uni-muenchen.de

**Abstract:** Isoflavone-rich legumes, including soy, are used for food production, as dietary supplements and in traditional medicine. Soy consumption correlates negatively with benign prostatic hyperplasia (BPH) and voiding symptoms. However, isoflavone effects on the prostate are hardly known. Here, we examined the effects on human prostate smooth muscle contractions and stromal cell growth, which are driving factors of voiding symptoms in BPH. Smooth muscle contractions were induced in prostate tissues from radical prostatectomy. Growth-related functions were studied in cultured stromal cells (WPMY-1). Neurogenic, $\alpha_1$-adrenergic and non-adrenergic contractions were strongly inhibited with 50 μM and by around 50% with 10 μM genistein. Daidzein inhibited neurogenic contractions using 10 and 100 μM. Agonist-induced contractions were inhibited by 100 μM but not 10 μM daidzein. A combination of 6 μM genistein with 5 μM daidzein still inhibited neurogenic and agonist-induced contractions. Proliferation of WPMY-1 cells was inhibited by genistein (>50%) and daidzein (<50%). Genistein induced apoptosis and cell death (by seven-fold relative to controls), while daidzein induced cell death (6.4-fold) without apoptosis. Viability was reduced by genistein (maximum: 87%) and daidzein (62%). In conclusion, soy isoflavones exert sustained effects on prostate smooth muscle contractions and stromal cell growth, which may explain the inverse relationships between soy-rich nutrition, BPH and voiding symptoms.

**Keywords:** benign prostatic hyperplasia (BPH); lower urinary tract symptoms (LUTSs); voiding symptoms; soy; soy food; isoflavones; prostate smooth muscle contraction

**Citation:** Huang, R.; Liu, Y.; Hu, S.; Tamalunas, A.; Waidelich, R.; Strittmatter, F.; Stief, C.G.; Hennenberg, M. Inhibition of $\alpha_1$-Adrenergic, Non-Adrenergic and Neurogenic Human Prostate Smooth Muscle Contraction and of Stromal Cell Growth by the Isoflavones Genistein and Daidzein. *Nutrients* **2022**, *14*, 4943. https://doi.org/10.3390/nu14234943

Academic Editors: Stuart Johnson and Rewati Bhattarai

Received: 2 November 2022
Accepted: 18 November 2022
Published: 22 November 2022



## 1. Introduction

Genistein, daidzein and puerarin are isoflavones naturally occurring in plants of the *Fabaceae* family, including soybeans (*Glycine max*), red clover (*Trifolium pratense*) and different *Pueraria* species, the latter known as Kudzu plant in Western nations and as Gegen in China [1,2]. Based on observations from traditional medicine and of soy consumption for nutrition, as well as epidemiological, experimental and clinical studies, a plethora of biological effects have been attributed to isoflavones, including antitumorigenic, antidiabetic, neuroprotective, cardioprotective, estrogenic and hormone-disrupting effects, in addition to vasorelaxation and inhibition of smooth muscle contractions [2,3]. In soy and red clover, genistein and daidzein are predominant isoflavones and constitute major parts of the total isoflavone pool [1,4]. In *Pueraria* species, the predominant isoflavone is puerarin [1,4]. Isoflavone-rich preparations from soy and red clover and from genistein itself are available as dietary supplements and used against menopause symptoms or to reduce cholesterol levels [5–7]. Preparations from different *Pueraria* species are used in Western countries ("Kudzu") and in traditional medicines in China ("Gegen") and other Eastern Asian regions for the treatment of cardiovascular diseases, diabetes, liver injury and symptoms of diarrhea and fever, as well as to improve circulation and blood flow [8–11].

Soy-rich nutrition has been repeatedly associated with different health-promoting effects, including improvement of benign prostatic hyperplasia (BPH) and lower urinary tract symptoms (LUTSs). In epidemiological studies, consumption of soy food correlated negatively with the prevalence of BPH and LUTSs, which was supposed to account for the low prevalences of BPH in Eastern Asian regions with traditionally high soy consumption rates [12–14]. Increasing isoflavone exposure, resulting from the increased popularity and consumption of soy food that has been noted in Western countries, has unknown risks and benefits, raising the need for improved understanding of isoflavone effects [15].

Voiding symptoms in BPH are characterized by impaired urinary flow and bladder emptying due to urethral obstruction driven by increased prostate smooth muscle tone, enlargement of the prostate or both [16,17]. Accordingly, contraction and growth in the prostate are targets for medical treatment in BPH. Available drugs include $\alpha_1$-adrenoceptor antagonists ($\alpha_1$-blockers) and the phosphodiesterase 5 inhibitor tadalafil for rapid symptom improvements by smooth muscle relaxation and 5$\alpha$-reductase inhibitors to reduce prostate growth, progression and the risk of complications and surgery [18]. However, the average improvements by available medications do not exceed 50%, which level is not far from that of placebos and contributes to dissatisfaction, discontinuation and finally hospitalization and surgery [18,19]. Combination therapies are required for simultaneous symptom improvements and prevention of progression but are beset by discontinuation rates of up to 90% [20]. The limitations of $\alpha_1$-blockers, which are the first-line option for medical treatment of male LUTSs, have been explained by non-adrenergic contractions, which may raise maximum smooth muscle tone in the prostate and could maintain urethral obstruction in medication-refractory LUTSs [21,22].

While relationships between isoflavone-rich nutrition and BPH/LUTSs have been widely assumed and anticontractile effects have been repeatedly reported in different smooth-muscle-rich organs, isoflavone effects in the prostate are poorly known. Previous studies addressing their effects on prostate smooth muscle contraction have not included human tissues, being limited to non-human prostates. In parallel, attempts to address isoflavone effects on the growth of prostate cells have largely focused on glandular cells. Here, we speculated that leguminous isoflavones may inhibit agonist-induced and neurogenic contractions of human prostate smooth muscle and affect the growth-related functions of stromal cells. Thus, the aim of this study was to examine the effects of genistein, daidzein and puerarin on $\alpha_1$-adrenergic, non-adrenergic and neurogenic contractions of human prostate tissues and on growth-related functions of cultured stromal cells.

## 2. Materials and Methods

### 2.1. Human Prostate Tissues

Prostate tissues were obtained from patients undergoing radical prostatectomy for prostate cancer. This study was performed in accordance with the Declaration of Helsinki of the World Medical Association and was approved by the Ethics Committee of the Ludwig-Maximilians University, Munich, Germany. Informed consent was obtained from all patients. All samples and data were collected and analyzed anonymously. Accordingly, no patient data were collected, stored or analyzed in the context of this study, and samples were not grouped for pathological backgrounds or any other condition. Macroscopic inspection of prostates for tumor burden and subsequent sampling were performed by pathologists, approximately within 30 min following the final resection of prostates from patients. Organ bath experiments were started within 60 min following sampling. For transport and storage, organs and tissues were stored in Custodiol® solution (Köhler, Bensheim, Germany). For macroscopic examination and sampling, the prostate was opened by a single longitudinal cut from the capsule to the urethra. Subsequently, both intersections were checked macroscopically for any obvious tumor infiltration. Tissues were taken solely from the transitional, periurethral zone, while most prostate cancers arise in the peripheral zone [23,24]. In fact, tumor infiltration in the periurethral zone (where sampling was performed) was found in less than 1% of prostates. Prostates showing tumors in the periurethral zone upon macroscopic inspection were not subjected to sampling. BPH is present in ca. 80% of patients with prostate cancer [25,26]. The average age of patients undergoing radical prostatectomy in our department is 66 ± 7 years [27], where the prevalence of histological BPH may range between 60 and 70% [16]. Typically, tissues sampled under the same conditions in our previous studies showed prostate-characteristic architectures with compositions of glands and smooth-muscle-rich stroma [28], while tissues containing only stroma without glands were usually limited to anterior parts of the human prostate [29].

### 2.2. Organ Bath

Strips of prostate tissues (6 × 3 × 3 mm) were mounted in chambers of an organ bath system (myograph 720 M, Danish Myotechnology, Aarhus, Denmark). The device includes four chambers, each filled with 10 mL aerated (95% $O_2$ and 5% $CO_2$) Krebs–Henseleit solution (37 °C, pH 7.4) with the following composition: 118 mM NaCl, 4.7 mM KCl, 2.55 mM $CaCl_2$, 1.2 mM $KH_2PO_4$, 1.2 mM $MgSO_4$, 25 mM $NaHCO_3$, 7.5 mM glucose. After mounting, tissues were stretched to a tension of 4.9 mN and equilibrated for 45 min. Typically, pretensions decrease spontaneously at the beginning of the equilibration period. Therefore, tensions were adjusted three times during the equilibration period, until a stable resting tone of 4.9 mN was attained within 45 min. Subsequently, contractions induced by 80 mM KCl were assessed by the addition of a 2 M stock solution to assess highmolar KCl-induced contractions as a measure of smooth muscle content and condition and as a later reference for agonist-induced and neurogenic contractions. Highmolar KCl results in biphasic responses, including a phasic contraction reaching a peak within a few minutes after the addition of KCl, which was used for normalization, as described below, followed by a decline to a tonic, stable contraction [30]. After the peak of the phasic contraction was obtained and the tension started to decline to the tonic phase, the chambers were washed by replacing the Krebs–Henseleit solution three times, resulting in a stable resting tone close to the first baseline before KCl. Thereafter, genistein, daidzein and puerarin, or equivalent volumes of dimethylsulfoxide (DMSO, solvent for genistein and daidzein) or methanol (solvent for puerarin) as controls were added. Cumulative concentration–response curves for agonists or frequency response curves by electric field stimulation (EFS) were created 30 min after the addition of isoflavones or solvent.

EFS simulates action potentials, resulting in smooth muscle contractions of prostate tissues by release of endogenous neurotransmitters, including noradrenaline. Using tetrodotoxin, it has been demonstrated that neurotransmission accounts for two-thirds or more of EFS-induced contractions in the human prostate [31,32]. For EFS, tissue strips were placed between two parallel platinum electrodes connected to a CS4 stimulator (Danish Myotechnology, Denmark). Square pulses with durations of 1 ms were applied with a voltage of 20 V for a train duration of 10 s. EFS-induced contractile responses were studied at frequencies of 2, 4, 8, 16 and 32 Hz, with train intervals of 30 s between stimulations.

For a single independent experiment, all four chambers of one organ bath device were filled with tissues from the same prostate. Two of them were examined with isoflavone and the two others with solvent for the controls. Only one frequency response or concentration–response curve was recorded with each mounted tissue. Allocation of isoflavones and solvent to the different chambers was randomly changed between experiments. Independent experiments were repeated in indicated numbers (n), using tissues from n different patients, resulting in numbers of independent experiments as indicated for each series. Single experiments were based on double determinations wherever this was possible. From a total of 200 organ bath experiments, double determinations for both groups (i.e., control and drug groups, each with two tissues from the same prostate) were possible in 164 experiments. In the remaining experiments, the number of sampled tissues did not allow the filling of two channels for both groups. Thus, in 22 experiments, double determinations were possible in only one of the two groups, while 14 experiments were performed with only one tissue for both groups. However, each experiment contained at least one sample for both groups, resulting in paired samples.

For the calculation of agonist- or EFS-induced contractions, tensions (peak heights in EFS-induced contractions and maximum contractions following agonist exposure) were expressed as % of 80 mM KCl-induced contractions (maximum of phasic contraction). Normalization to KCl may correct variations in smooth muscle content, tissue composition and stromal/epithelial ratios, resulting from varying phenotypes and degrees of BPH, or from other individual heterogeneities between samples and patients. For calculation, the maximum contraction level at a given agonist concentration or given frequency was assessed, regardless of the pattern of contraction. In fact, contraction patterns in our samples differed between $\alpha_1$-adrenergic and non-adrenergic agonists and between tissues and patients [30]. In either case, however, the next higher concentration in the concentration–response curves was applied as soon as the plateau contraction was attained and no further substantial increase was expected. Contractions in concentration and frequency response curves describe the maximum contraction levels at each concentration and frequency in control and drug groups.

$E_{max}$ values, $EC_{50}$ values for contractile agonists and frequencies (f) inducing 50% of the maximum EFS-induced contraction ($Ef_{50}$) were calculated for each single experiment by curve fitting using GraphPad Prism 6 (Statcon, Witzenhausen, Germany) and analyzed as described below. As presentation of single values in scatter plots was intended, curve fitting was performed separately for each single experiment, resulting in separate values for each independent experiment. Curves were fitted without predefined constraints for bottom, top or $EC_{50}$ values, by ordinary fit, without weighting and without choosing automatic outlier elimination by non-linear regression. The resulting values were checked for plausibility, and settings were adapted as follows if error messages occurred, as recommended in the "GraphPad Curve Fitting Guide" (GraphPad Software Inc., San Diego, CA, USA). Thus, high concentrations were excluded in two experiments with endothelin-1 because the results were marked as "ambiguous", as full contractions were obtained already at low concentrations in one experiment or the curve could not be converged due to downhill parts at higher concentrations.

### 2.3. *Cell Culture*

WPMY-1 cells are an SV40 large-T antigen-immortalized cell line, obtained from the stroma of a human prostate without prostate cancer [33]. According to the typical composition of the prostate stroma, where smooth muscle cells are the predominant cell type, WPMY-1 cells show characteristics of myofibroblasts and prostate smooth muscle cells, including expression of vimentin, $\alpha$-smooth muscle actin, calponin and $\alpha_{1A}$-adrenoceptors, but lacking expression of cytokeratins and tyrosine hydroxylase [28,33,34]. WPMY-1 cells were purchased from the American Type Culture Collection (ATCC, Manassas, VA, USA) and cultured in RPMI 1640 (Gibco, Carlsbad, CA, USA) supplemented with 10% fetal calf serum (FCS) and 1% penicillin/streptomycin at 37 °C with 5% $CO_2$. Before the addition of isoflavones, DMSO (solvent for genistein and daidzein) or methanol (solvent for puerarin) to the cells, the medium was changed to an FCS-free medium.

### 2.4. *Proliferation Assay*

The proliferation rate of cells was assessed using the 5-ethynyl-2′-deoxyuridine (EdU)-based EdU-Click 555 proliferation assay kit (Baseclick, Tutzing, Germany), according to the manufacturer's instructions. In this assay, incorporation of EdU into the DNA of proliferating cells is assessed by detection with fluorescing 5-carboxytetramethylrhodamine (5-TAMRA). Thirty thousand cells were placed in each well of a 16-well chambered coverslip (Thermo Scientific, Waltham, MA, USA) and cultured for 24 h. Subsequently, the medium was replaced by 10 mM EdU solution in FCS-free smooth muscle cell medium containing isoflavones or solvent, and, 24 h later, the cells were fixed with ROTI® Histofix 4% solution (Roth, Karlsruhe, Germany). Counterstaining of all nuclei was performed with DAPI.

Finally, analysis was performed by fluorescence microscopy (excitation: 546 nm; emission: 479 nm) using a laser scanning microscope (Leica SP2, Wetzlar, Germany). Stainings were quantified using "ImageJ" (National Institutes of Health, Bethesda, AR, USA).

### *2.5. Cell Apoptosis and Cell Death Analysis*

A flow-cytometry-based annexin V allophycocyanin (APC) and 7-aminoactinomycin D (7-AAD) apoptosis detection kit (BD Biosciences, Franklin Lakes, NJ, USA) was used to detect apoptotic (annexin V-positive, 7-AAD-negative) and dead (annexin V-positive, 7-AAD-positive) cells. Cell death in annexin V-positive/7-AAD-positive cells may result from apoptosis or necrosis, which cannot be distinguished by this assay. Cells were grown in 6-well plates, and isoflavones or solvent were added 24 h before flow cytometry. The remaining medium of each well was collected and centrifuged to collect the cell debris. Subsequently, the cells were washed with PBS and resuspended (together with the corresponding debris collected) in annexin V binding buffer (BD Biosciences), followed by the addition of 5 μL APC annexin V and 5 μL 7-AAD reagent to every $10^5$ cells from each sample. After incubation in the dark for 15 min at room temperature, 400 μL binding buffer was added to each sample before analysis by flow cytometry.

### *2.6. Viability Assay*

Viability was assessed using the Cell Counting Kit-8 (CCK-8) (Sigma-Aldrich, Munich, Germany). Cells were seeded in 96-well plates (5000 cells/well) and cultured for 24 h. Subsequently, isoflavones or solvent in required amounts were added, and cells were cultured for further 24, 48 or 72 h until assessment. Finally, 10 μL of [2-(2-methoxy-4-nitrophenyl)-3-(4-nitrophenyl)-5-(2,4-disulfophenyl)-2H-tetrazolium monosodium salt (WST-8) from the kit were added, and absorbance in each well was measured at 450 nm after incubation for 2 h at 37 °C.

### *2.7. Drugs and Nomenclature*

Genistein (5,7-dihydroxy-3-(4-hydroxyphenyl)-4H-1-benzopyran-4-one), daidzein (7-hydroxy-3-(4-hydroxyphenyl)-4H-1-benzopyran-4-one) and puerarin (8-(β-D-glucopyranosyl-7-hydroxy-3-(4-hydroxyphenyl)-4H-1-benzopyran-4-on) are isoflavones naturally occurring in soybeans (*Glycine max*), red clover (*Trifolium pratense*) and Kudzu (*Pueraria* spp.). Synthetic genistein (G6649), synthetic daidzein (D7802) and puerarin (P5555, lot no. MKCG6104) were purchased from Sigma-Aldrich (Munich, Germany). Stock solutions (10 mM) of genistein and daidzein were prepared with DMSO and with puerarin using methanol (as recommended by the providers), and all were stored at −20 °C until use.

Phenylephrine ((R)-3-[-1-hydroxy-2-(methylamino)ethyl]phenol) and methoxamine (α-(1-aminoethyl)-2,5-dimethoxybenzyl alcohol) are $\alpha_1$-selective adrenoceptor agonists. Aqueous stock solutions (10 mM) of noradrenaline, phenylephrine and methoxamine were freshly prepared before each experiment. Aqueous stock solutions of endothelin-1 (0.4 mM) were stored at −20 °C as small aliquots, so that repeated freezing and thawing cycles were avoided. U46619 ((Z)-7-[(1S,4R,5R,6S)-5-[(E,3S)-3-hydroxyoct-1-enyl]-3-oxabicyclo [2.2.1]heptan-6-yl]hept-5-enoic acid) is an agonist of the thromboxane $A_2$ receptor and was dissolved in ethanol. Stock solutions (10 mM) were stored at −80 °C until use. U46619 and endothelin-1 were obtained from Enzo Life Sciences (Lörrach, Freiburg im Breisgau, Germany). Noradrenaline, phenylephrine and methoxamine were obtained from Sigma-Aldrich (Munich, Germany).

### *2.8. Statistical Analyses*

Data in the frequency and concentration–response curves are means ± standard deviations (SDs). $E_{max}$, $Ef_{50}$ and $EC_{50}$ values are presented as single values (i.e., means from double determinations, as described above) from all independent experiments in scatter plots. Effect sizes become obvious from concentration–response and frequency response curves and scatter plots. Changes in $EC_{50}$ values are additionally reported as mean differences (MDs) with 95% confidence intervals (Cis) in the text if the effects were obvious. Changes in $E_{max}$ values observed in contraction experiments and cell culture experiments are reported as percentage decreases or x-fold increases (means with 95% CIs) in the text as well and were calculated by referring isoflavone samples to the corresponding controls (set to 100% or to one) in each single experiment. Calculation of MDs with 95% CIs was performed using GraphPad Prism 6. Derived from our experimental design, the control and drug groups were paired in each series. Comparison of whole frequency/concentration–response curves, i.e., of both groups within a series, was performed by two-way analysis of variance (ANOVA). Post hoc analyses for comparisons at single-agonist concentrations or frequencies were not performed, as this is discouraged in the "GraphPad Statistics Guide" (GraphPad Software Inc., San Diego, CA, USA). $E_{max}$, $Ef_{50}$ and $EC_{50}$ values were compared by paired Student's *t*-tests. All statistical tests were performed using GraphPad Prism 6. $p$-values < 0.05 were considered significant.

The present study and analyses have an exploratory character, as important features of hypothesis-testing studies are lacking [35]. Owing to the exploratory study design, the *p*-values reported here need to be considered as descriptive; they are not hypothesis-testing [35]. In order to use *p*-values sparingly [35], no *p*-values are reported in the text and *p*-values of 0.05 or higher are not indicated.

The minimum number of independent experiments in each series was pre-planned as $n = 5$, as statistical analyses with group sizes <5 have been discouraged [36]. Data were analyzed after five experiments in a series were completed, with the exception of the flow cytometry experiments, where six experiments were performed for each series. Originally, we intended to decide whether a series was to be continued or not following this interim analysis. This procedure is possible, as our study was explorative and not designed to test a pre-specified statistical null hypothesis [35], and flexible group sizes have been recommended if data are characterized by large variations, which applies here [36]. However, all series in this study revealed conclusive findings after five independent experiments or six initial independent experiments in the case of flow cytometry. For technical reasons, six instead of five experiments were performed for flow cytometry. In all series, no experiments were excluded from the analyses, and no data were excluded, with the required exceptions for curve fitting in two experiments, as described above.

## 3. Results

### *3.1. Effects of Genistein on $\alpha_1$-Adrenergic Contractions*

Noradrenaline, phenylephrine and methoxamine induced concentration-dependent contractions which were inhibited by 50 µM and 10 µM genistein (Figure 1). Contractions by all three $\alpha_1$-adrenergic agonists were inhibited by two-thirds or more using 50 µM genistein and by approximately half using 10 µM genistein (Figure 1). Inhibitions seen in concentration–response curves were reflected in reduced $E_{max}$ values for agonists calculated by curve fitting (Figure 1). $E_{max}$ values for noradrenaline were reduced by 66% (52 to 80) with 50 µM genistein (Figure 1A) and by 42% (16 to 67) with 10 µM genistein (Figure 1B) if values under genistein were normalized to values for the corresponding control groups in each single experiment. $E_{max}$ values for phenylephrine were reduced by 68% (40 to 96) with 50 µM genistein (Figure 1C) and by 56% (23 to 89) with µM genistein (Figure 1D). $E_{max}$ values for methoxamine were reduced by 73% (55 to 90) with 50 µM genistein (Figure 1E) and by 19% (−93 to 130) with 10 µM genistein (Figure 1F). Average $EC_{50}$ values for noradrenaline and phenylephrine were elevated slightly, by less than half a magnitude, by 50 µM genistein, while average $EC_{50}$ values for methoxamine were elevated by more than half a magnitude by 50 µM genistein (MD (logM) (0.13 to 1.18)) (Figure 1). Using 10 µM genistein, average $EC_{50}$ values for all three $\alpha_1$-adrenergic agonists remained unchanged (Figure 1).

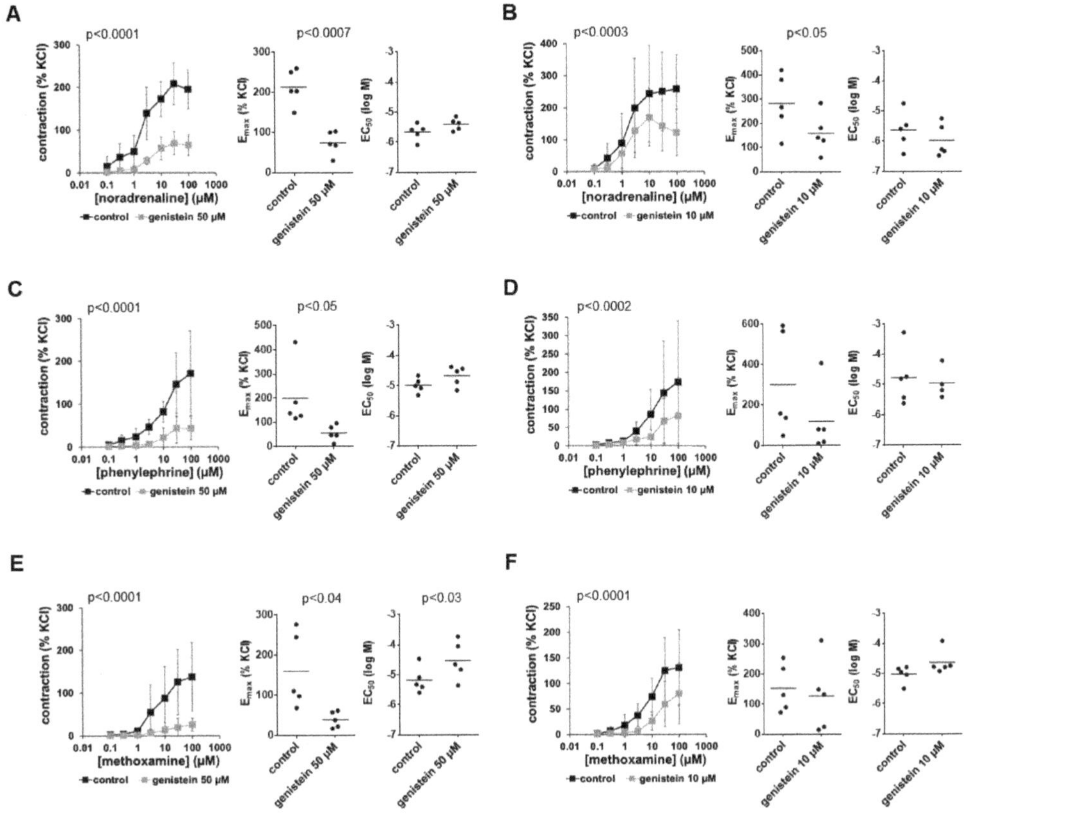

**Figure 1.** Effects of genistein on $\alpha_1$-adrenergic contractions of human prostate tissues. Contractions were induced by the $\alpha_1$-adrenergic agonists noradrenaline (**A**,**B**), phenylephrine (**C**,**D**) and methoxamine (**E**,**F**) in the presence of 50 µM genistein (**A**,**C**,**E**), 10 µM genistein (**B**,**D**,**F**) and corresponding amounts of solvent (DMSO) for the controls added 30 min before the construction of concentration–response curves. Data from experiments with tissues from $n$ = 5 patients are shown in each diagram. In each experiment, tissue from one patient was allocated to the control and genistein groups. Data are means ± SDs in concentration–response curves ($p$-values for whole groups from two-way ANOVA), and $E_{max}$ and $EC_{50}$ values are from all single experiments (calculated by curve fitting) in scatter plots ($p$-values from two-tailed $t$-tests).

### 3.2. Effects of Genistein on EFS-Induced Contractions

EFS induced frequency-dependent contractions which were inhibited by 50 µM and 10 µM genistein (Figure 2A,B). Inhibitions amounted to 75% or more at each frequency for 50 µM genistein and approximately half for 10 µM genistein (Figure 2A,B). Inhibitions seen in frequency response curves were paralleled by reduced $E_{max}$ values, calculated by curve fitting, which decreased by 84% (69 to 98) with 50 µM and by 44% (−26 to 115) with 10 µM genistein (Figure 2A,B). $Ef_{50}$ values remained unchanged by both concentrations of genistein (Figure 2A,B).

### 3.3. Effects of Genistein on Non-Adrenergic Contractions

Endothelin-1 and U46619 induced concentration-dependent contractions (Figure 2C–F). Contractions by both agonists were inhibited by two-thirds or more using 50 µM genistein and were still obvious using 10 µM genistein (Figure 2C–F). Inhibitions seen in concentration–response curves were reflected in reduced $E_{max}$ values for agonists calculated by curve fitting (Figure 2C–F). $E_{max}$ values for endothelin-1 were reduced by 65% (46 to 84) with 50 µM genistein (Figure 2C) and by 36% (18 to 54) with 10 µM genistein (Figure 2D). $E_{max}$ values for U46619 were reduced by 74% (55 to 74) with 50 µM genistein (Figure 2E) and by 41% (−12 to 93) with 10 µM genistein (Figure 2F). $EC_{50}$ values for endothelin-1 remained unchanged by both concentrations of genistein (Figure 2C,D). Average $EC_{50}$ values for U46619 were increased by both concentrations of genistein; there were, however, large variations in the series using 50 µM genistein which amounted to more than half or nearly one magnitude (MD (logM) 0.78 (−1.63 to 3.19) with 50 µM, 0.88 (0.06 to 1.69) with 10 µM).

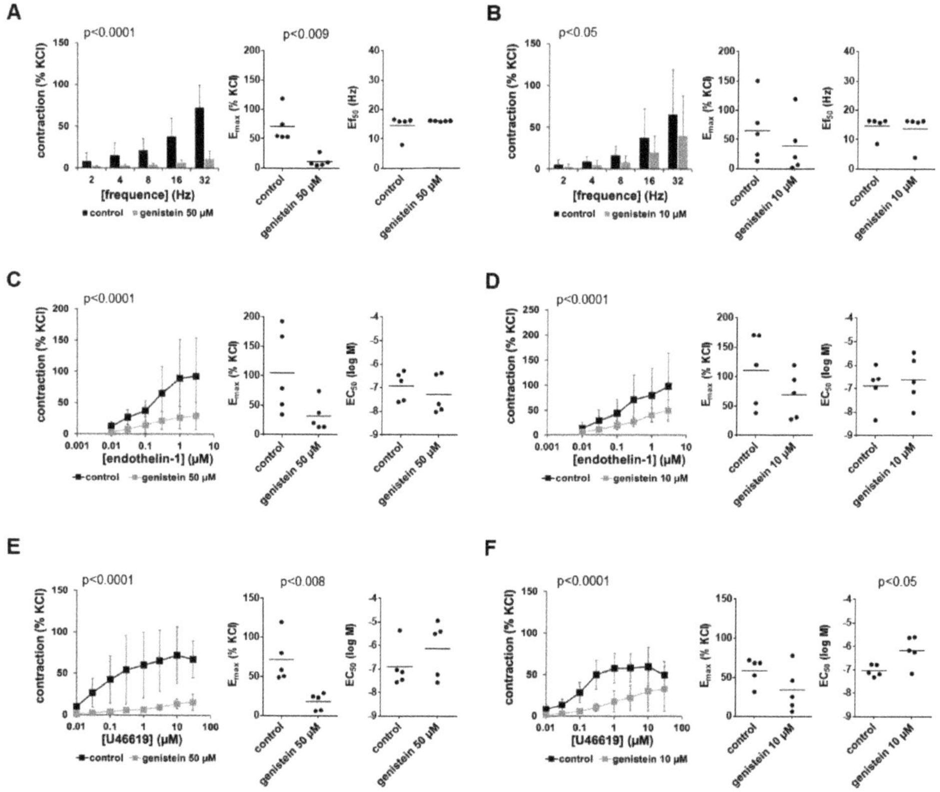


**Figure 2.** Effects of genistein on EFS-induced and non-adrenergic contractions of human prostate tissues. Contractions were induced by EFS (**A**,**B**), endothelin-1 (**C**,**D**) and U46619 (**E**,**F**) in the presence of 50 µM genistein (**A**,**C**,**E**), 10 µM genistein (**B**,**D**,**F**) and corresponding amounts of solvent (DMSO) for the controls added 30 min before the construction of frequency and concentration–response curves. Data from experiments with tissues from $n = 5$ patients are shown in each diagram. In each experiment, tissue from one patient was allocated to the control and genistein groups. Data are means ± SDs in frequency/concentration–response curves ($p$-values for whole groups from two-way ANOVA), and $E_{max}$, $Ef_{50}$ and $EC_{50}$ values are from all single experiments (calculated by curve fitting) in scatter plots ($p$-values from two-tailed $t$-tests).

### 3.4. *Effects of Daidzein on $\alpha_1$-Adrenergic Contractions*

Contractions induced by noradrenaline, phenylephrine and methoxamine were inhibited by 100 µM daidzein but not or to a neglectable extent by 10 µM daidzein (Figure 3). Inhibitions by 100 µM daidzein were largest at high agonist concentrations, where they amounted to around one-third for noradrenaline and phenylephrine, or around 50% for methoxamine (Figure 3). Inhibitions seen in concentration–response curves were paralleled by reduced $E_{max}$ values (Figure 3), which decreased by 35% (10 to 59) for noradrenaline, 44% (12 to 76) for phenylephrine and 45% (2 to 87) for methoxamine with 100 μM daidzein. Average $EC_{50}$ values were not consistently changed for all three agonists by 100 μM daidzein (Figure 3). Using 10 μM daidzein, $E_{max}$ and $EC_{50}$ values for all three $\alpha_1$-adrenergic agonists remained unchanged, apart from a decrease in $E_{max}$ values for noradrenaline-induced contractions (Figure 3).

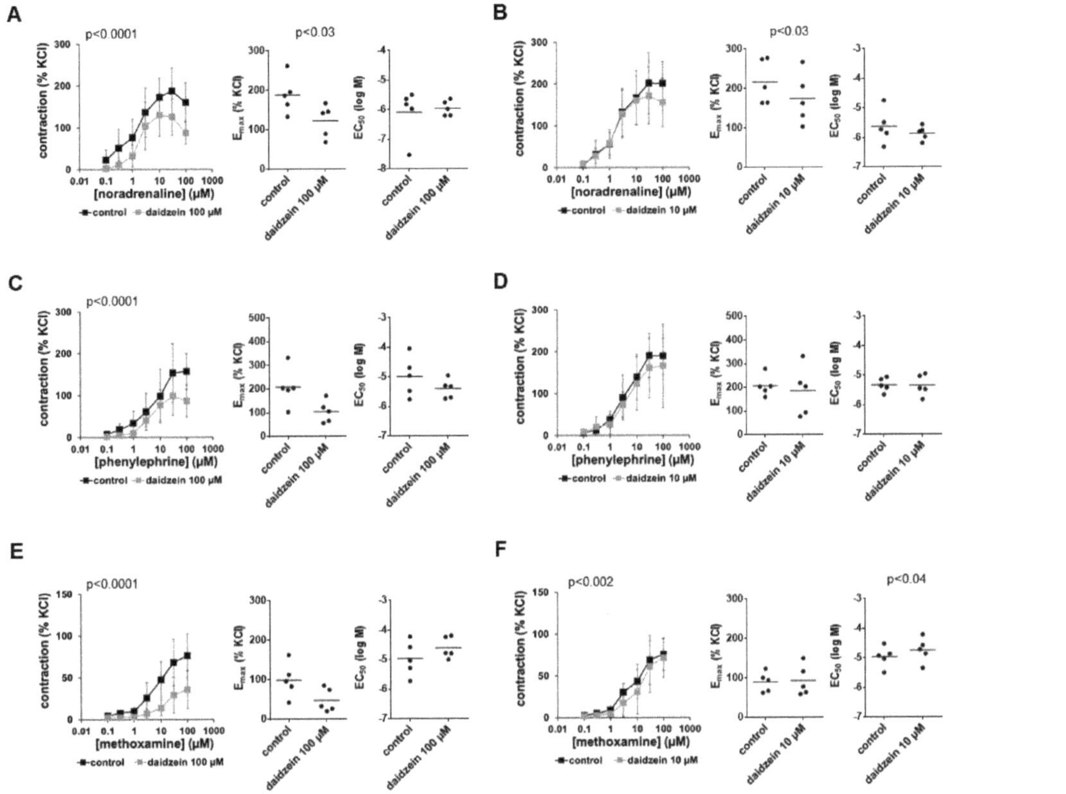


**Figure 3.** Effects of daidzein on $\alpha_1$-adrenergic contractions of human prostate tissues. Contractions were induced by the $\alpha_1$-adrenergic agonists noradrenaline (**A**,**B**), phenylephrine (**C**,**D**) and methoxamine (**E**,**F**) in the presence of 100 μM daidzein (**A**,**C**,**E**), 10 μM daidzein (**B**,**D**,**F**) and corresponding amounts of solvent (DMSO) added to controls 30 min before the construction of concentration–response curves. Data from experiments with tissues from $n = 5$ patients are shown in each diagram. In each experiment, tissue from one patient was allocated to the control and daidzein groups. Data are means ± SDs in concentration–response curves ($p$-values for whole groups from two-way ANOVA), and $E_{max}$ and $EC_{50}$ values are from all single experiments (calculated by curve fitting) in scatter plots ($p$-values from two-tailed $t$-tests).

### 3.5. *Effects of Daidzein on EFS-Induced Contractions*

EFS-induced contractions were inhibited by 100 μM and 10 μM daidzein (Figure 4A,B). Inhibitions amounted to more than two-thirds at each frequency with 100 μM daidzein and up to 50% using 10 μM daidzein (Figure 4A,B). Inhibitions seen in frequency response curves were paralleled by reduced $E_{max}$ values for EFS-induced contractions (Figure 4A,B), which were decreased by 70% (40 to 99) with 100 μM and by up to 50% (22 to 79) with 10 μM daidzein. $Ef_{50}$ values remained unchanged by both concentrations of daidzein (Figure 4A,B).

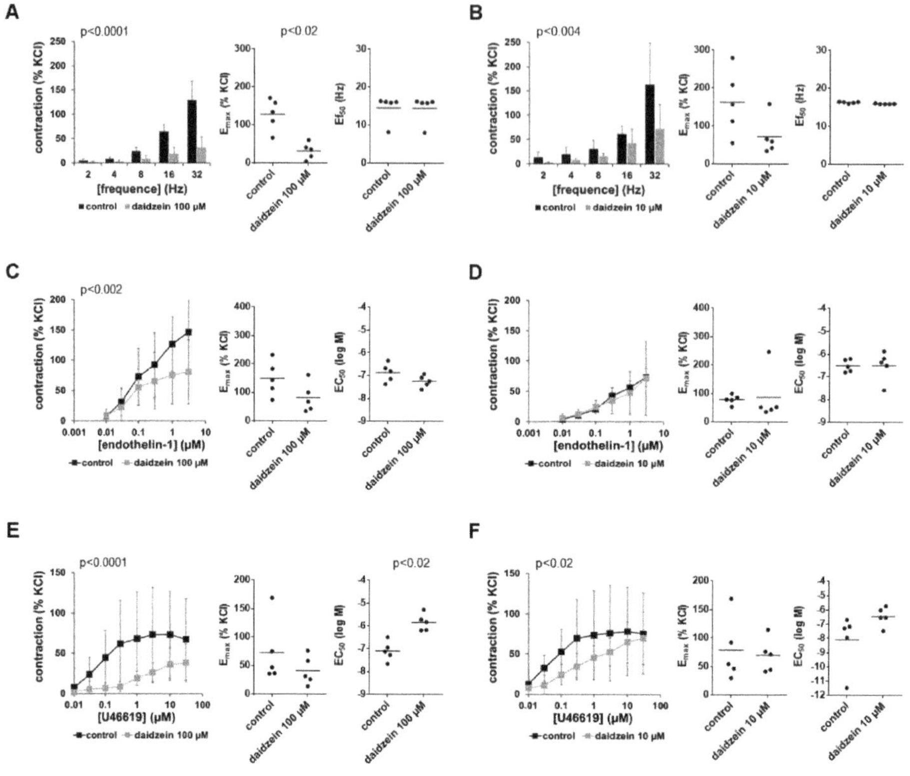


**Figure 4.** Effects of daidzein on EFS-induced and non-adrenergic contractions of human prostate tissues. Contractions were induced by EFS (**A**,**B**), endothelin-1 (**C**,**D**) and U46619 (**E**,**F**) in the presence of 100 µM daidzein (**A**,**C**,**E**), 10 µM daidzein (**B**,**D**,**F**) and corresponding amounts of solvent (DMSO) for the controls added 30 min before the construction of frequency and concentration–response curves. Data from experiments with tissues from $n = 5$ patients are shown in each diagram. In each experiment, tissue from one patient was allocated to the control and daidzein groups. Data are means ± SDs in frequency/concentration–response curves (*p*-values for whole groups from two-way ANOVA), and $E_{max}$, $Ef_{50}$ and $EC_{50}$ values are from all single experiments (calculated by curve fitting) in scatter plots (*p*-values from two-tailed *t*-tests).

### 3.6. Effects of Daidzein on Non-Adrenergic Contractions

Contractions induced by endothelin-1 were inhibited by 100 µM but not by 10 µM daidzein (Figure 4C,D). Inhibition was greatest at the two highest applied endothelin-1 concentrations, amounting to approximately 40% (Figure 4C). Inhibitions seen in concentration–response curves were paralleled by reduced $E_{max}$ values for endothelin-1 (Figure 4C), which were decreased by 43% (1 to 84) with 100 µM daidzein. $E_{max}$ values were not changed by 10 µM daidzein (Figure 4D), while $Ef_{50}$ values remained unchanged with both concentrations of daidzein (Figure 4C,D).

Contractions induced by U46619 were inhibited by 100 µM and 10 µM daidzein (Figure 4E,F). Inhibitions were between more than two-thirds to half across wide ranges of concentration–response curves but were smaller or lacking at the highest applied agonist concentrations using 10 µM daidzein (Figure 4E,F). Inhibitions seen in concentration–response curves were reflected by $E_{max}$ values for U46619, which were decreased by 39% (12 to 66) with 100 µM daidzein (Figure 4E) but remained unchanged with 10 µM daidzein (Figure 4F). Concentration–response curves for U46619 were right-shifted using both concentrations of daidzein, which was reflected in increased $EC_{50}$ values for U46619 with daidzein (Figure 4E,F). As $E_{max}$ values were not reduced with 10 µM daidzein, contractions in right-shifted concentration–response curves with daidzein fully recovered at high concentrations of U46619 (Figure 4F). $EC_{50}$ values for U46619 were increased by more than one magnitude with both concentrations of daidzein (MD (logM) 1.27 (0.38 to 2.16) with 100 µM, 1.66 (−0.72 to 4.04) with 10 µM) (Figure 4E,F).

### *3.7. Effects of Combined Genistein and Daidzein on Contractions*

By combined application of 6 µM genistein and 5 µM daidzein and compared to the corresponding controls, EFS-induced contractions were inhibited by more than half at each applied frequency (Figure 5A). Inhibitions seen in frequency response curves were paralleled by reduced $E_{max}$ values for EFS-induced contractions (Figure 5A), which were decreased by 57% (38 to 77). $Ef_{50}$ values remained unchanged by the combination (Figure 5A).

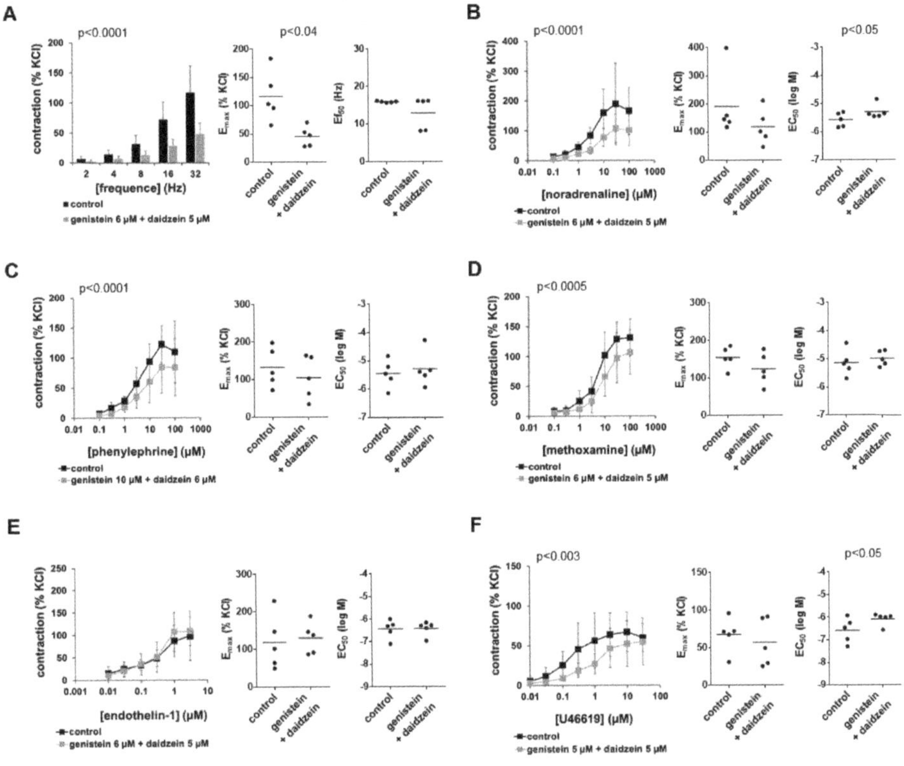


**Figure 5.** Effects of combined genistein and daidzein on EFS- and agonist-induced contractions of human prostate tissues. Contractions were induced by EFS (**A**), noradrenaline (**B**), phenylephrine (**C**), methoxamine (**D**), endothelin-1 (**E**) and U46619 (**F**) in the presence of 6 µM genistein plus 5 µM daidzein and corresponding amounts of solvent (DMSO) for the controls added 30 min before the construction of frequency and concentration–response curves. Data from experiments with tissues from $n = 5$ patients are shown in each diagram. In each experiment, tissue from one patient was allocated to the control and genistein + daidzein groups. Data are means ± SDs in frequency/concentration–response curves ($p$-values for whole groups from two-way ANOVA), and $E_{max}$, $Ef_{50}$ and $EC_{50}$ values are from all single experiments (calculated by curve fitting) in scatter plots ($p$-values from two-tailed $t$-tests).

Contractions by noradrenaline, phenylephrine and methoxamine were inhibited consistently but not by more than half by the combination of genistein with daidzein (Figure 5B–D). Inhibitions seen in concentration–response curves were paralleled by reduced $E_{max}$ values for agonists, which were decreased by 32% (−17 to 80) for noradrenaline,

23% (−6 to 53) for phenylephrine and 19% (−15 to 54) for methoxamine (Figure 5B–D). $EC_{50}$ values were increased slightly, i.e., less than half a magnitude for noradrenaline, or remained unaffected for phenylephrine and methoxamine (Figure 5B–D).

Contractions by endothelin-1 were not affected by the combination of 6 μM genistein and 5 μM daidzein (Figure 5E). U46619-induced contractions were inhibited partly, i.e., by around half at lower concentrations, but not or to a neglectable extent at high agonist concentrations (Figure 5F). Average $E_{max}$ values for U46619 remained unaffected by the combination, while $EC_{50}$ values were increased by around half a magnitude (MD (logM) 0.498 (0.03 to 0.97)) (Figure 5F).

### 3.8. *Effects of Puerarin on $\alpha_1$-Adrenergic Contractions*

Contractions induced by noradrenaline, phenylephrine and methoxamine were reduced marginally by 100 μM puerarin but not 10 μM puerarin (Figure 6). $E_{max}$ values were neither substantially nor consistently changed by 100 μM puerarin and remained unaffected by 10 μM puerarin (Figure 6). $EC_{50}$ values for all three $\alpha_1$-adrenergic agonists remained unchanged by both concentrations of puerarin (Figure 6).

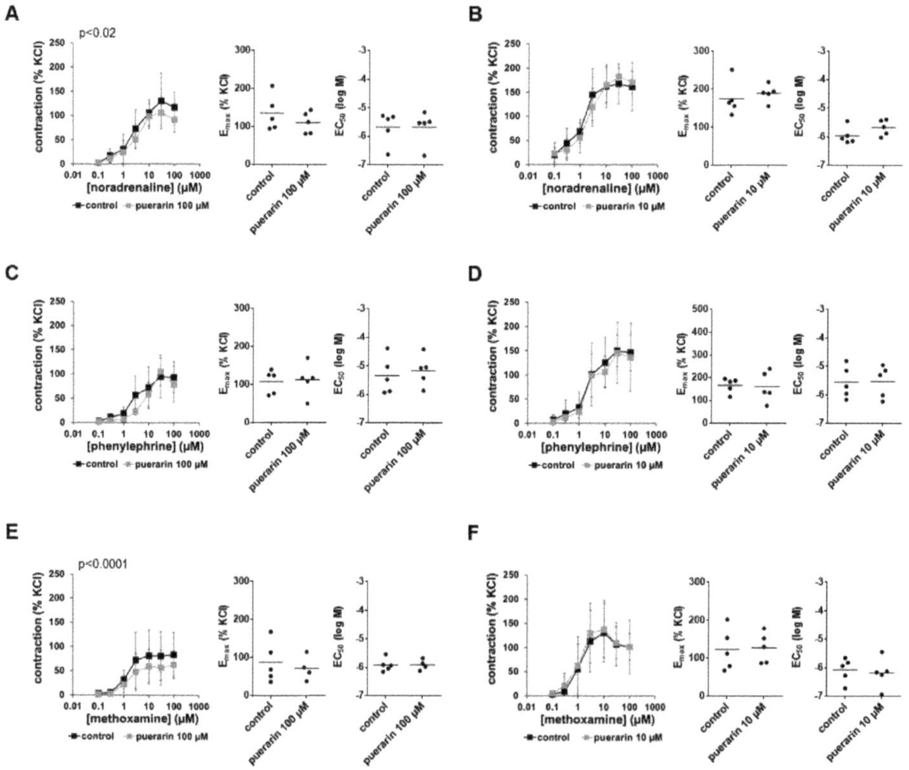


**Figure 6.** Effects of puerarin on $\alpha_1$-adrenergic contractions of human prostate tissues. Contractions were induced by the $\alpha_1$-adrenergic agonists noradrenaline (**A**,**B**), phenylephrine (**C**,**D**) and methoxamine (**E**,**F**) in the presence of 100 μM puerarin (**A**,**C**,**E**), 10 μM puerarin (**B**,**D**,**F**) and corresponding amounts of solvent (methanol) for the controls added 30 min before the construction of concentration–response curves. Data from experiments with tissues from $n = 5$ patients are shown in each diagram. In each experiment, tissue from one patient was allocated to the control and puerarin groups. Data are means ± SDs in concentration–response curves ($p$-values for whole groups from two-way ANOVA), and $E_{max}$ and $EC_{50}$ values are from all single experiments (calculated by curve fitting) in scatter plots ($p$-values from two-tailed $t$-tests).

### 3.9. Effects of Puerarin on EFS-Induced Contractions

Puerarin did not affect EFS-induced contractions using a concentration of 100 µM (Figure 7A), so effects at a lower concentration were not examined. According to the lack of effects in frequency response curves, no effects were seen in $E_{max}$ and $Ef_{50}$ values (Figure 7A).

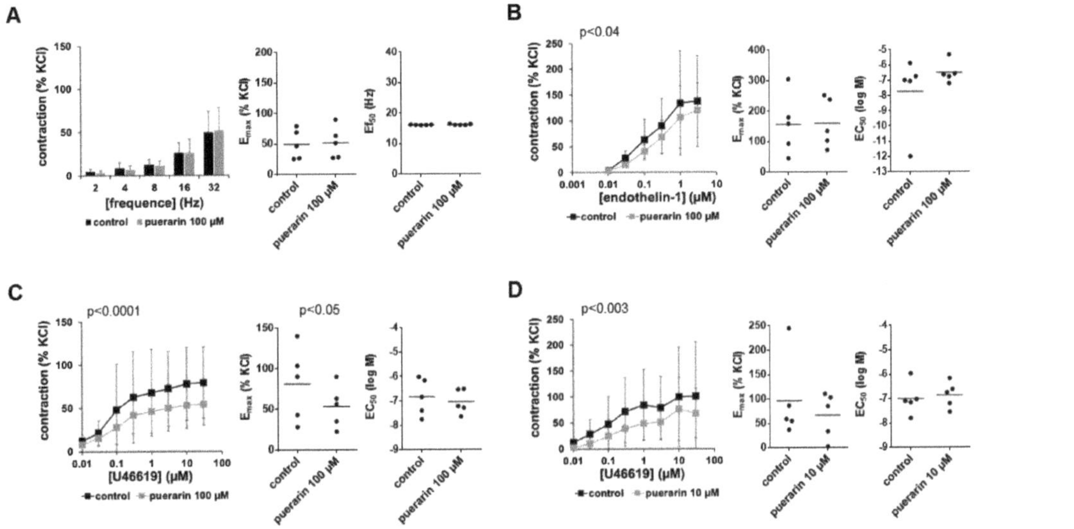


**Figure 7.** Effects of puerarin on EFS-induced and non-adrenergic contractions of human prostate tissues. Contractions were induced by EFS (**A**), endothelin-1 (**B**) and U46619 (**C**,**D**), in the presence of 100 µM puerarin (**A**–**C**) 10 µM puerarin (**D**) and corresponding amounts of solvent (methanol) for the controls added 30 min before the construction of frequency and concentration–response curves. Data from experiments with tissues from $n = 5$ patients are shown in each diagram. In each experiment, tissue from one patient was allocated to the control and puerarin groups. Data are means ± SDs in frequency/concentration–response curves ($p$-values for whole groups from two-way ANOVA), and $E_{max}$, $Ef_{50}$ and $EC_{50}$ values are from all single experiments (calculated by curve fitting) in scatter plots ($p$-values from two-tailed $t$-tests).

### 3.10. Effects of Puerarin on Non-Adrenergic Contractions

Puerarin did not affect endothelin-1-induced contractions using a concentration of 100 µM (Figure 7B), so effects at a lower concentration were not examined. According to the lack of effects in concentration–response curves, no effects of puerarin were seen in $E_{max}$ and $EC_{50}$ values (Figure 7B).

Contractions induced by U46619 were inhibited by puerarin to similar extents at concentrations of 100 µM and 10 µM, with maximum decreases in contractions of around one-third (Figure 7C,D). Inhibitions were obvious in concentration–response curves and reflected in $E_{max}$ values for U46619-induced contractions, which were decreased by 30% (16 to 44) with 100 µM puerarin and by 23% (−50 to 97) with 10 µM puerarin (Figure 7C,D). $EC_{50}$ values for U46619 were not changed by puerarin (Figure 7C,D).

### 3.11. Effects of Genistein, Daidzein and Puerarin on the Proliferation of Stromal Cells

Proliferation of WPMY-1 cells, assessed by EdU assays, was concentration-dependently decreased by genistein (Figure 8A). Decreases in proliferation rates amounted to 9% (4 to 15) with 5 µM, 36% (29 to 43) with 10 µM and 57% (48 to 67) with 50 µM genistein if proliferation rates with genistein were referred to corresponding controls without genistein in each single experiment. Daidzein reduced the proliferation of WPMY-1 cells as well and to similar extents at the two highest applied concentrations (Figure 8B). Decreases in proliferation rates amounted to 3% (0 to 6) with 5 µM, 29% (21 to 37) with 10 µM daidzein and 27% (22 to 32) with 50 µM daidzein. Puerarin decreased the proliferation of WPMY-1 cells only slightly and not concentration-dependently (Figure 8C), resulting in decreases in proliferation of 13% (6 to 21) with 5 µM, 11% (4 to 17) with 10 µM and 11% (3 to 19) with 50 µM puerarin.

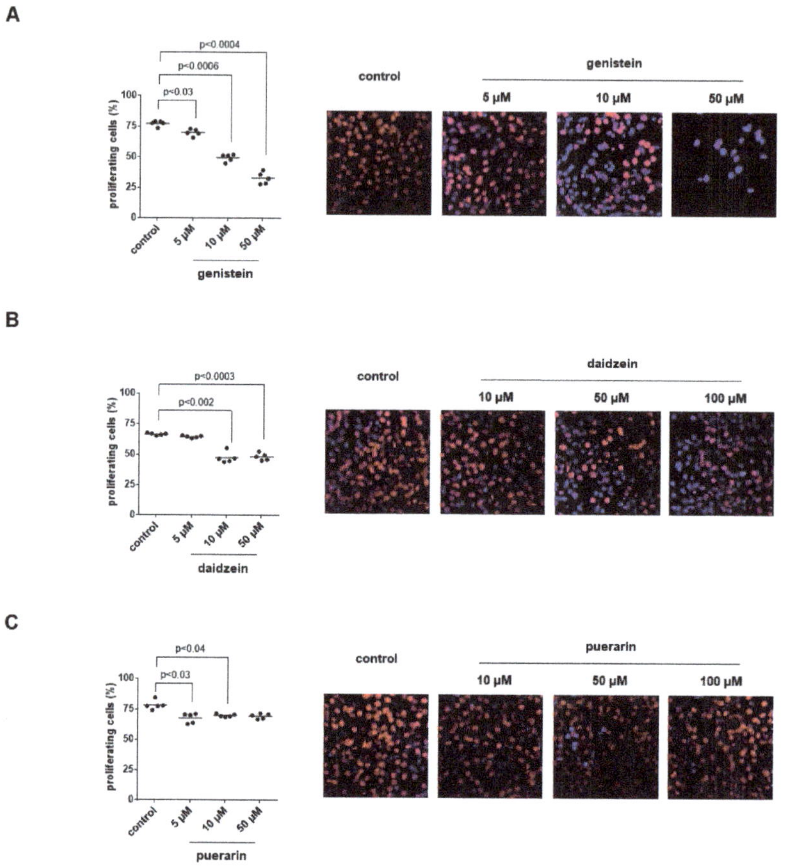


**Figure 8.** Effects of genistein, daidzein and puerarin on the proliferation of WPMY-1 cells. Proliferation was assessed by EdU assays (red, proliferating cells; blue, non-proliferating cells) after the cells were cultured for 24 h with genistein (**A**), daidzein (**B**) and puerarin (**C**) at the indicated concentrations or with solvent (DMSO for genistein and daidzein, methanol for puerarin) for the controls. Shown are all single values together with means from the quantification of proliferating cells in five independent experiments, along with representative images for each series, and *p*-values from comparisons with values for the corresponding DMSO group determined by one-way ANOVA with Dunnett's tests.

### *3.12. Effects of Genistein, Daidzein and Puerarin on Apoptosis and Cell Death of Stromal Cells*

Genistein increased the numbers of apoptotic and dead WPMY-1 cells, which were assessed by flow cytometric analysis of 7-AAD and annexin V (Figure 9A). Increases in the numbers of dead cells but not numbers of apoptotic cells by genistein were concentration-dependent (Figure 9A). The numbers of apoptotic cells were 2.1-fold higher (1.2 to 3.0) than for the corresponding controls with 25 µM, 1.9-fold higher (1.3 to 2.5) with 50 µM and 1.7-fold higher (1.3 to 2.2) with 100 µM genistein if the numbers with genistein were referred to the corresponding controls without genistein in each single experiment. The numbers of dead cells were 4.0-fold higher (2.8 to 5.2) than for the corresponding controls with 25 µM, 4.9-fold higher (4.3 to 5.4) with 50 µM and 7.0-fold higher (5.3 to 8.7) with 100 µM genistein.

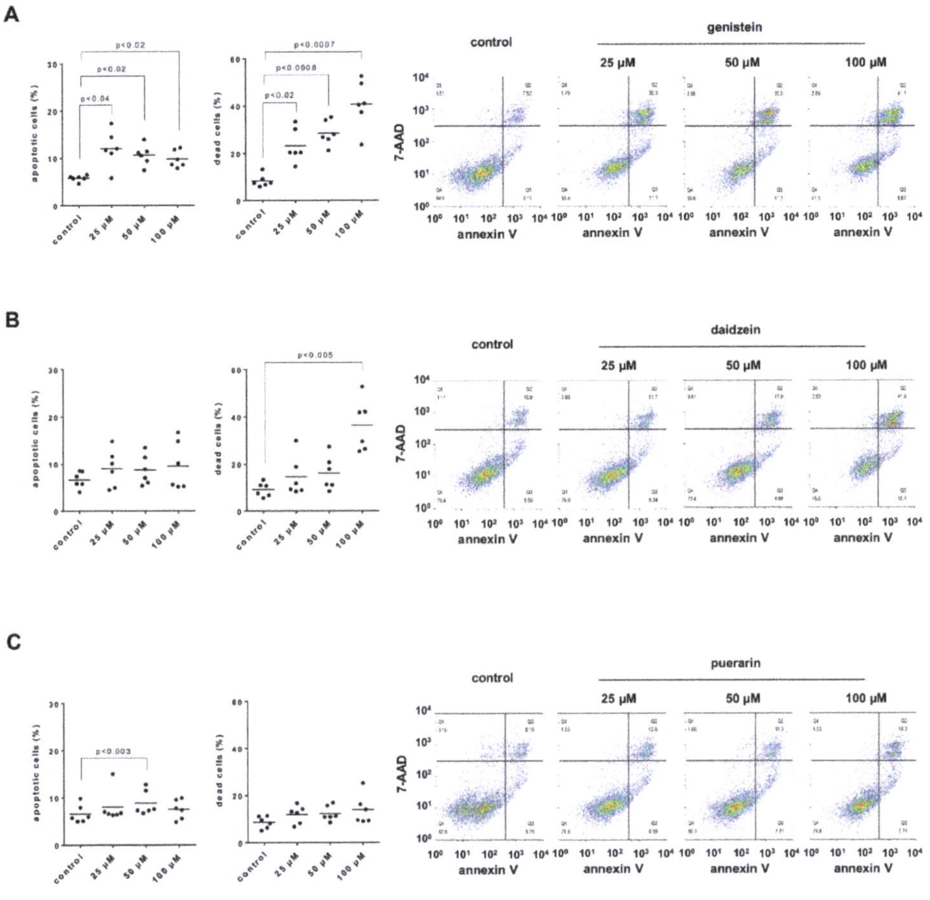


**Figure 9.** Effects of genistein, daidzein and puerarin on apoptosis and cell death in WPMY-1 cells. Numbers of cells in apoptosis (annexin V-positive, 7-AAD-negative) and of dead cells (resulting from apoptosis and/or necrosis; annexin V-positive, 7-AAD-positive) were assessed by flow cytometry. Flow cytometry was performed following culture with genistein (**A**), daidzein (**B**) and puerarin (**C**) at the indicated concentrations or solvent (DMSO for genistein and daidzein, methanol for puerarin) for the controls for 24 h. Shown are all single values together with means from the quantification of five independent experiments and representative experiments for each series, along with *p*-values from one-way ANOVA with Dunnett's tests.

Daidzein increased the numbers of apoptotic WPMY-1 cells without a concentration-dependent relationship and concentration-dependently increased the numbers of dead cells, which were most pronounced with the highest applied concentration (Figure 9B). The numbers of apoptotic cells were 1.5-fold higher (0.9 to 2.2) than for the corresponding controls with 25 µM, 1.5-fold higher (1.0 to 1.9) with 50 µM and 1.6-fold higher (0.8 to 2.4) with 100 µM daidzein. The numbers of dead cells were 2.6-fold higher (0.6 to 4.6) than for the corresponding controls with 25 µM, 2.8-fold higher (1.2 to 4.5) with 50 µM and 6.4-fold higher (3.7 to 9.0) with 100 µM daidzein.

Puerarin increased the numbers of apoptotic and dead WPMY-1 cells only slightly or to a neglectable extent, without concentration-dependent relationships (Figure 9C). The numbers of apoptotic cells were 1.4-fold higher (0.9 to 1.9) than for the corresponding controls with 25 µM, 1.6-fold higher (0.9 to 2.2) with 50 µM and 1.3-fold higher (0.9 to 1.7) with 100 µM puerarin. The numbers of dead cells were 2.1-fold higher (1.3 to 2.8) than for the corresponding controls with 25 µM, 2.2-fold higher (1.4 to 2.9) with 50 µM and 2.5-fold higher (0.9 to 4.0) with 100 µM puerarin.

### 3.13. *Effects of Genistein, Daidzein and Puerarin on the Viability of Stromal Cells*

Genistein reduced the viability of WPMY-1 cells, which was assessed by CCK-8 assays 24 h, 48 h and 72 after the addition of isoflavones (Figure 10A). The viability, as well as the effects of genistein, increased over time (Figure 10A). Decreases in viability started with 5 µM, without exceeding decreases of 50% up to 10 µM but approaching decreases of 59% (51 to 67) after 24 h and of 87% (86 to 89) after 72 h with 100 µM genistein if values with genistein were referred to the corresponding controls without genistein in each single experiment (Figure 10A).

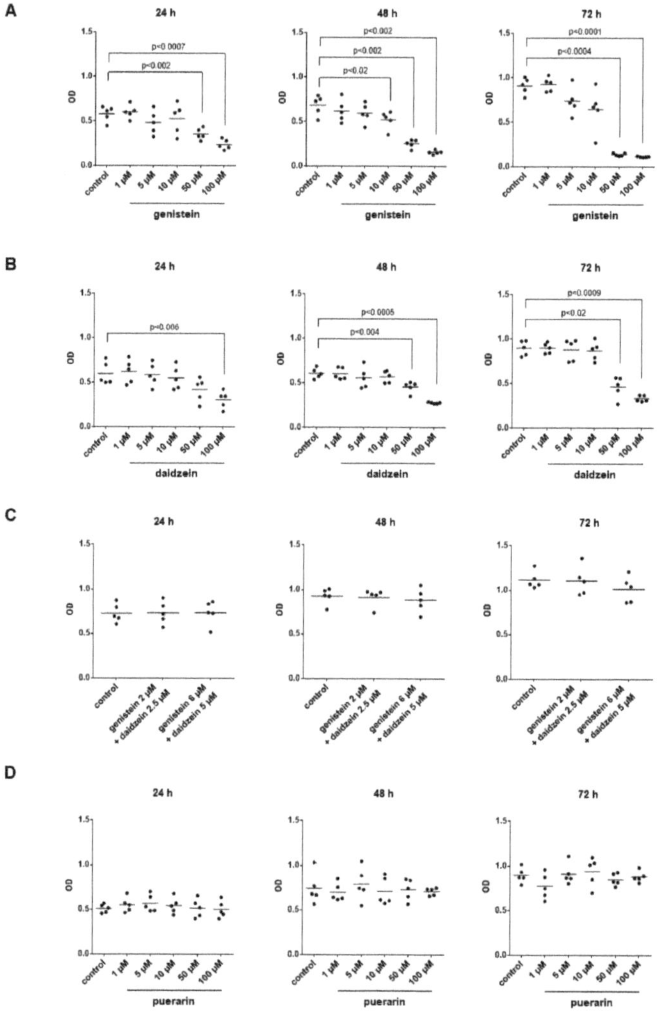


**Figure 10.** Effects of genistein, daidzein and puerarin on the viability of WPMY-1 cells. Viability was assessed by CCK-8 assays, following culture with genistein (**A**), daidzein (**B**), combinations of genistein with daidzein (**C**) and puerarin (**D**) at the indicated concentrations or solvent (DMSO for genistein and daidzein, methanol for puerarin) for the controls and for indicated periods. Shown are all single values together with means from quantification of five independent experiments and representative experiments for each series, along with *p*-values from one-way ANOVA with Dunnett's tests.

Daidzein decreased the viability of WPMY-1 cells, which was concentration-dependent but started with a concentration of 50 µM, while no certain decreases occurred with lower concentrations (Figure 10B). Maximum decreases amounted to 49% (33 to 64) after 24 h, 54% (50 to 58) after 48 h and 62% (55 to 70) after 72 h with 100 µM daidzein (Figure 10B).

The combination of 2 µM genistein with 2.5 µM daidzein did not affect the viability of WPMY-1 cells after 24–72 h (Figure 10C). The combination of 6 µM genistein with 5 µM daidzein did not affect the viability after 24–48 h, while the viability was reduced by 9% (−0.3 to 19) after 72 h with this combination (Figure 10C).

Puerarin did not affect the viability of WPMY-1 cells across the examined concentration range, with a maximum concentration of 100 µM puerarin (Figure 10D).

## 4. Discussion

High isoflavone exposure due to soy-rich nutrition correlated with low prevalence of BPH and voiding symptoms in epidemiological studies [12–14]. Clinical observations confirmed possible effects of isoflavones on BPH/LUTSs, as preparations from isoflavone-rich plants partly reduced symptoms and the risk for symptomatic BPH [37]. However, previous experimental studies addressing isoflavone impacts in the prostate focused on estrogenic effects, hormone-disrupting properties, epithelial prostate cells, animal models and prostate cancer [38–43]. Consequently, the effects of isoflavones on human prostate smooth muscle contraction and on the growth of stromal cells are mostly unknown, and directed comparisons to identify the most active isoflavones in the prostate are still pending. Our present findings suggest that genistein inhibited human prostate smooth muscle contractions to a remarkable extent and had marked effects on growth-related functions of stromal cells, whereas the effects of daidzein were weaker and the effects of puerarin appeared overall limited. Our findings provide a new basis for understanding isoflavone effects on voiding symptoms and BPH, which are relevant in light of the expanding popularity of soy food seen in many regions and the availability of isoflavone-rich plant preparations as dietary supplements.

We observed inhibition of neurogenic-, $\alpha_1$-adrenergic-, endothelin-1- and U46619-induced contractions of human prostate tissues by genistein. Inhibitions with 50 µM of genistein approached the effects of $\alpha_1$-adrenoceptor antagonists on EFS-induced and $\alpha_1$-adrenergic contractions of human prostate tissues previously observed by us under similar conditions [44]. Using 10 µM, the effects were still obvious, although weaker. In addition, genistein inhibited contractions induced by U46619 and endothelin-1 to a similar degree as $\alpha_1$-adrenergic contractions. These non-adrenergic contractions are resistant to $\alpha_1$-adrenergic antagonists but may induce a full, maximum prostate smooth muscle tone [21]. In BPH, increased prostate smooth muscle tone may contribute to urethral obstruction and finally to voiding symptoms [16,18]. $\alpha_1$-blockers are the first line option for medical treatment and are supposed to improve symptoms by inhibition of $\alpha_1$-adrenergic prostate smooth muscle contractions [16,18]. However, their efficacy to improve voiding symptoms is limited, probably due to contractions by non-adrenergic mediators [21,22]. Non-adrenergic contractions may keep prostate tone elevated and account for medication-refractory LUTSs, so that compounds showing inhibitory effects on non-adrenergic contractions are potentially of high clinical interest [21,22]. To the best of our knowledge, previous findings addressing the effects of genistein on prostate smooth muscle contractions are limited to one series, demonstrating the inhibition of EFS-induced contractions in rat prostate tissues with 100 µM genistein [41], while data for agonist-induced contractions, lower concentrations or human tissues are not available.

In cultured stromal cells, we observed concentration-dependent inhibition of proliferation, increases in cell death and decreases in viability with genistein, paralleled by concentration-independent increases in apoptosis. Cell death occurred probably independently of apoptosis, as the rate of apoptotic cells was insufficient to explain higher percentages of dead cells. Thus, genistein may act on different growth-related functions of stromal cells, including proliferation, apoptosis and apoptosis-independent survival, resulting in the observed decreases in viability starting at genistein concentrations of 5–10 μM. Previous studies addressing effects of genistein on stromal cells are limited to one series, where a single concentration of 10 μM showed no effect on survival in the presence of androgen [45]. For glandular epithelial cells of the prostate, antiproliferative and apoptotic effects of genistein are well-documented [46–48]. In BPH, both cell types may contribute to epithelial, stromal or mixed hyperplasia and thus to prostate growth and urethral obstruction [17]. Our current findings, together with previous findings from epithelial cells, may suggest that genistein could reduce prostate growth in BPH in vivo, provided adequate concentrations occur in the prostate. In fact, genistein inhibited prostate growth in rodent models of testosterone-induced BPH and high-fat-induced prostate growth [38,40,43], as well as ex vivo growth in prostate tissues from patients with BPH [49].

Similar to genistein, isoflavone-rich plant preparations reduced experimentally induced and age-related prostate growth in in vivo studies on rodents, including extracts from red clover, black soybeans, Kudzu (*Pueraria montana*) and *Pueraria mirifica* [38,40,42,50,51], as did a soy-based, phytoestrogen-rich diet [39]. These and similar preparations contain several different isoflavones, with compositions and predominant compounds varying between plant species. In soybeans, genistein and daidzein belong to the predominant isoflavones, with the content of genistein ranging 1.5–2-fold higher than daidzein [15,39]. Similar to animal models, preparations from soy and red clover reduced prostate size and slightly improved voiding symptoms in humans [37,52], which may explain the widely assumed benefits of soy-rich nutrition in the context of BPH and voiding symptoms. Clinically, voiding symptoms in BPH are expressed by the maximum urinary flow rate ($Q_{max}$) and the international prostate symptoms score (IPSS) assessed by standardized questionnaires, while postvoid residual urine volume (PVR) allows estimation of the risk for progression. In patients with BPH with a $Q_{max}$ < 15 mL/s, an IPSS > 8 and a PVR > 150 mL, daily treatment with a soy-based preparation for one year containing 40 mg isoflavones increased $Q_{max}$ by 1.7 mL/s, reduced IPSS by 4 points and decreased PVR by 49 mL [37]. However, these were not different from placebo effects: increases in $Q_{max}$ by 1.3 mL/s, decreases in IPSS by 2.9 points and decreases in PVR by 40 mL [37]. In another single-arm trial without a placebo group, patients with BPH received a red clover extract for one year equivalent to 60 mg of isoflavones per day, resulting in a decrease in prostate volume of 10% and a decrease in IPSS of 1.2 points [52]. In general, improvements by medical interventions require decreases in IPSS by 3 or more points in order to be perceptible by patients [53]. No threshold values have been defined for clinically relevant improvements in $Q_{max}$, as this highly depends on the baseline $Q_{max}$ before treatment initiation. In large-scale trials, $\alpha_1$-blockers enhanced $Q_{max}$ by 0.7–2.5 mL/s, which has been considered clinically relevant [54,55].

Consequently, the available clinical data and the effects of these preparations are both limited but may reflect a higher effectiveness of soy compared to red clover. The composition of isoflavone pools may differ between species. Genistein and daidzein are predominant isoflavones in soybeans and red clover and constitute most parts of their total isoflavone content [1,4,15]. However, the availability and content of genistein and daidzein in soy-based food may vary and depend on processing, including fermentation and washing procedures, even though they are high overall [15]. The predominant isoflavone in *Pueraria* roots is puerarin, with levels exceeding those of genistein plus daidzein together by around five-fold and contributing the largest part of the total isoflavone content [1,4,8]. In contrast, puerarin was undetectable in isoflavone preparations from soybeans and red clover [1,4]. Our findings suggest that genistein may be the isoflavone contributing most to the beneficial effects of mixed isoflavone preparations for BPH and LUTSs. Even though daidzein caused obvious inhibitions of EFS- and agonist-induced contractions and showed growth-inhibiting effects in stromal cells, these effects were weaker compared to 50 μM genistein and were minor or lacking at a concentration of 10 μM. Puerarin, the most abundant isoflavone in *Pueraria* roots, did not inhibit smooth muscle contractions of human tissues, apart from partial inhibition of U46619-induced contractions, while growth-related functions of stromal cells were widely unaffected. Regarding the cardiovascular system, effects of vasodilation and relaxation of vascular smooth muscle due to puerarin are well-known [11], and antihypertensive effects of the roots of *Pueraria* species used in Eastern Asian medicine have been considered [56]. In line with the lack of effects observed for puerarin in our study, improvements in voiding symptoms by *Pueraria* preparations have never been noted, despite their widespread use in traditional medicine in Eastern Asia and as dietary supplements in Western nations [8,9].

Isoflavone concentrations in the prostate have been assessed in patients receiving isoflavone preparations in several trials. Concentrations in the prostate were found to be higher compared to the plasma following the use of isoflavone preparations. In patients with BPH receiving 112.5 mg/d isoflavones, composed of 65.7% genistein and 31.7% daidzein in commercially available capsules, for three days, genistein concentrations amounted to 0.58 nmol/g prostate tissue but 0.78 μM in the plasma, and daidzein concentrations were 0.47 nmol/g prostate tissue but 0.66 μM in the plasma [57]. Assuming this could be simply extrapolated, this may correspond to 5.8 μM genistein or 4.7 μM daidzein in the prostate. Similar values were reported for patients undergoing radical prostatectomy for prostate cancer who received 27.2 mg of isoflavones (containing 10.6 mg genistein and 13.3 mg daidzein) per day in a soy-based preparation for 12–14 days, resulting in genistein concentrations of 2.2 μmol/kg in prostate tissue but 0.45 μM in the plasma, and daidzein concentrations of 2.4 μmol/kg in tissue but 0.28 μM in the plasma [58]. In men consuming a soymilk-based beverage for seven days containing 42–60 mg isoflavones, both compounds were detectable in the prostate fluid (consisting mostly of prostate secretion), with concentrations between 0.2 and 1 μM for genistein and between 0.2 and 30 μM for daidzein [48]. Consequently, we examined the effects of a combination of 6 μM genistein and 5 μM daidzein, representing a composition and concentration which may be expected from isoflavone-rich preparations in the prostate. This combination still inhibited $\alpha_1$-adrenergic-, EFS- and U46619-induced contractions. Whether these effects are sufficient for urodynamic effects in vivo cannot be concluded from our findings but it appears possible that they are. It could be speculated that even slight inhibitions of adrenergic and non-adrenergic contractions and slight growth-inhibiting effects may add up, which cannot be expected from $\alpha_1$-blockers.

While isoflavone concentrations in the prostate obviously exceed the plasma concentrations after ingestion of isoflavone preparations, the bioavailability and activity of isoflavones depend on glycosylation, conjugation and metabolization, cofactors and other nutrients, duration and form of ingestion, and other variables [6,59,60]. The naturally occurring form of isoflavones, predominating in plants and ingested with soy food are glucosides, which are poorly absorbed [6,59]. The absorbed forms are aglycones, resulting from hydrolysis of the sugar moiety by bacterial and intestinal mucosal β-glucosidases [6,59]. Accordingly, the bioavailability is higher with fermented soy products, for which the content of aglycones is higher, compared to non-fermented soy products [6]. Furthermore, deglycosylation and absorption are critically enhanced by fiber-rich, matrix food components [59,60]. Following absorption, aglycosylated isoflavones undergo conjugation, mainly including glucuronidation but also sulfation, other conjugations and further metabolization [6,59]. Conjugation occurs mostly in enterocytes, and glucuronides are the major forms of isoflavones in the plasma, where the free, unconjugated forms represent only a small portion [6].

Metabolites of isoflavones may be bioactive as well, with equol resulting from the conversion of daidzein as a prominent example [59]. However, our study focused on non-conjugated isoflavone compounds and did not include metabolites or modified forms. Apart from food matrix components, absorption and bioavailability but also bioactivity in vitro may depend on further cofactors and nutrients. In cultured prostate cancer cells, isoflavones show antiproliferative and antigenotoxic effects, which are enhanced by metal ions, in particular iron, copper and zinc ions [61]. In fact, the risk for prostate cancer incidence decreases with soy food consumption, which is not related to soy protein but has been referred to isoflavones [62]. The mortality from prostate cancer, however, is not reduced by isoflavones or may even increase [62,63], reflecting the need for an improved understanding of still incompletely understood isoflavone effects in the prostate and of their relationships with cofactors and other nutrients.

Considering that genistein showed the largest effects of three different isoflavones on prostate functions, the use of genistein rather than mixed isoflavone preparations for the treatment of voiding symptoms in BPH could be discussed. In contrast to available drugs, genistein may theoretically address contraction and growth in the prostate at once, as a single compound. Currently, this requires combination therapies [18]. Several compounds were recently identified as inhibiting prostate smooth muscle contraction and stromal cell growth at once, similar to genistein. However, these were inhibitors for kinases of GTPases, with limited translational value due to expectable side effects. Considering that genistein is a plant constituent and that pure genistein is available without prescription as a dietary supplement, it may be easily assumed that genistein is better tolerated than most small molecule inhibitors, though this may be in fact a dangerous oversimplification [64]. Genistein belongs to the group of "pan-assay interference compounds" (PAINS), together with other natural phytochemicals [65], showing low specificity, multiple and uncertain targets and a plethora of effects in preclinical studies. In fact, genistein is applied as an "phytoestrogenic" to reduce menopause symptoms [6]. In experimental medicine, it has been used as an inhibitor for non-receptor, Src family tyrosine kinases, so that it appears surprising that the compound is freely available as a dietary supplement. While it is apparently safe in women using it against menopause symptoms, tolerability is unknown in elderly men with BPH and voiding symptoms, and side effects should not be excluded. Isoflavones are potent inhibitors of cytochrome P450 isoforms and modulate activities of various other metabolic enzymes and transporters. As a result, they may not only influence their own bioavailability but also strongly influence the pharmacokinetics and dispositions of co-administered drugs. Clinically relevant and even life-threating drug–herb, drug–diet and drug–drug interactions have been reported due to the use of isoflavones [64]. Thus, the safety of genistein and isoflavones needs to be assessed before any application in BPH, where polypharmacy is common.

## 5. Conclusions

The major soy isoflavones, genistein and daidzein, exert sustained effects on prostate smooth muscle contractions and stromal cell growth, which may explain the inverse relationships between soy-rich nutrition, BPH and voiding symptoms. Improvements in voiding symptoms due to isoflavone-rich preparations with soy or red clover appear possible.

**Author Contributions:** Conceptualization, R.H., C.G.S. and M.H.; formal analysis, R.H. and M.H.; investigation, R.H., Y.L., S.H., A.T., R.W., F.S., C.G.S. and M.H.; resources, R.W. and C.G.S.; data curation, R.H. and M.H.; writing—original draft preparation, R.H. and M.H.; writing—review and editing Y.L., S.H., A.T., R.W., F.S. and C.G.S.; supervision, C.G.S. and M.H.; project administration, M.H.; funding acquisition, R.H., R.W., C.G.S. and M.H. All authors have read and agreed to the published version of the manuscript.

**Funding:** This research was funded by the Deutsche Forschungsgemeinschaft (DFG), grant number HE 5825/9-1, and the China Scholarship Council (CSC), grant number 202008080134.

**Institutional Review Board Statement:** The study was conducted in accordance with the Declaration of Helsinki and approved by the Ethics Committee of the Ludwig-Maximilians University, Munich, Germany (17-106, 16 March 2017).

**Informed Consent Statement:** Informed consent was obtained from all subjects involved in the study.

**Data Availability Statement:** The datasets supporting the conclusions of this article are included within the article.

**Acknowledgments:** We thank T. Kirchner, Frederick Klauschen and their coworkers (Institute of Pathology, Ludwig-Maximilians University, Munich) for the asservation of tissue samples from human prostates. We thank Maximilian M. Saller and his colleagues (Department of Orthopaedics and Trauma Surgery, Musculoskeletal University Center Munich (MUM), University Hospital, LMU Munich) for their support with fluorescence microscopy.

**Conflicts of Interest:** The authors declare no conflict of interest.

## References

1. Burdette, C.Q.; Marcus, R.K. Determination of isoflavone content in soy, red clover, and kudzu dietary supplement materials by liquid chromatography-particle beam/electron ionization mass spectrometry. *J. AOAC Int.* **2013**, *96*, 925–932. [CrossRef] [PubMed]
2. Aboushanab, S.A.; Khedr, S.M.; Gette, I.F.; Danilova, I.G.; Kolberg, N.A.; Ravishankar, G.A.; Ambati, R.R.; Kovaleva, E.G. Isoflavones derived from plant raw materials: Bioavailability, anti-cancer, anti-aging potentials, and microbiome modulation. *Crit. Rev. Food Sci. Nutr.* **2021**, 1–27. [CrossRef]
3. Silva, H. The Vascular Effects of Isolated Isoflavones-A Focus on the Determinants of Blood Pressure Regulation. *Biology* **2021**, *10*, 49. [CrossRef] [PubMed]
4. Delmonte, P.; Perry, J.; Rader, J.I. Determination of isoflavones in dietary supplements containing soy, Red Clover and kudzu: Extraction followed by basic or acid hydrolysis. *J. Chromatogr. A* **2006**, *1107*, 59–69. [CrossRef] [PubMed]
5. Mamagkaki, A.; Bouris, I.; Parsonidis, P.; Vlachou, I.; Gougousi, M.; Papasotiriou, I. Genistein as a dietary supplement; formulation, analysis and pharmacokinetics study. *PLoS ONE* **2021**, *16*, e0250599. [CrossRef]
6. Abdi, F.; Rahnemaei, F.A.; Roozbeh, N.; Pakzad, R. Impact of phytoestrogens on treatment of urogenital menopause symptoms: A systematic review of randomized clinical trials. *Eur. J. Obs. Gynecol. Reprod. Biol.* **2021**, *261*, 222–235. [CrossRef]
7. Chen, L.R.; Chen, K.H. Utilization of Isoflavones in Soybeans for Women with Menopausal Syndrome: An Overview. *Int. J. Mol. Sci.* **2021**, *22*, 3212. [CrossRef] [PubMed]
8. Lu, J.Z.; Ye, D.; Ma, B.L. Constituents, Pharmacokinetics, and Pharmacology of Gegen-Qinlian Decoction. *Front. Pharm.* **2021**, *12*, 668418. [CrossRef]
9. Wong, K.H.; Li, G.Q.; Li, K.M.; Razmovski-Naumovski, V.; Chan, K. Kudzu root: Traditional uses and potential medicinal benefits in diabetes and cardiovascular diseases. *J. Ethnopharmacol.* **2011**, *134*, 584–607. [CrossRef]
10. Zhang, Z.; Lam, T.N.; Zuo, Z. Radix Puerariae: An overview of its chemistry, pharmacology, pharmacokinetics, and clinical use. *J. Clin. Pharm.* **2013**, *53*, 787–811. [CrossRef]
11. Zhou, Y.X.; Zhang, H.; Peng, C. Puerarin: A review of pharmacological effects. *Phytother. Res.* **2014**, *28*, 961–975. [CrossRef] [PubMed]
12. Ambrosini, G.L.; de Klerk, N.H.; Mackerras, D.; Leavy, J.; Fritschi, L. Dietary patterns and surgically treated benign prostatic hyperplasia: A case control study in Western Australia. *BJU Int.* **2008**, *101*, 853–860. [CrossRef]
13. Denis, L.; Morton, M.S.; Griffiths, K. Diet and its preventive role in prostatic disease. *Eur. Urol.* **1999**, *35*, 377–387. [CrossRef] [PubMed]
14. Wong, S.Y.; Lau, W.W.; Leung, P.C.; Leung, J.C.; Woo, J. The association between isoflavone and lower urinary tract symptoms in elderly men. *Br. J. Nutr.* **2007**, *98*, 1237–1242. [CrossRef] [PubMed]
15. Bennetau-Pelissero, C. Risks and benefits of phytoestrogens: Where are we now? *Curr. Opin. Clin. Nutr. Metab. Care* **2016**, *19*, 477–483. [CrossRef]
16. Lepor, H. Pathophysiology, epidemiology, and natural history of benign prostatic hyperplasia. *Rev. Urol.* **2004**, *6* (Suppl. S9), S3–S10. [PubMed]
17. Strand, D.W.; Costa, D.N.; Francis, F.; Ricke, W.A.; Roehrborn, C.G. Targeting phenotypic heterogeneity in benign prostatic hyperplasia. *Differentiation* **2017**, *96*, 49–61. [CrossRef]
18. Oelke, M.; Bachmann, A.; Descazeaud, A.; Emberton, M.; Gravas, S.; Michel, M.C.; N'Dow, J.; Nordling, J.; de la Rosette, J.J.; European Association of Urology. EAU guidelines on the treatment and follow-up of non-neurogenic male lower urinary tract symptoms including benign prostatic obstruction. *Eur. Urol.* **2013**, *64*, 118–140. [CrossRef]
19. Cindolo, L.; Pirozzi, L.; Fanizza, C.; Romero, M.; Tubaro, A.; Autorino, R.; De Nunzio, C.; Schips, L. Drug adherence and clinical outcomes for patients under pharmacological therapy for lower urinary tract symptoms related to benign prostatic hyperplasia: Population-based cohort study. *Eur. Urol.* **2015**, *68*, 418–425. [CrossRef]
20. Cindolo, L.; Pirozzi, L.; Sountoulides, P.; Fanizza, C.; Romero, M.; Castellan, P.; Antonelli, A.; Simeone, C.; Tubaro, A.; de Nunzio, C.; et al. Patient's adherence on pharmacological therapy for benign prostatic hyperplasia (BPH)-associated lower urinary tract symptoms (LUTS) is different: Is combination therapy better than monotherapy? *BMC Urol.* **2015**, *15*, 96. [CrossRef]
21. Hennenberg, M.; Acevedo, A.; Wiemer, N.; Kan, A.; Tamalunas, A.; Wang, Y.; Yu, Q.; Rutz, B.; Ciotkowska, A.; Herlemann, A.; et al. Non-Adrenergic, Tamsulosin-Insensitive Smooth Muscle Contraction is Sufficient to Replace alpha1 -Adrenergic Tension in the Human Prostate. *Prostate* **2017**, *77*, 697–707. [CrossRef]
22. Hennenberg, M.; Stief, C.G.; Gratzke, C. Prostatic alpha1-adrenoceptors: New concepts of function, regulation, and intracellular signaling. *Neurourol. Urodyn.* **2014**, *33*, 1074–1085. [CrossRef] [PubMed]
23. Pradidarcheep, W.; Wallner, C.; Dabhoiwala, N.F.; Lamers, W.H. Anatomy and histology of the lower urinary tract. *Handb. Exp. Pharm.* **2011**, 117–148. [CrossRef]

24. Shaikhibrahim, Z.; Lindstrot, A.; Ellinger, J.; Rogenhofer, S.; Buettner, R.; Perner, S.; Wernert, N. The peripheral zone of the prostate is more prone to tumor development than the transitional zone: Is the ETS family the key? *Mol. Med. Rep.* **2012**, *5*, 313–316. [CrossRef]
25. Alcaraz, A.; Hammerer, P.; Tubaro, A.; Schroder, F.H.; Castro, R. Is there evidence of a relationship between benign prostatic hyperplasia and prostate cancer? Findings of a literature review. *Eur. Urol.* **2009**, *55*, 864–873. [CrossRef]
26. Orsted, D.D.; Bojesen, S.E. The link between benign prostatic hyperplasia and prostate cancer. *Nat. Rev. Urol.* **2013**, *10*, 49–54. [CrossRef]
27. Grabbert, M.; Buchner, A.; Butler-Ransohoff, C.; Kretschmer, A.; Stief, C.G.; Bauer, R.M. Long-term functional outcome analysis in a large cohort of patients after radical prostatectomy. *Neurourol. Urodyn.* **2018**, *37*, 2263–2270. [CrossRef]
28. Wang, Y.; Kunit, T.; Ciotkowska, A.; Rutz, B.; Schreiber, A.; Strittmatter, F.; Waidelich, R.; Liu, C.; Stief, C.G.; Gratzke, C.; et al. Inhibition of prostate smooth muscle contraction and prostate stromal cell growth by the inhibitors of Rac, NSC23766 and EHT1864. *Br. J. Pharm.* **2015**, *172*, 2905–2917. [CrossRef]
29. Chakrabarty, B.; Lee, S.; Exintaris, B. Generation and Regulation of Spontaneous Contractions in the Prostate. *Adv. Exp. Med. Biol.* **2019**, *1124*, 195–215. [CrossRef]
30. Huang, R.; Liu, Y.; Li, B.; Wang, R.; Tamalunas, A.; Waidelich, R.; Strittmatter, F.; Stief, C.G.; Hennenberg, M. Inhibition of human prostate smooth muscle contraction by the inhibitors of protein kinase C, GF109203X, and Go6983. *Prostate* **2022**, *82*, 59–77. [CrossRef]
31. Angulo, J.; Cuevas, P.; Fernandez, A.; La Fuente, J.M.; Allona, A.; Moncada, I.; Saenz de Tejada, I. Tadalafil enhances the inhibitory effects of tamsulosin on neurogenic contractions of human prostate and bladder neck. *J. Sex. Med.* **2012**, *9*, 2293–2306. [CrossRef] [PubMed]
32. Spek, A.; Li, B.; Rutz, B.; Ciotkowska, A.; Huang, R.; Liu, Y.; Wang, R.; Strittmatter, F.; Waidelich, R.; Stief, C.G.; et al. Purinergic smooth muscle contractions in the human prostate: Estimation of relevance and characterization of different agonists. *Naunyn. Schmiedebergs. Arch. Pharm.* **2021**, *394*, 1113–1131. [CrossRef]
33. Webber, M.M.; Trakul, N.; Thraves, P.S.; Bello-DeOcampo, D.; Chu, W.W.; Storto, P.D.; Huard, T.K.; Rhim, J.S.; Williams, D.E. A human prostatic stromal myofibroblast cell line WPMY-1: A model for stromal-epithelial interactions in prostatic neoplasia. *Carcinogenesis* **1999**, *20*, 1185–1192. [CrossRef] [PubMed]
34. Wang, Y.; Gratzke, C.; Tamalunas, A.; Wiemer, N.; Ciotkowska, A.; Rutz, B.; Waidelich, R.; Strittmatter, F.; Liu, C.; Stief, C.G.; et al. P21-Activated Kinase Inhibitors FRAX486 and IPA3: Inhibition of Prostate Stromal Cell Growth and Effects on Smooth Muscle Contraction in the Human Prostate. *PLoS ONE* **2016**, *11*, e0153312. [CrossRef]
35. Michel, M.C.; Murphy, T.J.; Motulsky, H.J. New Author Guidelines for Displaying Data and Reporting Data Analysis and Statistical Methods in Experimental Biology. *Mol. Pharm.* **2020**, *97*, 49–60. [CrossRef]
36. Curtis, M.J.; Bond, R.A.; Spina, D.; Ahluwalia, A.; Alexander, S.P.; Giembycz, M.A.; Gilchrist, A.; Hoyer, D.; Insel, P.A.; Izzo, A.A.; et al. Experimental design and analysis and their reporting: New guidance for publication in BJP. *Br. J. Pharm.* **2015**, *172*, 3461–3471. [CrossRef]
37. Wong, W.C.; Wong, E.L.; Li, H.; You, J.H.; Ho, S.; Woo, J.; Hui, E. Isoflavones in treating watchful waiting benign prostate hyperplasia: A double-blinded, randomized controlled trial. *J. Altern. Complement. Med.* **2012**, *18*, 54–60. [CrossRef]
38. Mohamad, J.; Masrudin, S.S.; Alias, Z.; Muhamad, N.A. The effects of Pueraria mirifica extract, diadzein and genistein in testosterone-induced prostate hyperplasia in male Sprague Dawley rats. *Mol. Biol. Rep.* **2019**, *46*, 1855–1871. [CrossRef]
39. Weber, K.S.; Setchell, K.D.; Stocco, D.M.; Lephart, E.D. Dietary soy-phytoestrogens decrease testosterone levels and prostate weight without altering LH, prostate 5alpha-reductase or testicular steroidogenic acute regulatory peptide levels in adult male Sprague-Dawley rats. *J. Endocrinol.* **2001**, *170*, 591–599. [CrossRef]
40. Masrudin, S.S.; Mohamad, J. Preventive effect of Pueraria mirifica on testosterone-induced prostatic hyperplasia in Sprague Dawley rats. *Andrologia* **2015**, *47*, 1153–1159. [CrossRef]
41. Brandli, A.; Simpson, J.S.; Ventura, S. Isoflavones isolated from red clover (Trifolium pratense) inhibit smooth muscle contraction of the isolated rat prostate gland. *Phytomedicine* **2010**, *17*, 895–901. [CrossRef] [PubMed]
42. Jang, H.; Bae, W.J.; Yuk, S.M.; Han, D.S.; Ha, U.S.; Hwang, S.Y.; Yoon, S.H.; Kim, S.W.; Han, C.H. Seoritae extract reduces prostate weight and suppresses prostate cell proliferation in a rat model of benign prostate hyperplasia. *Evid. Based. Complement. Altern. Med.* **2014**, *2014*, 475876. [CrossRef] [PubMed]
43. Cai, L.Q.; Cai, J.; Wu, W.; Zhu, Y.S. 17alpha-Estradiol and genistein inhibit high fat diet induced prostate gene expression and prostate growth in the rat. *J. Urol.* **2011**, *186*, 1489–1496. [CrossRef] [PubMed]
44. Wang, X.; Li, B.; Ciotkowska, A.; Rutz, B.; Erlander, M.G.; Ridinger, M.; Wang, R.; Tamalunas, A.; Waidelich, R.; Stief, C.G.; et al. Onvansertib, a polo-like kinase 1 inhibitor, inhibits prostate stromal cell growth and prostate smooth muscle contraction, which is additive to inhibition by alpha1-blockers. *Eur. J. Pharm.* **2020**, *873*, 172985. [CrossRef]
45. Nomura, H.; Kawashima, H.; Masaki, S.; Hosono, T.Y.; Matsumura, K.; Tamada, S.; Tanaka, T.; Nakatani, T. Effect of selective estrogen receptor modulators on cell proliferation and estrogen receptor activities in normal human prostate stromal and epithelial cells. *Prostate. Cancer Prostatic. Dis.* **2009**, *12*, 375–381. [CrossRef]
46. Rao, A.; Woodruff, R.D.; Wade, W.N.; Kute, T.E.; Cramer, S.D. Genistein and vitamin D synergistically inhibit human prostatic epithelial cell growth. *J. Nutr.* **2002**, *132*, 3191–3194. [CrossRef]
47. Wang, X.; Clubbs, E.A.; Bomser, J.A. Genistein modulates prostate epithelial cell proliferation via estrogen- and extracellular signal-regulated kinase-dependent pathways. *J. Nutr. Biochem.* **2006**, *17*, 204–210. [CrossRef]

48. Hedlund, T.E.; van Bokhoven, A.; Johannes, W.U.; Nordeen, S.K.; Ogden, L.G. Prostatic fluid concentrations of isoflavonoids in soy consumers are sufficient to inhibit growth of benign and malignant prostatic epithelial cells in vitro. *Prostate* **2006**, *66*, 557–566. [CrossRef]
49. Geller, J.; Sionit, L.; Partido, C.; Li, L.; Tan, X.; Youngkin, T.; Nachtsheim, D.; Hoffman, R.M. Genistein inhibits the growth of human-patient BPH and prostate cancer in histoculture. *Prostate* **1998**, *34*, 75–79. [CrossRef]
50. Risbridger, G.P.; Wang, H.; Frydenberg, M.; Husband, A. The in vivo effect of red clover diet on ventral prostate growth in adult male mice. *Reprod. Fertil. Dev.* **2001**, *13*, 325–329. [CrossRef]
51. Semenov, A.L.; Gubareva, E.A.; Ermakova, E.D.; Dorofeeva, A.A.; Tumanyan, I.A.; Radetskaya, E.A.; Yurova, M.N.; Aboushanab, S.A.; Kanwugu, O.N.; Fedoros, E.I.; et al. Astaxantin and Isoflavones Inhibit Benign Prostatic Hyperplasia in Rats by Reducing Oxidative Stress and Normalizing Ca/Mg Balance. *Plants* **2021**, *10*, 2735. [CrossRef] [PubMed]
52. Engelhardt, P.F.; Riedl, C.R. Effects of one-year treatment with isoflavone extract from red clover on prostate, liver function, sexual function, and quality of life in men with elevated PSA levels and negative prostate biopsy findings. *Urology* **2008**, *71*, 185–190; discussion 190. [CrossRef] [PubMed]
53. Barry, M.J.; Williford, W.O.; Chang, Y.; Machi, M.; Jones, K.M.; Walker-Corkery, E.; Lepor, H. Benign prostatic hyperplasia specific health status measures in clinical research: How much change in the American Urological Association symptom index and the benign prostatic hyperplasia impact index is perceptible to patients? *J. Urol.* **1995**, *154*, 1770–1774. [CrossRef] [PubMed]
54. McConnell, J.D.; Roehrborn, C.G.; Bautista, O.M.; Andriole, G.L., Jr.; Dixon, C.M.; Kusek, J.W.; Lepor, H.; McVary, K.T.; Nyberg, L.M., Jr.; Clarke, H.S.; et al. The long-term effect of doxazosin, finasteride, and combination therapy on the clinical progression of benign prostatic hyperplasia. *N. Engl. J. Med.* **2003**, *349*, 2387–2398. [CrossRef]
55. Roehrborn, C.G.; Siami, P.; Barkin, J.; Damiao, R.; Major-Walker, K.; Nandy, I.; Morrill, B.B.; Gagnier, R.P.; Montorsi, F.; Comb, A.T.S.G. The effects of combination therapy with dutasteride and tamsulosin on clinical outcomes in men with symptomatic benign prostatic hyperplasia: 4-year results from the CombAT study. *Eur. Urol.* **2010**, *57*, 123–131. [CrossRef] [PubMed]
56. Cao, Y.; Xie, L.; Liu, K.; Liang, Y.; Dai, X.; Wang, X.; Lu, J.; Zhang, X.; Li, X. The antihypertensive potential of flavonoids from Chinese Herbal Medicine: A review. *Pharm. Res.* **2021**, *174*, 105919. [CrossRef]
57. Guy, L.; Vedrine, N.; Urpi-Sarda, M.; Gil-Izquierdo, A.; Al-Maharik, N.; Boiteux, J.P.; Scalbert, A.; Remesy, C.; Botting, N.P.; Manach, C. Orally administered isoflavones are present as glucuronides in the human prostate. *Nutr. Cancer* **2008**, *60*, 461–468. [CrossRef]
58. Gardner, C.D.; Oelrich, B.; Liu, J.P.; Feldman, D.; Franke, A.A.; Brooks, J.D. Prostatic soy isoflavone concentrations exceed serum levels after dietary supplementation. *Prostate* **2009**, *69*, 719–726. [CrossRef]
59. Vitale, D.C.; Piazza, C.; Melilli, B.; Drago, F.; Salomone, S. Isoflavones: Estrogenic activity, biological effect and bioavailability. *Eur. J. Drug Metab. Pharm.* **2013**, *38*, 15–25. [CrossRef]
60. Nielsen, I.L.; Williamson, G. Review of the factors affecting bioavailability of soy isoflavones in humans. *Nutr. Cancer* **2007**, *57*, 1–10. [CrossRef]
61. Stanislawska, I.J.; Figat, R.; Kiss, A.K.; Bobrowska-Korczak, B. Essential Elements and Isoflavonoids in the Prevention of Prostate Cancer. *Nutrients* **2022**, *14*, 1225. [CrossRef] [PubMed]
62. Fan, Y.; Wang, M.; Li, Z.; Jiang, H.; Shi, J.; Shi, X.; Liu, S.; Zhao, J.; Kong, L.; Zhang, W.; et al. Intake of Soy, Soy Isoflavones and Soy Protein and Risk of Cancer Incidence and Mortality. *Front. Nutr.* **2022**, *9*, 847421. [CrossRef] [PubMed]
63. Sawada, N.; Iwasaki, M.; Yamaji, T.; Shimazu, T.; Inoue, M.; Tsugane, S.; Japan Public Health Center-based Prospective Study Group. Soy and isoflavone consumption and subsequent risk of prostate cancer mortality: The Japan Public Health Center-based Prospective Study. *Int. J. Epidemiol.* **2020**, *49*, 1553–1561. [CrossRef] [PubMed]
64. Wahajuddin; Taneja, I.; Arora, S.; Raju, K.S.; Siddiqui, N. Disposition of pharmacologically active dietary isoflavones in biological systems. *Curr. Drug Metab.* **2013**, *14*, 369–380. [CrossRef] [PubMed]
65. Baell, J.; Walters, M.A. Chemistry: Chemical con artists foil drug discovery. *Nature* **2014**, *513*, 481–483. [CrossRef]

*nutrients*

MDPI

*Article*

# In Vitro Digestion and Colonic Fermentation of UHT Treated Faba Protein Emulsions: Effects of Enzymatic Hydrolysis and Thermal Processing on Proteins and Phenolics

Jingyu Gu [1], Minhao Li [1], Malik Adil Nawaz [2,*], Regine Stockmann [2,*], Roman Buckow [3] and Hafiz A. R. Suleria [1]

[1] School of Agriculture and Food, Faculty of Veterinary and Agricultural Sciences, The University of Melbourne, Parkville, VIC 3010, Australia
[2] Commonwealth Scientific and Industrial Research Organisation (CSIRO), Agriculture and Food, Werribee, VIC 3030, Australia
[3] Centre for Advanced Food Engineering, School of Chemical and Biomolecular Engineering, The University of Sydney, Darlington, NSW 2006, Australia
* Correspondence: malik.nawaz@csiro.au (M.A.N.); regine.stockmann@csiro.au (R.S.)

**Abstract:** Faba bean (*Vicia faba* L.) protein is a new plant protein alternative source with high nutrient content especially protein and phenolic compounds. The present study investigated physicochemical properties, phenolic content, antioxidant potential, and short chain fatty acids (SCFAs) production during in vitro digestion and colonic fermentation of faba bean hydrolysates and oil-in-water (O/W) emulsions. Results indicate that the enzymic hydrolysates of faba proteins exhibited higher protein solubility, increased electronegativity, and decreased surface hydrophobicity than native faba protein. O/W emulsions showed improved colloidal stability for the faba protein hydrolysates after ultra-high temperature processing (UHT). Furthermore, UHT processing preserved total phenolic content, DPPH and ABTS radical scavenging abilities while decreasing total flavonoid content and ferric reducing power. Besides, the release of phenolic compounds in faba bean hydrolysates (FBH) and emulsions (FBE) improved after intestinal digestion by 0.44 mg GAE/g and 0.55 mg GAE/g, respectively. For colonic fermentation, FBH demonstrated an approximately 10 mg TE/g higher ABTS value than FBE (106.45 mg TE/g). Total SCFAs production of both FBH and FBE was only 0.03 mM. The treatment of FBH with 30 min enzymatic hydrolysis displayed relatively higher antioxidant capacities and SCFAs production, indicating its potential to bring more benefits to gut health. Overall, this study showed that enzymic hydrolysis of faba proteins not only improved the colloidal emulsion stability, but also released antioxidant capacity during in vitro digestibility and colonic fermentation. Colonic fermentation metabolites (SCFAs) were related to the degree of hydrolysis for both FBH and FBE. Additional studies are required to further elucidate and differentiate the role of phenolics during faba protein processing and digestion stages in comparison to contributions of peptides, amino acids and microelements to digestion rates, antioxidant capacities and colonial SCFA production.

**Keywords:** faba proteins; enzymatic hydrolysis; UHT processing; emulsion; phenolic compounds; in vitro digestion; colonic fermentation

**Citation:** Gu, J.; Li, M.; Nawaz, M.A.; Stockmann, R.; Buckow, R.; Suleria, H.A.R. In Vitro Digestion and Colonic Fermentation of UHT Treated Faba Protein Emulsions: Effects of Enzymatic Hydrolysis and Thermal Processing on Proteins and Phenolics. *Nutrients* **2023**, *15*, 89. https://doi.org/10.3390/nu15010089

Academic Editors: Stuart Johnson and Rewati Bhattarai

Received: 12 November 2022
Revised: 17 December 2022
Accepted: 20 December 2022
Published: 24 December 2022



## 1. Introduction

The Faba bean (*Vicia faba* L.) is one of the most widely grown winter season legume crops [1]. As a member of the Fabaceae family, it is typically named fava bean or broad bean, and usually cultivated as food and fodder [2–4]. Like other legumes, faba bean is a good source of quality protein (27–34%) [5] and phenolics, especially condensed tannins [6]. In addition, a high proportion of other minor compounds are also present in faba beans, such as levodopa as well as phytic acids which may aid in treating Parkinson's disease and have anti-cancer properties [7,8]. Recently, faba protein along with other legume proteins has gained popularity due to lower environmental impacts and lower cost of production compared to animal proteins [9]. However, application of faba proteins in plant based foods has not yet been fully explored and faba proteins are still considered an emerging legume protein source.

Phenolic-protein interactions may result in nutrition losses due to protein precipitation and the inactivation of digestive enzymes [10]. Moreover, complexes between phenolics and proteins may affect the bio-accessibility and bioavailability of proteins and phenolics by protecting them from oxidizing as they pass through the GI tract [11]. To improve the bioavailability of phenolics and protein digestibility, protein hydrolysis may be commercially implemented by using protease enzymes such as Alcalase to help decompose protein into smaller peptides and free amino acids, thereby promoting antioxidant capacity [12,13]. Several processing techniques (such as ultra-high temperature processing, high shear strength mixing, and homogenization) also assist in enhancing the stability of protein emulsions by generating pressure and heating impacts [14], which further provide specific antioxidant potential due to the formation of some Maillard reaction products [15].

Earlier studies reported that protein hydrolysates derived from Alcalase showed more resistance to digestive enzymes and higher antioxidant activities than hydrolysates generated from other proteases [16,17]. However, to the best of our knowledge, the influence of Alcalase hydrolysis on phenolic compounds contents and activity in hydrolysates of faba bean is not understood. Furthermore, the knowledge on how UHT, an important commercial processing operation to stabilize plant-based drinks, transforms the phenolic contents and antioxidant properties of faba bean protein emulsions is limited. Lastly, there is a lack of comprehensive understanding of how in vitro gastrointestinal digestion influences the release of phenolic compounds in hydrolysates and emulsions. Therefore, the objective of the present study was to determine how physicochemical properties, phenolics contents, and antioxidant activities, changed during in vitro gastrointestinal digestion and colonic fermentation of oil-in-water (O/W) emulsions formed with functional faba been protein hydrolysates with subsequent UHT processing treatment. The production of SCFAs, of FBH and FBE that is generated during colonic fermentation in a faecal model, was also examined.

## 2. Materials and Methods

### 2.1. Chemicals and Reagents

All chemicals and reagents used were of analytical grade and purchased from Sigma-Aldrich (St. Louis, MO, USA) unless otherwise stated.

### 2.2. Sample Preparation

Commercial dehulled faba beans were locally purchased, ground into powders and defatted with n-hexane and stored at room temperature (Experimental design of the study is presented in Figure 1).

#### 2.2.1. Extraction of Faba Bean Phenolics (FBP)

Ethanolic extraction of faba bean was performed according to the previously published method with slight modifications [18]. Briefly, slurry of faba bean powder (5 g) was prepared in 20 mL ethanol (70% $v/v$) by homogenising with Ultra-Turrax T25 Homogenizer (IKA Werke GmbH & Co. KG, Staufen, Germany) for 30 s at 10,000 rpm. Homogenised samples were incubated in a shaking incubator (ZWYR-240 incubator shaker, Labwit, Ashwood, VIC, Australia) for 12 h at 4 °C for 120 rpm. Subsequently, incubated samples were centrifuged by Hettich Refrigerated Centrifuge (ROTINA380R, Tuttlingen, Germany) at 5000 rpm for 15 min at 4 °C.

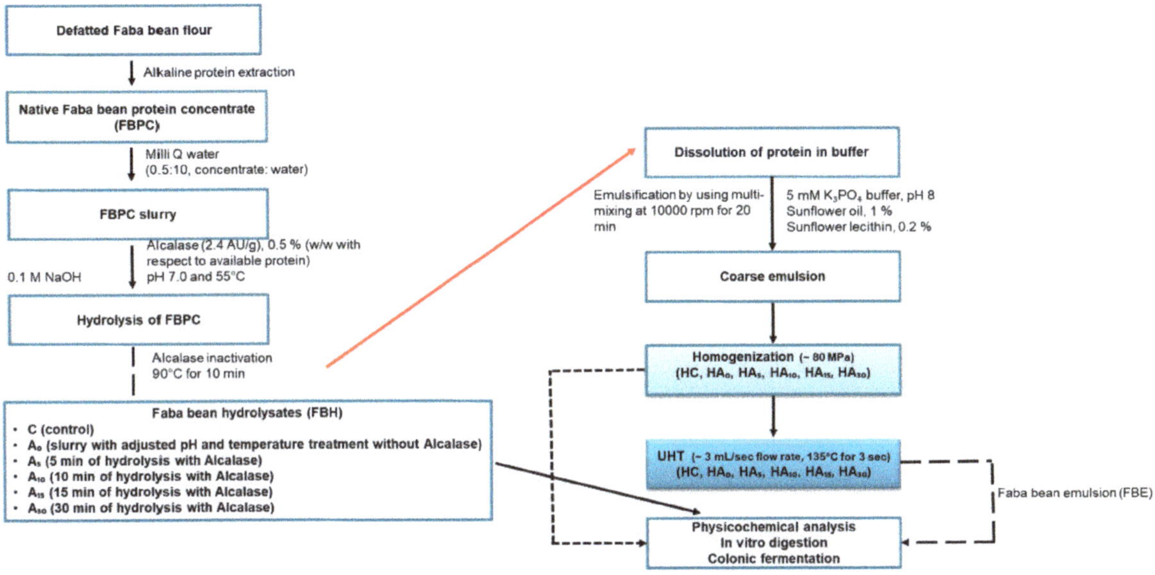


**Figure 1.** Experimental design.

2.2.2. Preparation of Faba Bean Protein Concentrate (FBPC)

Faba bean flour was defatted using n-hexane. Defatted flour was mixed with distilled water (1:10) to make a slurry and pH was adjusted to 9.0 using 2 N NaOH. The slurry was then mixed for 2 h at room temperature using an overhead stirrer, followed by centrifugation at 10,000× *g* for 15 min at 10 °C and collection of the supernatant. The pH of the supernatant was adjusted to 4.3 and centrifuged at 10,000× *g* for 15 min 10 °C. The pellets were separated from the supernatant, and pH of the pellets was adjusted to 7.0, followed by freeze drying. The composition and caloric value of the concentrate (per 100 g) were carbohydrate (27 g), total fat (3 g), protein content (63 g), total ash (4 g), moisture content (3 g), and calculated calories (410 kcal).

2.2.3. Preparation of Faba Bean Hydrolysates (FBH)

FBPC was hydrolysed by Alcalase according to the method described by Ghribi et al. [19]. Briefly, 10% (*w*/*w*) FBPC dispersions pre-equilibrated at pH 7.0 were hydrolysed using 0.5% (*v*/*v*) of Alcalase (2.4 AU/g of protein) (Alcalase®, Novozymes Australia Pty. Ltd., North Ride, NSW, Australia) at 50 °C for various time periods viz., 5 min ($A_5$), 10 min ($A_{10}$), 15 min ($A_{15}$), and 30 min ($A_{30}$). After the target time of reaction, hydrolysis was stopped by inactivating Alcalase at 85 °C for 20 min. Control sample (slurry without pH adjustment and Alcalase) and $A_0$ sample (slurry with adjusted pH 8 and temperature treatment without Alcalase) were also prepared.

2.2.4. Preparation of Faba Bean Emulsions (FBE)

O/W emulsions (approximately 3000 g) were prepared by adding 5% (*w*/*w*) of native FBPC (C) or hydrolysed FBPC ($A_0$, $A_5$, $A_{10}$, $A_{15}$, and $A_{30}$), 1% (*w*/*w*) sunflower oil and 0.2% (*w*/*w*) sunflower lecithin (as a co-emulsifier) in Milli-Q grade water (pH 7.0). The dispersions were mixed with a multi-mixer at 1000 rpm for 10 min. Following mixing, the coarse emulsions were prepared by homogenising the lipidic and aqueous phases using a Silverson high-shear laboratory mixer (Silverson®, Artarmon, NSW, Australia) at 10,000 rpm for 20 min. The coarse emulsions were homogenised by two passes in an EmulsiFlex-C5 (Avestin, Ottawa, Canada) operating at ~80 MPa followed by UHT processing using a bench-top UHT unit (FT74XTS: UHT/HTST System, Armfield Ltd.,

Hampshire, UK), pre-heated to 105 °C for 3 s and operated at high heat-treatment at 135 °C for 3 s with a sample flow rate of 3 mL/s.

### 2.3. In Vitro Gastrointestinal Digestion

FBH and FBE were digested in three stages, including oral, gastric and intestinal, followed by the harmonized INFOGEST 2.0 protocol of static in vitro GID conditions, including enzymes, $CaCl_2$ and the simulated oral (SOF), gastric (SGF) and intestinal (SIF) fluids described by Sánchez-Velázquez et al. [20] with some modifications. All the samples were first stirred in water (1:2, *w/v*) and an aliquot of 5 mL was taken as a non-digested (ND) aliquot. Samples were then dissolved in SOF (1:1, *v/v*) at pH 7.0 to add 75 U of α-amylase/mL and stirred at 37 °C for 2 min. 5 mL was then taken as an oral-phase aliquot. To simulate the gastric phase (GP), oral-phase bolus was mixed with SGF (1:1, *v/v*) at 37 °C and adjusted to pH 3 before adding 2000 U of porcine pepsin/mL. Gastric phase (GP) aliquots were removed after incubation at 37 °C for 2 h. Gastric-digested samples were further digested by dissolving in SIF (1:1, *v/v*), adjusted pH to 7.0, trypsin (100 U/mL), and bile salt (10 mM) were added to the intestinal phase (IP). Intestinal phase aliquots were taken after incubation at 37 °C for 2 h.

### 2.4. In Vitro Colonic Fermentation

The in vitro colonic fermentation procedure was conducted by using a modified method of Gu et al. [21]. A pig faecal model considered a good predictor for human colonic fermentation was selected. Briefly, Pig faecal samples were used as gut microbiome source as human faeces substitutes since pigs and humans are primarily colonic fermenters sharing a comparable gut microbiome. Ten mixed male and female large landrace grower white pigs (around 50 kg live weight) and raised in animal house of Diamond Valley Pork (Laverton North, VIC, Australia), with the standard grower diets for two weeks. The faeces were taken immediately after the pigs defecated and were put into an anaerobic chamber. The medium (20% faeces) was prepared by mixing 20 g faeces with 80 g 0.1 M sterilized pre-nitrogen flushed phosphate buffer (pH 7.0) in a stomacher mixer (MiniMix® Lab Blender, Thomas Scientific, Swedesboro, NJ, USA) for 5 min followed by the filtration through a sterile muslin cloth. Sediments from the small intestinal digestion (FBH and FBE) were prepared after centrifugation at 10,000× *g* for 10 min. Then, aliquots of 5 mL faecal slurry were added into six sets of tubes with the sediments and 5 mL of the basal media were added. These six sets of tubes were first flushed with nitrogen and then incubated with shaking at 120 rpm for 0 h, 2 h, 4 h, 8 h, 16 h, and 24 h, respectively in darkness. Afterwards, the tubes were centrifuged at 10,000× *g* and 5 °C for 10 min and the supernatant was taken from the sediment. Supernatants were stored at −80 °C for analysis of phytochemical bioactivity and short chain fatty acid analysis (SCFA) production.

### 2.5. Physicochemical Properties of Hydrolysates and Emulsions

#### 2.5.1. Degree of Hydrolysis (DH)

DH of all FBE was calculated by quantifying of α-amino groups released during hydrolysis by o-phthaldialdehyde (OPA) as described by Church et al. [22]. Briefly, on a 96-well fluorescence plate, 5 μL of sample (glycine standards or hydrolysate) was mixed with 200 μL of OPA reagent (97.5 mL of 100 mM sodium teteraborate (pH 9.9), 0.5 mL of 20% SDS, 1 mL of 5 mg/mL methylated OPA, 1 mL of 5 mg/mL aqueous dithiothreitol) for 5 min at 25 °C, followed by incubation at 37 °C for 20 min. Fluorescence emission (excitation λ 340 nm, emission λ 450 nm) of incubated samples was measured by a plate reader (Varioskan Flash, Thermo Scientific, Waltham, MA, USA). DH was calculated by using Equation (1):

$$DH\ (\%) = \frac{C_S \times D_{F1} \times D_{F2} \times 100}{m_3 \times H_{tot}} \quad (1)$$

where, $C_S$ (concentration of sample (mmol/L)), $D_{F1}$ (dilution factor prior to OPA reaction), $D_{F2}$ (dilution factor during OPA reaction), $m_3$ (mass of protein per litre), $H_{tot}$ (total peptide bonds in protein substrate (assumed as 7.8 mmol/g)).

#### 2.5.2. Protein Solubility

Hydrolysate suspensions (1% *w*/*v*) in 5 mM potassium phosphate buffer (pH 8, prepared in Milli Q water) were prepared by mixing for 10 min. Suspensions were then centrifuged (12,000× *g*, 20 °C for 20 min), followed by estimation of the total nitrogen contents of supernatant using Dumas combustion method with LECO Trumac® N (LECO Corporation, St. Joseph, MA, USA). Total protein contents in the samples were calculated using a conversion factor of 6.25 [23]. The protein solubility was expressed as a percentage of supernatant protein over total protein.

#### 2.5.3. ζ-Potential

The ζ-potential of prediluted (~10-folds) samples FBH and FBE samples were measured using Malvern Zetasizer at 25 °C at a refractive index of 1.33 [24].

#### 2.5.4. Surface Hydrophobicity ($S_0$)

The $S_0$ of FBH and FBE were measured using 8-anilino-1-naphthalene sulfonic acid (ANS) as a fluorescent probe on a spectrofluorometer as described by Nwachukwu and Aluko [25]. Briefly, four different serial dilutions (0.006, 0.003, 0.0012 and 0.0009%) of every sample were prepared followed by sub-division of each dilution into two batches (with and without ANS solution). 20 µL of ANS (8.0 mM prepared in 5 mM phosphate buffer) was added to 5 mL of the first batch while the second batch was used as a blank. The ANS-protein conjugation was measured at 370 nm (excitation) and 490 nm (emission) wavelengths using a plate reader (Varioskan Flash, Thermo Scientific, Waltham, MA, USA). The $S_0$ index was obtained through linear regression analysis of the plot between fluorescence intensity (FI) and protein concentrations (as the slope of fluorescence intensity-protein concentration).

#### 2.5.5. Particle Size

The average particle size distribution of all FBE was determined using a Malvern Mastersizer 3000 (Malvern Instruments Ltd., Worcestershire, UK). The continuous phase was Milli-Q grade water (Refractive Index: 1.33), while the dispersed phase was sunflower oil and plant protein. Measurements for particle size was obtained using ~12.5% laser obscuration and the droplet size distribution proportion at 10% is d(0.1), 50% is d(0.5) and 90% is d(0.9), respectively. The uniformity, surface weighted mean, $D^{3,2}$ and volume-weighted mean $D^{4,3}$ were also achieved from the particle size distribution graphs [26]. Then, the polydispersity index (PDI) was also calculated according to Equation (2).

$$\text{PDI} = \text{d}(0.9) - \text{d}(0.1)/\text{d}(0.5) \quad (2)$$

### *2.6. Estimation of Phenolics and Their Antioxidant Potential*

Estimations were performed according to the methods of Akhtar, Wu, Ponnampalam, Cottrell, Dunshea and Suleria [18] using Multiskan Go microplate photometer (Thermo Fisher Scientific Inc., Waltham, MA, USA).

#### 2.6.1. Determination of Total Phenolic Content (TPC)

Folin–Ciocalteu's method was used to estimate the TPC of FB, FBPC, FBH, FBE and their corresponding digests. 25 µL extract, 25 µL Folin–Ciocalteu's reagent solution (1:3 diluted with water) and 200 µL water were added into the 96-well plate (Costar, Corning, NY, USA) and incubated for 5 min at room temperature (25 °C). Afterwards, 25 µL of 10% (*w*/*w*) sodium carbonate was added and incubated for 60 min at 25 °C. Absorbance was then measured at 765 nm using a spectrophotometer (Thermo Fisher

Scientific, Waltham, MA, USA). Gallic acid standard curve with concentrations ranging from 0 to 200 μg/mL was used to determine TPC content and expressed in mg of gallic acid equivalents per gram of extract (mg GAE/g).

#### 2.6.2. Determination of Total Flavonoid Content (TFC)

TFC of FB, FBPC, FBH, and FBE and their corresponding digests were quantified by using aluminium chloride. Extract (80 μL), 80 μL of 2% aluminium chloride and 120 μL of 50 g/L sodium acetate solution were added into the 96-well plate and incubated in dark for 2.5 h at room temperature. Absorbance was measured at 440 nm. Quercetin standard curve with concentrations ranging from 0–50 μg/mL was used to determine TFC and expressed in mg quercetin equivalents per gram of extract (mg QE/g).

#### 2.6.3. Determination of Total Condensed Tannin (TCT)

TCT of FB, FBPC, FBH, and FBE and their corresponding digests were measured using vanillin sulphuric acid method. 25 μL of 32% sulphuric acid, 25 μL of sample extract and followed by150 μL of 4% vanillin solution were added to 96-well plate which was covered and incubated for 15 min. The absorbance was measured at 500 nm. Catechin standard curve with concentrations ranging from 0 to 1 mg/mL was used for estimation of TCT and expressed in mg catechin equivalents (CE) per gram of extract (mg CE/g).

#### 2.6.4. 2,2′-Diphenyl-1-Picrylhydrazyl (DPPH) Assay

DPPH (4 mg) was dissolved in methanol (100 mL) to prepare DPPH radical solution. 40 μL of extract and 260 μL of DPPH solution were first added to 96-well plate and then vigorously shaken in the dark for 30 min at 25 °C. The absorbance was measured at 517 nm. A Trolox standard curve with concentrations ranging from 0 to 200 μg/mL was used to evaluate the DPPH radical scavenging activity and expressed in mg of Trolox equivalent per gram (mg TE/g) of extract.

#### 2.6.5. Ferric Reducing Antioxidant Power (FRAP) Assay

Sodium acetate solution (300 mM), TPTZ solution (10 mM) and Fe [III] solution (20 mM) were mixed at the ratio of 10:1:1, to prepare the FRAP solution. Then, 20 μL of the extract and 280 μL prepared dye solution were added to a 96-well plate and incubated for 10 min at 37 °C. The absorbance was measured at 593 nm. Trolox standard curve with concentrations ranging from 0 to 200 μg/mL was used to evaluate the FRAP values and expressed in mg of Trolox equivalent per gram of extract (mg TE/g).

#### 2.6.6. 2,2′-Azino-Bis-3-Ethylbenzothiazoline-6-Sulfonic Acid (ABTS) Assay

$ABTS^+$ stock solution was prepared by mixing 5 mL of 7 mM ABTS solution and 88 μL of 140 mM potassium persulfate. Then, the reaction mixture was incubated in dark for 16 h. 10 μL of the extract and 290 μL dye solution were added to the 96-well plate and incubated for 6 min at 25 °C. The absorbance was measured at 734 nm. The antioxidant activity was calculated using the standard curve of Trolox with concentrations ranging from 0 to 500 μg/mL. Results were expressed in mg of Trolox equivalent per gram of extract (mg TE/g).

### *2.7. Short Chain Fatty Acids (SCFAs) Production (GC-FID) Analysis*

SCFAs production of the colonic digesta (FBH and FBE) was assessed according to the protocol described by Gu, Suleria, Dunshea and Howell [21]. Briefly, the acidified post-fermented faeces samples were extracted with water/formic acid (99:1, *v*/*v*) and analysed using gas chromatography (7890B Agilent, Santa Clara, CA, USA) equipped with a flame ionisation detector (FID) using capillary column (SGE BP21, 12 × 0.53 mm internal diameter (ID) with 0.5 μm film thickness, SGE International, Ringwood, VIC, Australia, P/N 054473). The injection volume was 1 μL and 4-methyl-valeric acid was used as the internal standard. Concentration of detected acetic, propionic, and butyric acids were expressed as μmol/L.

*2.8. Statistical Analysis*

Results are presented in means ± standard deviations. All the experiments were performed in triplicates. Experimental data was statistically analysed using one-way ANOVA on Minitab 19 (Minitab® for Windows Release 19, Minitab Inc, Chicago, IL, USA). Tukey's HSD was calculated to test the significant difference among samples at the level of $p \leq 0.05$, at a 5 % level of significance.

## 3. Results and Discussions

*3.1. Changes in the Phenolic Contents and Antioxidant Properties during FBPC Preparation*

The effects of concentrating the faba bean protein on the phenolic estimations and antioxidant properties are shown in Table 1. The total phenolic and flavonoid contents of FBPC were 2.79 mg GAE/g and 394.37 µg QE/g, respectively, which were significantly higher than those of FB. Meanwhile, no significant difference was observed in total condensed tannin after making protein concentrate. That might be due to the initial alkaline conditions during production of the concentrate, which could promote the oxidation of condensed tannins [27]. FBPC exhibited significantly higher values in all three antioxidant assays compared with FB. The increased DPPH radical scavenging capacity after processing faba bean into powders with high protein concentration is in agreement with the study of Nnamezie et al. [28] which also showed improved free radical scavenging abilities resulting from greater phenolics content in the extracts with higher concentrations. The higher the phenolic compound concentrations are, the more available hydroxyl groups within the reaction media and, therefore, the more chances for free radicals to donate hydrogen atoms [29].

**Table 1.** The estimations of phenolic content and antioxidant capacities of faba bean (FB), protein concentrate (FBPC), hydrolysates (FBH) and emulsions (FBE) *.

| Treatments | Samples | TPC (mg GAE/g) | TFC (µg QE/g) | TCT (mg CE/g) | DPPH (mg TE/g) | FRAP (µg TE/g) | ABTS (mg TE/g) |
|---|---|---|---|---|---|---|---|
| Faba bean flour | FB | $1.78 \pm 0.01^{b}$ | $119.72 \pm 13.44^{b}$ | $1.48 \pm 0.18^{a}$ | $7.18 \pm 0.32^{b}$ | $1650.74 \pm 0.01^{b}$ | $28.06 \pm 1.08^{b}$ |
| Faba bean protein concentrate | FBPC | $2.79 \pm 0.14^{a}$ | $394.37 \pm 28.27^{a}$ | $1.74 \pm 0.05^{a}$ | $15.58 \pm 0.55^{a}$ | $3271.86 \pm 0.19^{a}$ | $41.50 \pm 0.07^{a}$ |
| Hydrolysates (FBH) | HC | $0.23 \pm 0.01^{cd}$ | $37.65 \pm 0.92^{c}$ | $0.09 \pm 0.01^{g}$ | $0.86 \pm 0.05^{a}$ | $249.11 \pm 3.93^{b}$ | $5.39 \pm 0.61^{a}$ |
| | $HA_0$ | $0.23 \pm 0.01^{d}$ | $50.33 \pm 3.98^{b}$ | $0.09 \pm 0.01^{fg}$ | $0.61 \pm 0.02^{bc}$ | $242.01 \pm 9.87^{b}$ | $4.90 \pm 0.53^{ab}$ |
| | $HA_5$ | $0.25 \pm 0.01^{bcd}$ | $63.90 \pm 1.13^{a}$ | $0.14 \pm 0.02^{efg}$ | $0.65 \pm 0.01^{bc}$ | $146.46 \pm 2.90^{de}$ | $3.72 \pm 0.25^{c}$ |
| | $HA_{10}$ | $0.24 \pm 0.02^{cd}$ | $13.71 \pm 0.22^{ef}$ | $0.17 \pm 0.02^{cde}$ | $0.64 \pm 0.01^{bc}$ | $152.61 \pm 2.63^{d}$ | $3.92 \pm 0.49^{bc}$ |
| | $HA_{15}$ | $0.24 \pm 0.01^{cd}$ | $15.45 \pm 0.41^{ef}$ | $0.15 \pm 0.02^{def}$ | $0.51 \pm 0.01^{d}$ | $269.37 \pm 2.75^{a}$ | $4.32 \pm 0.25^{abc}$ |
| | $HA_{30}$ | $0.29 \pm 0.01^{a}$ | $38.69 \pm 5.46^{c}$ | $0.23 \pm 0.02^{b}$ | $0.89 \pm 0.08^{a}$ | $197.51 \pm 8.09^{c}$ | $5.36 \pm 0.61^{a}$ |
| Emulsions (FBE) | EC | $0.23 \pm 0.02^{d}$ | $30.56 \pm 4.15^{cd}$ | $0.15 \pm 0.01^{de}$ | $0.69 \pm 0.02^{b}$ | $137.63 \pm 3.64^{ef}$ | $4.60 \pm 0.37^{abc}$ |
| | $EA_0$ | $0.22 \pm 0.00^{d}$ | $19.34 \pm 3.91^{e}$ | $0.21 \pm 0.01^{bc}$ | $0.65 \pm 0.01^{bc}$ | $143.77 \pm 1.88^{def}$ | $4.48 \pm 0.12^{abc}$ |
| | $EA_5$ | $0.25 \pm 0.01^{bcd}$ | $35.12 \pm 4.73^{c}$ | $0.20 \pm 0.01^{bcd}$ | $0.59 \pm 0.01^{cd}$ | $138.67 \pm 2.50^{ef}$ | $4.58 \pm 0.15^{abc}$ |
| | $EA_{10}$ | $0.23 \pm 0.02^{cd}$ | $23.52 \pm 2.61^{de}$ | $0.20 \pm 0.01^{bcd}$ | $0.60 \pm 0.01^{cd}$ | $130.35 \pm 1.56^{f}$ | $4.31 \pm 0.26^{abc}$ |
| | $EA_{15}$ | $0.26 \pm 0.01^{abc}$ | $35.02 \pm 3.93^{c}$ | $0.23 \pm 0.02^{b}$ | $0.58 \pm 0.01^{cd}$ | $135.72 \pm 5.90^{ef}$ | $4.39 \pm 0.36^{abc}$ |
| | $EA_{30}$ | $0.28 \pm 0.01^{ab}$ | $39.15 \pm 4.11^{c}$ | $0.29 \pm 0.05^{a}$ | $0.66 \pm 0.03^{bc}$ | $146.55 \pm 2.48^{de}$ | $4.37 \pm 0.24^{abc}$ |

* Values (mean ± SD) illustrated in this table are in triplicates (n = 3). For each assay, different superscript letters indicated the significant differences among the faba bean (FB) and protein concentrate (FBPC); and hydrolysates (FBH), and emulsions (FBE), respectively, within a column at 95% confidence level ($p < 0.05$). C, native FBPC protein; $A_0$, FBPC protein hydrolysates viz., slurry with adjusted pH and temperature treatment without Alcalase; $A_5$, slurry with 5 min hydrolysis with Alcalase; $A_{10}$, slurry with 10 min hydrolysis with Alcalase; $A_{15}$, slurry with 15 min hydrolysis with Alcalase; $A_{30}$, slurry with 30 min hydrolysis with Alcalase; TPC, Total phenolic content; TFC, total flavonoid content; TCT, total condensed tannin; DPPH, 2,2′-diphenyl-1-picrylhydrazyl; FRAP, ferric reducing antioxidant power; ABTS, 2,2′-azinobis-3-ethylbenzo-thiazoline-6-sulfonic acid; GAE, gallic acid equivalents; QE, quercetin equivalents; CE, catechin equivalents; TE, Trolox equivalents.

*3.2. Effects of Enzymatic Hydrolysis on the Phenolic Contents and Antioxidant Properties*

The phenolic estimations and antioxidant potential determinations of all the hydrolysates are shown in Table 1. Total phenolic content and total condensed tannin were significantly enhanced after enzymatic hydrolysis for 30 min. Puspita et al. [30] also reported an increase in the total phenolic content of Alcalase extracts of Japanese wireweed (*Sargassum muticum*). Similar improved phenolic contents in the enzymatic extracts of various fruits (kiwi, pear, green apple, raspberry, blackberry, strawberry and blueberry) and vegetables (pumpkin, green and red pepper) were also observed by Álvarez et al. [31]. At the same time, total flavonoids slightly increased during the first 5 min of hydrolysis but were reduced after longer periods of hydrolysis. This reduction is likely due to anthocyanins, common flavonoidsin faba beans, being highly unstable under alkaline conditions of the Alcalase treatment, and they are decolourised within a short time due to hydration at the 2-position of the anthocyanidin skeleton [32,33]. Ferric reducing antioxidant power of $HA_{15}$, the one subject to hydrolysis for 15 min, was the highest, up to 269.37 µg TE/g, significantly larger than that of the non-hydrolysed one ($HA_0$). Interestingly, after 15 min hydrolysis, the FRAP value sharply declined to 197.51 µg TE/g. The increase in FRAP may be ascribed to the breakdown of the peptide bonds and consequent higher hydrogen ion availability, which helps proton donation occur at particular side-chain groups [34]. Shahi et al. [35] added that the enhanced ferric reducing power could result from the cleaved peptide chains releasing amino acids like lysine and tryptophan with antioxidant activities. The strongest DPPH radical scavenging ability was observed in $HA_{30}$, treated with Alcalase for 30 min, up to 0.89 mg TE/g, which was significantly differed from other hydrolysates. Li et al. [36] found that the Alcalase hydrolysates of *Camellia oleifera* seed cake protein showed greater DPPH than papain and trypsin hydrolysates. Similarly, $HA_{30}$ also represented the highest ABTS value (5.36 mg TE/g) but was not significantly differed from the non-hydrolysed one. At the same time, Ali [2] observed a significant increase in the ABTS values of faba hydrolysates after treating with pepsin for 180 min.

Sbroggio, Montilha, FIGUEIREDO, Georgetti and Kurozawa [34] mentioned that the differences between DPPH and ABTS free radical scavenging modes might be related to the peptides produced having different structures. Extensive hydrolysis could result in forming shorter peptides, including dipeptides and tripeptides, and free amino acids, and therefore peptides got higher hydrophilic properties, easier to approach the ABTS free radical that is water-soluble [37]. Chen et al. [38] also confirmed that soybean peptides, due to their hydrophilicity, have difficulties in having interactions with hydrophobic peroxy radicals like DPPH. This could be a reason why the ABTS value was much higher than that of DPPH. Besides, Samaei et al. [39] added that the ABTS assay has higher sensitivity than the DPPH method since the ABTS radical has preferential interactions with the hydroxylated aromatic compounds within the sequences of peptides, which is associated with the protein composition and hydrophobic properties of the hydrolysates, leading to the peptides of hydrolysates having differences in scavenging ABTS and DPPH radicals.

The effect of hydrolysis time on the DPPH was correlated with degree of hydrolysis as shown in SDS-PAGE (Figure 2). According to SDS-PAGE patterns of hydrolysates viz., $A_5$ to $A_{30}$ showed that the Alcalase effectively cleaved proteins in hydrolysates and size was significantly reduced with the increase in reaction time. During the initial hydrolysis stage, both enzyme activity and substrate concentration were comparatively higher, resulting in the breakdown of peptide bonds and proteolysis occurring at higher rates [13]. This may lead to the initial increase in DPPH free radical scavenging capacity, and then further interactions of peptides or functional groups of amino acids may reduce DPPH values. Vasconcellos et al. [40] pointed out that glycinin peptides displayed higher antioxidant effects than conglycinin peptides emphasising the importance of protein structure and amino acid composition at constant degree of hydrolysis. At the end of the hydrolysis, $A_{30}$ resulted in bands under 20 kDa corresponding to glycinin, and that may cause the improved DPPH at $A_{30}$.

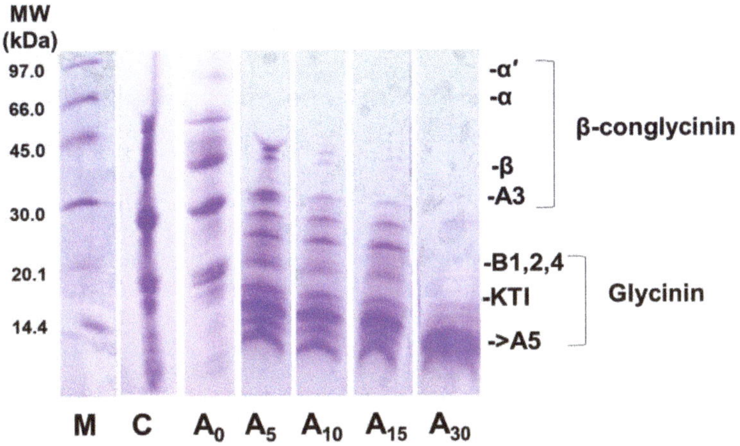


**Figure 2.** SDS-PAGE profiles of native FBPC protein (C), and FBPC protein hydrolysates viz., slurry with adjusted pH and temperature treatment without Alcalase ($A_0$), 5 min hydrolysis with Alcalase ($A_5$), 10 min hydrolysis with Alcalase ($A_{10}$), 15 min hydrolysis with Alcalase ($A_{15}$), and 30 min hydrolysis with Alcalase ($A_{30}$). (M) molecular weight markers.

The results shown in Table 1 revealed that the enzymatic hydrolysis using Alcalase for 30 min is the most effective one, which significantly increased the total phenolic and condensed tannin content and retained or improved DPPH and ABTS radical scavenging capacities but decreased the total flavonoid content and FRAP. The effects of enzymatic hydrolysis were also greatly influenced by the hydrolysis treatment time. Generally, prolonging the hydrolysis time could increase the phenolic content and antioxidant activities of all the hydrolysates to some extent.

### 3.3. *Effects of UHT Processing on the Phenolic Content and Antioxidant Properties of Faba Bean Concentrate*

The phenolic estimations and antioxidant activities of $HA_0$, the slurry of FBPC with adjusted pH and temperature treatment without Alcalase treatment, and $EA_0$, the hydrolysate $A_0$ with UHT treatment, are shown in Table 1. By comparing $HA_0$ and emulsion $EA_0$, no significant difference was observed in total phenolic content. However, Xu and Chang [41] found reduced TPC values in yellow and black soybean milk after UHT processing, and Dai et al. [42] reported that the TPC values of UHT-treated strawberry samples were significantly lower than the raw ones. Meanwhile, the total flavonoid content of $EA_0$ was noticeably lower than that of $HA_0$. Anthocyanins would degrade into various degradation products due to their thermal instability, particularly in a neutral environment [32]. That might be a reason why TFC declined after UHT treatment. While the improved TFC value of UHT-treated yellow soybean milk was attributed to the releasement of free phenolic compounds from lignin within cell walls [43]. It was observed that $HA_0$ exhibited a significantly higher total condensed tannin content (0.21 mg CE/g) compared with $EA_0$ (0.09 mg CE/g), which indicated UHT greatly decreased? the content of condensed tannins.

In terms of DPPH and ABTS values, $HA_0$ and $EA_0$ did not significantly differ, revealing that UHT processing retained both DPPH and ABTS radical scavenging abilities. Xu and Chang [41] reported that the DPPH value of UHT soymilk produced from soybeans of Proto variety was considerably higher than the raw soymilk. Besides, heat treatments like cooking and pasteurization may lead to the breakdown of large polymerized structural substances present in the cell wall and the releasement of antioxidant compounds, increasing the antioxidant activity [44]. The FRAP value noticeably decreased from 0.24 μg TE/g ($HA_0$) to 0.14 μg TE/g ($EA_0$) after UHT processing, resulting from losing natural antioxidants during the thermal processing. However, Nicoli et al. [45] stated that heat treatments might induce compounds having new antioxidant characteristics to form. Dias et al. [46] added that thermal processing would not only result in phenolic compound degradation but also unfold and form heat-induced antioxidants like melanoidins that have great antioxidant potential. These studies agree with the retained DPPH and ABTS values after the UHT processing in this research.

### *3.4. Effects of Enzymatic Hydrolysis Combined with UHT Processing on the Phenolic Contents and Antioxidant Properties*

The results shown in Table 1 indicated that the enzymatic hydrolysis using Alcalase for 30 min combined with UHT processing is the most effective one. There was a considerable increase in total phenolic content after this combination processing. The total flavonoid content of the emulsion treated with 30 min hydrolysis and UHT processing showed a significant improvement. Similarly, total condensed tannin noticeably increased to 0.29 mg CE/g after the combination treatment.

Meanwhile, there was no significant difference in both DPPH and ABTS radical scavenging abilities observed after enzymatic hydrolysis for 30 min following the UHT processing. Likewise, ferric reducing power did not show any remarkable change after the combination processing. These indicated that combing enzymatic hydrolysis using Alcalase for 30 min with UHT treatment could retain all the antioxidant activities, while a significant increase in the DPPH values of soy protein isolate hydrolysates, treated with a combination of high-temperature treatment under 121 °C for 10 min and enzymatic hydrolysis using Protamex, was found by Yoo and Chang [47]. Similarly, Voss et al. [48] reported that the okara hydrolysates produced by autoclave processing at 121 °C for 15 min and Alcalase hydrolysis showed a great increase in ABTS value.

### *3.5. Molecular Weight Distribution and Physicochemical Properties of Hydrolysates*

Partial enzymatic hydrolysis caused significant changes in the structure of FBPC revealed by SDS-PAGE profiles (Figure 2). C (native FBPC) exhibited four major bands with molecular weights of ~50, ~37, ~30, and ~21 kDa. These bands correspond to the $\alpha$-subunits, $\beta$-subunits and their intermediate subunits of 11S (legumin-like) globulins [49]. The pattern $A_0$ showed was similar to C but much clearer, revealing that temperature without enzyme had no effect on protein patterns. SDS-PAGE patterns of FBPC hydrolysates viz., $A_5$ to $A_{30}$ showed that the Alcalase effectively cleaved proteins in FBPC and size was significantly reduced with the increase in reaction time. $A_5$ to $A_{15}$ remarkably reduced the four distinctive bands and was accompanied by the increased occurrence of protein bands whose molecular weight was less than 20 kDa. Further hydrolysis viz., $A_{30}$ resulted in bands under 15 kDa. SDS-PAGE results agree with the degree of hydrolysis (Table 2). OPA spectrophotometric assay showed increased amounts of $\alpha$-amino groups with the increased reaction time of Alcalase resulting in ~1% for $A_5$, ~2% for $A_{10}$, ~9% for $A_{15}$, and ~16% for $A_{30}$.

The physicochemical properties of hydrolysates were also assessed (Table 2). Results showed that prolonging the Alcalase hydrolysis periods increased the solubility of FBPC by 7–23%, with the highest solubility shown in $A_{30}$ at pH 7.0. This improved protein solubility might be attributed to the smaller peptides produced by prolonged hydrolysis. These smaller peptides possibly became more soluble due to their formation of stronger hydrogen bonds with water [50]. The electronegativity of FBPC was also enhanced by prolonged hydrolysis at pH 7.0, possibly due to smaller peptides exposed and increased numbers of ionizable amino acids [51]. Besides, Akbari et al. [52] pointed out that enzymatic hydrolysis could dissociate carboxylic groups and release carboxylate ions ($COO^-$), resulting in increased electronegativity. In addition, the net negative charge increased with the increase of hydrolysis, which in turn improved protein solubility because of the larger intermolecular repulsive electrostatic forces [5]. Interestingly, the increase in Alcalase hydrolysis also led to a noticeable decrease in surface hydrophobicity ($S_0$). The hydrophobic areas were enzymatic cleaved, and the hydrophobic residues exposed were refolded, which might be why $S_0$ was reduced [53]. Overall, enzymatic hydrolysis by Alcalase with increased time generally resulted in hydrolysates having better functional properties.

**Table 2.** Physicochemical properties (Degree of hydrolysis ~ DH %, Protein Solubility (%), ζ-potential, Hydrophobicity index ~ $S_0$, Droplet size ~ $D^{3,2}$, and Polydispersity index ~ PDI) of faba bean hydrolysates (FBH) and faba bean emulsions (FBE) *.

| Treatment | DH % | Protein Solubility (%) | ζ-Potential (mV) | $S_0$ | $D^{3,2}$ (μm) | PDI (%) |
|---|---|---|---|---|---|---|
| $HA_0$ | 0 | 60.95 ± 0.90 [a] | −30.93 ± 0.45 [a] | 236.78 ± 3.00 [a] | - | - |
| $HA_5$ | 1 | 67.78 ± 1.57 [b] | −34.67 ± 0.41 [b] | 205.91 ± 7.46 [ab] | - | - |
| $HA_{10}$ | 2 | 72.06 ± 0.90 [c] | −37.53 ± 0.46 [c] | 193.32 ± 2.83 [ab] | - | - |
| $HA_{15}$ | 9 | 77.14 ± 0.90 [d] | −39.67 ± 0.36 [d] | 167.23 ± 7.10 [ab] | - | - |
| $HA_{30}$ | 16 | 83.65 ± 0.67 [e] | −43.75 ± 0.47 [e] | 122.64 ± 3.52 [b] | - | - |
| EC | - | - | −23.10 ± 0.96 [a] | 173.07 ± 1.65 [b] | 14.10 ± 0.46 [a] | 41.55 ± 3.99 [a] |
| $EA_0$ | - | - | −21.47 ± 1.10 [a] | 244.59 ± 13.80 [d] | 14.73 ± 0.35 [a] | 30.16 ± 2.17 [b] |
| $EA_5$ | - | - | −28.45 ± 0.39 [b] | 167.12 ± 5.30 [b] | 0.36 ± 0.06 [b] | 36.99 ± 2.48 [a] |
| $EA_{10}$ | - | - | −30.65 ± 0.34 [c] | 147.90 ± 2.26 [a] | 0.14 ± 0.01 [b] | 17.73 ± 0.78 [c] |
| $EA_{15}$ | - | - | −33.56 ± 0.40 [d] | 145.71 ± 1.94 [a] | 0.10 ± 0.01 [b] | 16.31 ± 0.79 [c] |
| $EA_{30}$ | - | - | −37.60 ± 0.25 [e] | 135.55 ± 2.69 [a] | 0.10 ± 0.01 [b] | 14.31 ± 0.53 [c] |

* Values (mean ± SD) illustrated in this table are in triplicates (n = 3). For each parameter, different superscript letters indicated the significant differences among FBPC protein hydrolysates viz., slurry with adjusted pH and temperature treatment without Alcalase ($HA_0$), 5 min hydrolysis with Alcalase ($HA_5$), 10 min hydrolysis with Alcalase ($HA_{10}$), 15 min hydrolysis with Alcalase ($HA_{15}$), and 30 min hydrolysis with Alcalase ($HA_{30}$), and native FBPC protein (EC), FBPC protein hydrolysates viz., slurry with adjusted pH and temperature treatment without Alcalase after UHT treatment ($EA_0$), 5 min hydrolysis with Alcalase after UHT treatment ($EA_5$), 10 min hydrolysis with Alcalase after UHT treatment ($EA_{10}$), 15 min hydrolysis with Alcalase after UHT treatment ($EA_{15}$), and 30 min hydrolysis with Alcalase after UHT treatment ($EA_{30}$) within a column at 95% confidence level ($p < 0.05$).

### 3.6. Physicochemical Properties of Emulsions

The physicochemical properties of emulsions were assessed (Table 2). The droplet size of $EA_0$ was larger than that of EC, which indicated that UHT processing could increase the particle size. [54] stated that UHT treatment at 135 °C would cause heat-induced protein aggregations on the droplet surface, leading to increased particle size. Other globules like proteins also denature 135 °C. Therefore, in addition to the fat globules, the particle size analysis may include several protein aggregates and protein-fat aggregates as well, resulting in increased particle size [55]. However, the particle size showed a significant reduction after UHT operation with increased Alcalase hydrolysis ($EA_5$ to $EA_{30}$). This may be attributed to the smaller peptides with greater emulsifying properties produced by prolonged hydrolysis. With the increase in Alcalase hydrolysis, peptides would have higher electronegativity, protein solubility and lower surface hydrophobicity which helps avoid the aggregation and facilitate small particles to form [16]. Besides, the polydispersity index [56], non-uniformity degree of distribution, was also calculated, and similar trends were observed in PDI, prolonging the Alcalase hydrolysis decreased PDI values.

ζ-potential is measured for predicting the stability of the emulsion. Generally, emulsions with low electronegativity are unstable since storage easily makes them flocculate or coagulate. In contrast, emulsions with higher electronegativity are relatively stable owing to their less attractive forces compared to repulsive forces [57]. The ζ-potential value of $EA_0$ (−21.47 mV) was less negative than that of EC (−23.10 mV), which revealed that UHT processing led to lower electronegativity. Protein aggregates would form primarily at high temperatures, leading to decreased protein solubility, which may account for lower electronegativity [58]. With the increase in Alcalase hydrolysis, the number of smaller peptides with more negative charge increased and that might result in the improvement in electronegativity ($EA_5$ to $EA_{30}$). The enhanced surface hydrophobicity after UHT processing would also be attributed to the aggregated protein since attractive hydrophobic interactions occur between them, forming the oil droplets coated by protein [57]. Prolonging the hydrolysis caused a significant decrease in the hydrophobicity of UHT treated emulsions, which is consistent with Zang et al. [53]. Overall, emulsions treated with Alcalase hydrolysis were stable after UHT treatment.

### 3.7. Phenolic Estimations across In Vitro Digestion

The phenolic estimations of FBH and FBE across three digestion phases are shown in Table 3. FBE showed no significant difference with FBH in total phenolic content after in vitro digestion. The TPC values of all the samples significantly increased across in vitro digestion. This observation is also consistent with the study of Ribeiro et al. [59] that the total phenolic content of juçara-based smoothies treated with pasteurization and sonication both had a noticeable increase after intestinal digestion. However, Ma et al. [60] reported an increase of 1.64% and a reduction of 19.97% in the TPC values of digested bamboo leaves soup at gastric and intestinal phases, respectively.

**Table 3.** Phenolic estimations of faba bean hydrolysates (FBH) and emulsions (FBE) across in vitro digestion *.

| Sample Types | Samples | Phases | TPC (mg GAE/g) | TFC (mg QE/g) | TCT (mg CE/g) | DPPH (mg TE/g) | FRAP (mg TE/g) | ABTS (mg TE/g) |
|---|---|---|---|---|---|---|---|---|
| Hydrolysates (FBH) | $HA_0$ | Oral | $0.53 \pm 0.03^{c}$ | $0.02 \pm 0.00^{a}$ | $0.41 \pm 0.03^{a}$ | $0.29 \pm 0.02^{b}$ | $0.43 \pm 0.04^{b}$ | $5.12 \pm 0.64^{b}$ |
| | | Gastric | $0.59 \pm 0.03^{b}$ | - | - | $0.45 \pm 0.06^{a}$ | $0.32 \pm 0.03^{c}$ | $7.61 \pm 0.31^{b}$ |
| | | Intestinal | $1.21 \pm 0.15^{a}$ | - | - | $0.13 \pm 0.01^{c}$ | $0.77 \pm 0.02^{a}$ | $66.36 \pm 3.82^{a}$ |
| | $HA_5$ | Oral | $0.40 \pm 0.01^{b}$ | - | - | $0.16 \pm 0.02^{b}$ | $0.41 \pm 0.01^{a}$ | $5.94 \pm 0.21^{c}$ |
| | | Gastric | $0.63 \pm 0.02^{a}$ | - | - | $0.48 \pm 0.08^{a}$ | $0.10 \pm 0.01^{c}$ | $7.90 \pm 0.12^{b}$ |
| | | Intestinal | $0.39 \pm 0.04^{b}$ | - | - | $0.07 \pm 0.01^{c}$ | $0.24 \pm 0.00^{b}$ | $22.46 \pm 0.39^{a}$ |
| | $HA_{10}$ | Oral | $0.46 \pm 0.04^{c}$ | - | $1.16 \pm 0.05^{b}$ | $0.11 \pm 0.00^{b}$ | $0.53 \pm 0.03^{a}$ | $10.52 \pm 0.07^{b}$ |
| | | Gastric | $0.61 \pm 0.03^{b}$ | - | $1.29 \pm 0.02^{a}$ | $0.39 \pm 0.05^{a}$ | $0.10 \pm 0.01^{c}$ | $8.22 \pm 0.68^{c}$ |
| | | Intestinal | $0.92 \pm 0.14^{a}$ | - | - | - | $0.50 \pm 0.01^{a}$ | $27.72 \pm 0.30^{a}$ |
| | $HA_{30}$ | Oral | $0.53 \pm 0.03^{c}$ | - | $0.54 \pm 0.08^{a}$ | $0.37 \pm 0.01^{b}$ | $1.10 \pm 0.05^{a}$ | $12.25 \pm 0.09^{b}$ |
| | | Gastric | $0.79 \pm 0.02^{b}$ | - | - | $0.73 \pm 0.06^{a}$ | $0.45 \pm 0.05^{c}$ | $9.62 \pm 0.15^{b}$ |
| | | Intestinal | $1.16 \pm 0.11^{a}$ | - | - | - | $0.58 \pm 0.01^{b}$ | $79.79 \pm 0.95^{a}$ |
| Emulsions (FBE) | $EA_0$ | Oral | $0.49 \pm 0.03^{c}$ | $0.04 \pm 0.00^{b}$ | $2.04 \pm 0.02^{a}$ | $0.28 \pm 0.02^{b}$ | $1.04 \pm 0.04^{b}$ | $12.79 \pm 0.41^{b}$ |
| | | Gastric | $0.69 \pm 0.02^{b}$ | - | - | $0.42 \pm 0.05^{a}$ | $0.90 \pm 0.00^{c}$ | $9.10 \pm 0.15^{b}$ |
| | | Intestinal | $1.15 \pm 0.13^{a}$ | $1.40 \pm 0.23^{a}$ | - | $0.19 \pm 0.01^{c}$ | $1.35 \pm 0.01^{a}$ | $116.13 \pm 8.91^{a}$ |
| | $EA_5$ | Oral | $0.51 \pm 0.02^{c}$ | - | - | $0.24 \pm 0.02^{a}$ | $0.99 \pm 0.04^{b}$ | $13.95 \pm 0.43^{b}$ |
| | | Gastric | $0.80 \pm 0.05^{b}$ | - | - | $0.27 \pm 0.03^{a}$ | $0.96 \pm 0.07^{b}$ | $9.94 \pm 0.53^{c}$ |
| | | Intestinal | $0.65 \pm 0.09^{a}$ | $0.80 \pm 0.11^{a}$ | - | $0.07 \pm 0.01^{b}$ | $1.10 \pm 0.02^{a}$ | $22.06 \pm 0.88^{a}$ |
| | $EA_{10}$ | Oral | $0.50 \pm 0.04^{c}$ | - | $2.28 \pm 0.02^{a}$ | $0.29 \pm 0.00^{b}$ | $1.11 \pm 0.02^{b}$ | $14.80 \pm 0.76^{b}$ |
| | | Gastric | $0.79 \pm 0.01^{b}$ | - | - | $0.40 \pm 0.02^{a}$ | $0.93 \pm 0.04^{c}$ | $9.47 \pm 0.43^{b}$ |
| | | Intestinal | $1.17 \pm 0.08^{a}$ | $2.61 \pm 0.10^{a}$ | - | $0.14 \pm 0.01^{c}$ | $2.06 \pm 0.09^{a}$ | $78.66 \pm 6.38^{a}$ |
| | $EA_{30}$ | Oral | $0.45 \pm 0.04^{c}$ | - | - | $0.26 \pm 0.02^{b}$ | $1.00 \pm 0.03^{b}$ | $14.98 \pm 0.86^{b}$ |
| | | Gastric | $0.86 \pm 0.05^{b}$ | - | - | $0.43 \pm 0.06^{a}$ | $1.01 \pm 0.05^{b}$ | $10.00 \pm 0.27^{b}$ |
| | | Intestinal | $1.17 \pm 0.07^{a}$ | $0.99 \pm 0.14^{a}$ | - | $0.15 \pm 0.01^{c}$ | $1.22 \pm 0.05^{a}$ | $67.46 \pm 1.44^{a}$ |

* Values (mean ± SD) illustrated in this table are in triplicates (n = 3). The data of all the samples have been subtracted the control values. For each sample of each assay, different superscript letters indicated the significant differences among oral, gastric and intestinal phases within three rows of a column at a 95% confidence level ($p < 0.05$). FBH, faba bean hydrolysates; FBE, faba bean emulsions; $A_0$, FBPC protein hydrolysates viz., slurry with adjusted pH and temperature treatment without Alcalase; $A_5$, slurry with 5 min hydrolysis with Alcalase; $A_{10}$, slurry with 10 min hydrolysis with Alcalase; $A_{15}$, slurry with 15 min hydrolysis with Alcalase; $A_{30}$, slurry with 30 min hydrolysis with Alcalase; TPC, Total phenolic content; TFC, total flavonoid content; TCT, total condensed tannin; GAE, gallic acid equivalents; QE, quercetin equivalents; CE, catechin equivalents. DPPH, 2,2′-diphenyl-1-picrylhydrazyl; FRAP, ferric reducing antioxidant power; ABTS, 2,2′-azinobis-3-ethylbenzo-thiazoline-6-sulfonic acid; TE, Trolox equivalent.

The enzymes in the intestinal phase would have actions on the food residual substrate, which helps phenolics release and thus enhance the total phenolic content [61]. Besides, gallic or *p*-coumaric acid and quercetin present in faba beans may increase during digestion since they may convert into other compounds [62]. Zeng et al. [63] added that the phenolics in faba beans are primarily present in their covalent bound form, which may be why they successfully survive the gastric and intestinal digestion to arrive at the colon.

In terms of total flavonoid content, FBE exhibited great total flavonoid contents after intestinal digestion, ranging from 0.80 to 2.61 (mg QE/g) but with almost no detection in the other two phases. Flavonoids are marginally more sensitive to gastric digestion as there were losses observed in all the anthocyanins and the majority of flavonols during the pepsin treatment, particularly higher related to the anthocyanins [64]. This may be responsible for no detection of flavonoids in the gastric phase for all FBE and FBH. On the contrary, Maduwanthi and Marapana [65] found that the gastric digestion stage displayed significantly higher total flavonoid content in bananas (*Musa acuminata*, AAB) treated with ripening agents ethephon and acetylene than the intestinal phase did. The step-wise release of total flavonoids in digests of four different apple varieties was shown from the gastric phase to the intestinal phase in the study of Bouayed et al. [66].

Regarding TCT, the total condensed tannin of some FBH and FBE sharply decreased after oral digestion and even no detection after gastrointestinal digestion. The interactions between tannins and pancreatic enzymes along with the slightly alkaline pH at the intestinal digestion could result in the degradation of tannins [67,68]. Besides, Wojtunik-Kulesza et al. [61] demonstrated that tannins as high molecular weight phenolic compounds could have strong interaction with proteins and get precipitated by the hydrogen bonding and hydrophobic effect. The total condensed tannin of this research was analysed through the filtered supernatant of the digests, which would also be responsible for no detection of TCT for most samples after gastrointestinal digestion.

### 3.8. Antioxidant Activities across the Stages of In Vitro Digestion

The antioxidant activities of FBH and FBE across three digestion phases are shown in Table 3. FBH and FBE performed the highest DPPH in the gastric phase and remarkably decreased after intestinal digestion. It is agreed with the previous studies [59,60,66]. Generally, phenolic compounds like quercetin and caffeic acid are highly stable under an acidic pH environment, like the pH of the gastric phase [69]. However, the intestinal condition with slightly alkaline pH would induce chemical conversions, forming quinone intermediates and other oxidizing compounds which were unstable [70]. Additionally, impair the structures of aglycones, changing the chemical characteristics of antioxidants and thus decreasing their antioxidant potential [71]. Chen et al. [72] also mentioned that the alkaline pH of the pancreatic digestion stage would degrade the antioxidant compounds and thereby causing a reduction in antioxidant activities.

The gastric phase of both FBH and FBE exhibited the lowest FRAP values among the three phases, and this might result from the peptides released in the gastric digestion having limited biological activity [73]. Besides, FBE showed significant higher ferric reducing power compared with FBH across in vitro digestion. This could be attributed to the generation of a brown polymer called melanoidins through the Maillard Reaction and other antioxidative products when emulsions are subjected to UHT treatment at 135 °C [15]. Additionally, the ferric reducing power of FBH and FBE was highest in the intestinal phase. It is consistent with the findings of Chen et al. [74] that five varieties of sesame seeds owned the strongest ferric reducing power at the intestinal phase. However, the DPPH values of bamboo leave soup were comparatively higher at the gastric phase and decreased after intestinal digestion [60].

ABTS values of all the samples significantly improved after gastrointestinal digestion. The ABTS radical scavenging ability was primarily released at the intestinal digestion stage. Changing the environment from acidic to alkaline would result in the deprotonation of the hydroxyl groups on the phenolic aromatic rings and thus improve its antioxidant potential [75]. This supports the study of Bouayed, Hoffmann and Bohn [66] that aglycone form exhibits stronger antioxidant activities than that of the glycoside. The bioactive peptides originating from pancreatin hydrolysis with antioxidant characteristics would also contribute to the improved ABTS radical scavenging ability at the intestinal phase [71]. The increasing tendency across in vitro digestion is agreed by Chen, Lin, Lin, Zheng and Chen [74] and Chotphruethipong et al. [76]. However, a considerable reduction was observed in ABTS values of the red grape pomace after intestinal digestion.

### 3.9. Phenolic Estimations across Colonic Fermentation

The variations of total phenolic compounds, flavonoids, and tannins content in faba bean hydrolysates (FBH) and faba bean emulsions (FBE) during colonic fermentation are displayed in Table 4. It was noted that, during colonic fermentation, total phenolic compounds significantly increased at the first 2 h, except for $HA_{30}$. Nearly all the TPC values at 8 h of fermentation were the highest among the FBH and FBE digesta, while $HA_{10}$ showed the highest TPC value at 4.67 mg GAE/g. However, for $EA_{30}$, it gained its peak at 2 h (3.80 mg GAE/g). Besides, TPC values significantly decreased after 8 h and then got the lowest levels at 24 h of fecal reaction. In terms of the total flavonoids, all FBH and FBE samples illustrated nearly no flavonoids at 0 h. TFC values gradually increased after 2 h of fermentation and significantly dropped after 4 h, except for $EA_5$. The emulsions with 5 min enzymatic hydrolysis significantly increased from 4 h, peaked at 8 h (2.61 mg QE/g), and later reduced and gained no detection at 16 h fermentation. As for the total tannin content, $HA_5$ was detected with the highest TCT value at 0 h (5.44 mg CE/g), but most FBH and FBE were found to have nearly 0 tannins within the first 8 h of fermentation. All FBH and FBE significantly rose after 8 h, especially for the FBE digesta, getting the highest values at 24 h of fecal reaction.

**Table 4.** Phenolic estimations of faba bean (FB), protein concentrate (FBPC), hydrolysates (FBH) and emulsions (FBE) across colonic fermentation *.

| Sample Types | Samples | Phases | TPC (mg GAE/g) | TFC (mg QE/g) | TCT (mg CE/g) | DPPH (mg TE/g) | FRAP (mg TE/g) | ABTS (mg TE/g) |
|---|---|---|---|---|---|---|---|---|
| Hydrolysates (FBH) | $HA_0$ | 0 h | $3.39 \pm 0.04^{b}$ | - | - | $2.36 \pm 0.01^{f}$ | $2.91 \pm 0.07^{d}$ | $26.55 \pm 1.37^{d}$ |
| | | 2 h | $3.44 \pm 0.03^{b}$ | $0.14 \pm 0.00^{c}$ | - | $3.88 \pm 0.00^{d}$ | $3.99 \pm 0.02^{b}$ | $16.15 \pm 0.01^{f}$ |
| | | 4 h | $2.40 \pm 0.07^{e}$ | $1.06 \pm 0.08^{a}$ | - | $3.45 \pm 0.03^{e}$ | $4.14 \pm 0.00^{a}$ | $33.89 \pm 0.12^{c}$ |
| | | 8 h | $3.63 \pm 0.06^{a}$ | $0.45 \pm 0.01^{b}$ | - | $5.78 \pm 0.03^{c}$ | $2.77 \pm 0.00^{e}$ | $19.25 \pm 0.00^{e}$ |
| | | 16 h | $2.80 \pm 0.00^{c}$ | - | - | $6.40 \pm 0.01^{b}$ | $3.76 \pm 0.00^{c}$ | $92.52 \pm 0.06^{b}$ |
| | | 24 h | $2.52 \pm 0.01^{d}$ | - | - | $6.60 \pm 0.03^{a}$ | $4.09 \pm 0.03^{a}$ | $116.94 \pm 0.05^{a}$ |
| | $HA_5$ | 0 h | $3.17 \pm 0.00^{c}$ | $0.03 \pm 0.01^{d}$ | $5.44 \pm 0.13^{a}$ | $1.13 \pm 0.02^{f}$ | $3.30 \pm 0.08^{d}$ | $33.98 \pm 0.35^{d}$ |
| | | 2 h | $3.64 \pm 0.01^{b}$ | $0.17 \pm 0.00^{c}$ | - | $4.32 \pm 0.00^{d}$ | $3.45 \pm 0.00^{c}$ | $16.18 \pm 0.01^{f}$ |
| | | 4 h | $2.94 \pm 0.04^{d}$ | $0.26 \pm 0.01^{b}$ | - | $3.66 \pm 0.00^{e}$ | $4.61 \pm 0.01^{a}$ | $36.17 \pm 0.06^{c}$ |
| | | 8 h | $4.07 \pm 0.03^{a}$ | $0.60 \pm 0.02^{a}$ | - | $6.08 \pm 0.01^{c}$ | $1.30 \pm 0.00^{f}$ | $19.35 \pm 0.00^{e}$ |
| | | 16 h | $2.52 \pm 0.02^{e}$ | - | - | $7.88 \pm 0.26^{b}$ | $3.05 \pm 0.00^{e}$ | $92.38 \pm 0.03^{b}$ |
| | | 24 h | $2.00 \pm 0.02^{f}$ | - | - | $8.48 \pm 0.09^{a}$ | $3.64 \pm 0.01^{b}$ | $116.72 \pm 0.06^{a}$ |
| | $HA_{10}$ | 0 h | $3.15 \pm 0.01^{c}$ | - | - | $4.27 \pm 0.09^{d}$ | $2.10 \pm 0.00^{f}$ | $23.59 \pm 0.37^{d}$ |
| | | 2 h | $3.93 \pm 0.00^{b}$ | $0.36 \pm 0.00^{c}$ | - | $5.14 \pm 0.01^{c}$ | $4.53 \pm 0.01^{b}$ | $15.92 \pm 0.03^{f}$ |
| | | 4 h | $2.81 \pm 0.04^{e}$ | $0.93 \pm 0.00^{b}$ | - | $3.64 \pm 0.09^{e}$ | $4.79 \pm 0.00^{a}$ | $34.52 \pm 0.22^{c}$ |
| | | 8 h | $4.67 \pm 0.02^{a}$ | $1.02 \pm 0.00^{a}$ | - | $7.43 \pm 0.05^{b}$ | $3.39 \pm 0.00^{c}$ | $19.16 \pm 0.00^{e}$ |
| | | 16 h | $2.95 \pm 0.05^{d}$ | - | $0.91 \pm 0.02^{b}$ | $7.61 \pm 0.10^{a}$ | $3.11 \pm 0.01^{d}$ | $94.60 \pm 0.03^{b}$ |
| | | 24 h | $2.38 \pm 0.02^{f}$ | - | $1.22 \pm 0.03^{a}$ | $7.67 \pm 0.06^{a}$ | $3.02 \pm 0.01^{e}$ | $119.75 \pm 0.05^{a}$ |
| | $HA_{30}$ | 0 h | $3.75 \pm 0.00^{a}$ | $0.20 \pm 0.00^{c}$ | - | $3.11 \pm 0.01^{f}$ | $2.22 \pm 0.00^{e}$ | $56.29 \pm 0.26^{c}$ |
| | | 2 h | $3.17 \pm 0.02^{d}$ | $0.20 \pm 0.00^{c}$ | - | $5.59 \pm 0.00^{c}$ | $4.26 \pm 0.04^{a}$ | $16.89 \pm 0.00^{f}$ |
| | | 4 h | $3.15 \pm 0.03^{d}$ | $0.89 \pm 0.01^{a}$ | - | $3.86 \pm 0.01^{e}$ | $4.13 \pm 0.00^{b}$ | $33.69 \pm 0.10^{d}$ |
| | | 8 h | $3.70 \pm 0.07^{a}$ | $0.29 \pm 0.01^{b}$ | - | $4.33 \pm 0.02^{d}$ | $3.12 \pm 0.00^{c}$ | $18.99 \pm 0.00^{e}$ |
| | | 16 h | $3.43 \pm 0.06^{b}$ | $0.15 \pm 0.02^{d}$ | $0.63 \pm 0.01^{b}$ | $7.19 \pm 0.09^{b}$ | $3.04 \pm 0.04^{d}$ | $97.71 \pm 0.04^{b}$ |
| | | 24 h | $3.33 \pm 0.03^{c}$ | $0.11 \pm 0.01^{e}$ | $0.83 \pm 0.02^{a}$ | $8.14 \pm 0.04^{a}$ | $3.01 \pm 0.02^{d}$ | $123.95 \pm 0.03^{a}$ |
| Emulsions (FBE) | $EA_0$ | 0 h | $3.17 \pm 0.02^{d}$ | - | - | $3.00 \pm 0.03^{f}$ | $3.52 \pm 0.05^{c}$ | $19.65 \pm 0.05^{d}$ |
| | | 2 h | $3.32 \pm 0.03^{c}$ | $0.38 \pm 0.00^{b}$ | - | $4.87 \pm 0.05^{d}$ | $3.72 \pm 0.00^{b}$ | $15.33 \pm 0.03^{f}$ |
| | | 4 h | $2.45 \pm 0.02^{e}$ | $0.96 \pm 0.02^{a}$ | - | $3.77 \pm 0.05^{e}$ | $4.38 \pm 0.00^{a}$ | $50.07 \pm 0.09^{c}$ |
| | | 8 h | $3.85 \pm 0.01^{a}$ | $0.29 \pm 0.00^{c}$ | - | $6.10 \pm 0.00^{c}$ | $2.51 \pm 0.00^{f}$ | $18.69 \pm 0.00^{e}$ |
| | | 16 h | $3.63 \pm 0.04^{b}$ | - | $5.51 \pm 0.00^{b}$ | $6.84 \pm 0.05^{b}$ | $3.15 \pm 0.06^{e}$ | $97.17 \pm 0.07^{b}$ |
| | | 24 h | $3.56 \pm 0.02^{b}$ | - | $7.35 \pm 0.02^{a}$ | $7.08 \pm 0.03^{a}$ | $3.36 \pm 0.04^{d}$ | $123.33 \pm 0.06^{a}$ |
| | $EA_5$ | 0 h | $2.31 \pm 0.01^{f}$ | $0.18 \pm 0.00^{c}$ | - | $3.31 \pm 0.02^{e}$ | $2.56 \pm 0.01^{e}$ | $61.22 \pm 0.01^{c}$ |
| | | 2 h | $3.71 \pm 0.01^{b}$ | $0.26 \pm 0.00^{b}$ | - | $5.01 \pm 0.00^{c}$ | $3.57 \pm 0.00^{b}$ | $15.88 \pm 0.01^{e}$ |
| | | 4 h | $2.83 \pm 0.13^{e}$ | $0.25 \pm 0.02^{b}$ | - | $3.45 \pm 0.04^{de}$ | $4.70 \pm 0.00^{a}$ | $57.03 \pm 0.19^{d}$ |
| | | 8 h | $4.20 \pm 0.01^{a}$ | $2.61 \pm 0.00^{a}$ | - | $3.58 \pm 0.03^{d}$ | $2.39 \pm 0.00^{f}$ | $61.11 \pm 0.01^{c}$ |
| | | 16 h | $3.25 \pm 0.01^{c}$ | - | $2.35 \pm 0.01^{b}$ | $5.99 \pm 0.13^{b}$ | $3.02 \pm 0.01^{d}$ | $88.35 \pm 0.01^{b}$ |
| | | 24 h | $2.93 \pm 0.02^{d}$ | - | $3.13 \pm 0.01^{a}$ | $6.80 \pm 0.05^{a}$ | $3.23 \pm 0.01^{c}$ | $97.43 \pm 0.08^{a}$ |
| | $EA_{10}$ | 0 h | $3.51 \pm 0.00^{b}$ | $0.10 \pm 0.00^{b}$ | - | $2.23 \pm 0.10^{e}$ | $2.32 \pm 0.01^{f}$ | $61.28 \pm 0.02^{c}$ |
| | | 2 h | $3.40 \pm 0.06^{c}$ | $0.06 \pm 0.00^{c}$ | - | $6.31 \pm 0.02^{a}$ | $4.09 \pm 0.00^{a}$ | $16.67 \pm 0.00^{f}$ |
| | | 4 h | $3.55 \pm 0.01^{b}$ | $0.22 \pm 0.02^{a}$ | - | $5.29 \pm 0.05^{d}$ | $3.94 \pm 0.00^{b}$ | $36.02 \pm 0.06^{d}$ |
| | | 8 h | $3.93 \pm 0.01^{a}$ | $0.10 \pm 0.00^{b}$ | - | $5.66 \pm 0.05^{c}$ | $2.42 \pm 0.00^{e}$ | $19.33 \pm 0.00^{e}$ |
| | | 16 h | $3.43 \pm 0.01^{c}$ | - | $2.20 \pm 0.01^{b}$ | $5.87 \pm 0.01^{b}$ | $2.86 \pm 0.03^{d}$ | $87.43 \pm 0.03^{b}$ |
| | | 24 h | $3.26 \pm 0.01^{d}$ | - | $2.93 \pm 0.02^{a}$ | $5.95 \pm 0.02^{b}$ | $3.01 \pm 0.02^{c}$ | $110.13 \pm 0.05^{a}$ |
| | $EA_{30}$ | 0 h | $2.12 \pm 0.02^{d}$ | - | $0.53 \pm 0.01^{c}$ | $1.85 \pm 0.01^{f}$ | $2.51 \pm 0.00^{e}$ | $76.29 \pm 1.26^{c}$ |
| | | 2 h | $3.80 \pm 0.03^{a}$ | $0.15 \pm 0.00^{b}$ | - | $5.92 \pm 0.00^{c}$ | $3.41 \pm 0.00^{b}$ | $17.02 \pm 0.00^{f}$ |
| | | 4 h | $3.19 \pm 0.02^{b}$ | $0.88 \pm 0.00^{a}$ | - | $4.46 \pm 0.05^{e}$ | $4.55 \pm 0.01^{a}$ | $32.61 \pm 0.24^{d}$ |
| | | 8 h | $2.90 \pm 0.06^{c}$ | - | - | $4.96 \pm 0.02^{d}$ | $2.41 \pm 0.00^{f}$ | $18.95 \pm 0.01^{e}$ |
| | | 16 h | $1.93 \pm 0.00^{e}$ | - | $4.13 \pm 0.01^{b}$ | $7.03 \pm 0.16^{b}$ | $2.70 \pm 0.00^{d}$ | $84.57 \pm 0.18^{b}$ |
| | | 24 h | $1.61 \pm 0.02^{f}$ | - | $5.51 \pm 0.02^{a}$ | $7.72 \pm 0.06^{a}$ | $2.79 \pm 0.02^{c}$ | $106.45 \pm 0.09^{a}$ |

* Values (mean ± SD) illustrated in this table are in triplicates (n = 3). The data of all the samples have been subtracted the control values. For each sample of each assay, different superscript letters indicated the significant differences among 0 h, 2 h, 4 h, 8 h, 16 h and 24 h phases within five rows of a column at a 95% confidence level ($p < 0.05$). FBH, faba bean hydrolysates; FBE, faba bean emulsions; $A_0$, FBPC protein hydrolysates viz., slurry with adjusted pH and temperature treatment without Alcalase; $A_5$, slurry with 5 min hydrolysis with Alcalase; $A_{10}$, slurry with 10 min hydrolysis with Alcalase; $A_{15}$, slurry with 15 min hydrolysis with Alcalase; $A_{30}$, slurry with 30 min hydrolysis with Alcalase; TPC, Total phenolic content; TFC, total flavonoid content; TCT, total condensed tannin; GAE, gallic acid equivalents; QE, quercetin equivalents; CE, catechin equivalents. DPPH, 2,2′-diphenyl-1-picrylhydrazyl; FRAP, ferric reducing antioxidant power; ABTS, 2,2′-azinobis-3-ethylbenzo-thiazoline-6-sulfonic acid; TE, Trolox equivalent.

These clearly demonstrated that phenolic compound release was enhanced throughout colonic fermentation, commonly after 8 h of fermentation, consistent with the study of Wang et al. [77]. Condensed tannins can be found in the plant in the form of oligomers as well as polymers of flavanols, which corresponds to extractable tannins as well as non-extractable tannins. According to Quatrin et al. [78], a large proportion of hydrolysable tannins can be converted quickly by gut microbiota within two hours. Meanwhile, tannins with non-extractable properties that are tightly combined with the protein or around sixty percent of the existing dietary fiber are hard to extract with aqueous acidic, organic solvents or digestive enzymes, however, as they approach the colon, they are metabolized by the gut flora and consequently contribute to the TPC results [79].

Flavonoids frequently form glycosidic bonds with other components. The rapid decrease in TFC values after 4 h of fermentation may be due to the combination of flavonoids with sulphates, glycosides and glucuronides after digestion in the upper digestive tract which are subsequently metabolized in the colon with the help of enzymes such as sulfatase, bacterial glycosidases and glucuronidases [80,81]. During gastrointestinal digestion, only a tiny portion of flavonoids (2–15%) was able to be completely digested and absorbed [80]. The flavonoid aglycon may continuously be metabolized in the presence of gut microbiota, resulting in fission products such as valerolactones and phenolic acids [82]. The degradation of condensed tannins contributes to a gradual rise in colonic TPC value [78]. However, because of some assay limitations, alterations shown in the content of condensed tannins during colonic fermentation are not able to be identified using an ultraviolet spectrophotometer. One of the possible explanations might be interactions among tannin metabolites, other compounds existing in the fermented residue, along with the chemicals applied in the assay, making the precipitate and finally falling within the limits of low determination [83].

### *3.10. Antioxidant Activities across Colonic Fermentation*

The variations in the antioxidant potential of faba beans during colonic fermentation were measured by DPPH, FRAP, and ABTS assays, as is displayed in Table 4. In general, a similar tendency in DPPH was found for the digested FBH and FBE, with the values rising and obtaining the highest level after 24 h of fermentation with fluctuations at 4 h and 8 h for some digesta. $HA_5$ was observed to have a comparatively higher DPPH value of 8.48 mg TE/g after 24 h, followed by $HA_{30}$ (8.14 mg TE/g). The reducing power of FBH and FBE could be observed through FRAP. FBH and FBE showed an increasing reducing power trend within the first 4 h. In contrast, a decreased trend was noted after 4 h but subsequently increased or sustained from 8 h of fecal reaction. $HA_{10}$ exhibited the highest reducing power after 4 h fermentation with a value of 4.79 mg TE/g. According to the ABTS results, the values of both FBH and FBE samples significantly reduced from 0 h, fluctuated between 2–8 h, but increased considerably after 8 h, and gained the highest ABTS values at 24 h of reaction (123.95 mg TE/g for $HA_{30}$). In addition, it seems that FBH demonstrated a relatively higher trend of ABTS values than the FBE group.

Cárdenas-Castro et al. [84] noted similar alterations in the DPPH value in tomatoes during colonic fermentation, and the DPPH value continued to build up after 24 h of fermentation, which is consistent with our results. This trend maintained upward tendency and might result from the existence of other bioactive compounds in faba beans that express antioxidant potential, such as vitamins [85,86]. Furthermore, the rising trends of antioxidant potential (DPPH and ABTS) that existed throughout colonic fermentation were considerably persistent with the improvement in total phenolic content. It was demonstrated that colonic fermentation could aid in releasing and breaking down phenolics from the residual of FBH and FBE digesta, while also enhancing its in vitro antioxidant potential.

Besides, melanoidins and dietary fiber in faba beans may cause microorganisms in the large intestine to produce compounds such as metabolites of tannins along with phenolic acids, enhancing DPPH, FRAP along with ABTS results [85]. Phenolic acids existing in faba beans are correlated with the in vitro expression of antioxidant capacity [87]. The modifications in antioxidant activities of faba beans at various stages of fermentation also implied that enzymes in the gastrointestinal tract could not completely neutralize biologically active molecules with antioxidant capacity, thus allowing the bioactive substances to enter the colonic site and release their bioactive effects [86,88,89].

### *3.11. Short Chain Fatty Acids Production*

Five types of short chain fatty acids, including acetic, propionic, iso-butyric, butyric and valeric acids in FBH and FBE were measured and displayed in Figure 3. The significant difference analysis result of Total SCFAs has been updated in the Supplemetary Materials (Table S1). Polysaccharides, oligosaccharides, and dietary fibers, containing pectin, beta-glucan, cellulose, gum, etc., will not be digested and absorbed in the upper digestive tract and subsequently, enter the colon. The flora present in the colon will then combine with them for further metabolism. Generally, there was a much lower SCFAs level in our fermented FBH and FBE samples compared to the regular faba bean powder illustrated in another study, in which the level of acetic, butyric and propionic acid after 24 h of reaction were around 41.58 mM, 11.05 mM, and 13.05 mM, respectively [90]. Nonetheless, acetic acid remained the major SCFA produced during colonic fermentation, followed by isobutyric and propionic acids. This finding was consistent with the studies of Gullón, Gullón Estévez, Tavaria, Vasconcelos and Gomes [90] and Wu, Liu, Lu, Barrow, Dunshea and Suleria [89]. However, regarding Çalışkantürk Karataş et al. [91], who measured the potential of faba bean gastrointestinal digesta to enhance gut microbiota fermentation, it is displayed that faba bean digesta facilitated the generation of short chain fatty acids, primarily acetic acid (56.9 μmol), followed by butyric acid (36.1 μmol), propionic acid (23.9 μmol), and valeric acid (8.8 μmol) per 100 mg residue. These slightly different patterns may be due to their use of anaerobic batch cultures to evaluate the impacts on metabolic products.

As shown in Figure 3, during colonic fermentation, the production of SCFAs was found an increasing trend after 4 h of fermentation and peaked after around 16 h, and then sustained or began decreasing steadily. Nevertheless, as noted by Çalışkantürk Karataş, Günay and Sayar [91], the formation of SCFAs still gradually rose after 12 h and showed a peak at 24 h for the whole faba bean. This slow formation pattern might result from the slow pace of fermentation of dietary fiber via gut flora that was picked from the human model, which existed differences from the present study. Periago et al. [92] also discovered that the generation of SCFAs was linked not only to the physical (electrostatic force) and chemical (hydrogen or ester bond) trapping configuration of dietary fiber but also to the species as well as quantities of microbiota.

In addition, the production of SCFAs among FBH and FBE with distinct periods of Alcalase hydrolysis was analogous. Still, it appeared a difference that the SCFAs level in $HA_{30}$ (enzymatic hydrolysis for 30 min) was generally higher than in other treatments. The lower production and the differences among various FBE along with FBH are most likely correlated to the degradation, Maillard reaction along with caramelization of certain carbohydrates when undergoing UHT treatment, which results in a loss of total polysaccharide contents for fermentation [93,94]. The gut microbiota could efficiently metabolize various Maillard reaction products, inducing the generation of short chain fatty acids. Except for $HA_{30}$, the similar level discovered in the fermented samples could be attributed to a substance named melanoidins. Melanoidins could form short chain fatty acids along with colonic fermentation, which finally makes up for the degradation loss caused by UHT treatment [15,94].

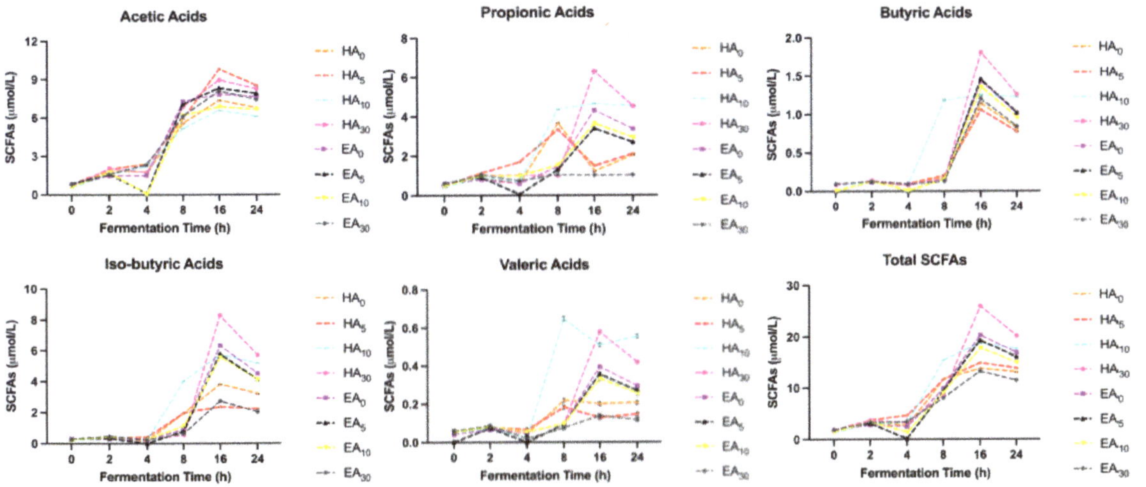


**Figure 3.** The production of five types of short chain fatty acids, including acetic, propionic, butyric, iso-butyric and valeric acids in FBH ($HA_0$, $HA_5$, $HA_{10}$, & $HA_{30}$), and FBE ($EA_0$, $EA_5$, $EA_{10}$, & $EA_{30}$) fermented with six-time variables (0 h, 2 h, 4 h, 8 h, 16 h, & 24 h).

## 4. Conclusions

In conclusion, enzymatic hydrolysis of faba bean proteins by Alcalase typically contributed to hydrolysates with better functional properties such as higher protein solubility, ζ-potential and lower surface hydrophobicity. In addition, emulsions treated with Alcalase were found stable after UHT treatment as indicated by $EA_0$ having the larger droplet size (14.73 μm), higher hydrophobicity index (244.59), and lower electronegativity (−21.47 mV). Both phenolic contents and antioxidant activities were significantly increased by about 36% and 54%, respectively, in the faba bean protein concentrate compared to the faba bean flour indicating that these compounds and activities partitioned with the proteins. Enzymatic hydrolysis with Alcalase for 30 min resulted in a comparatively higher total phenolic (0.29 mg GAE/g) and condensed tannin content (0.23 mg CE/g) with higher DPPH and ABTS value viz., 0.89 and 5.36 mg TE/g, respectively, but lower TFC and FRAP value (38.69 μg QE/g and 197.51 μg TE/g). UHT treatment maintained TPC, DPPH and ABTS but reduced the TFC by 30.99 μg QE/g and FRAP by 98.24 μg TE/g. Enzymatic hydrolysis combined with UHT significantly increased the content of total phenolics by 0.06 mg GAE/g and retained all the antioxidant activity values of DPPH, FRAP, and ABTS. The release of phenolic compounds in hydrolysates and UHT emulsions increased after intestinal digestion by 0.44 mg GAE/g and 0.55 mg GAE/g, respectively. Intestinal digestion was the stage when most phenolics were released with the highest antioxidant potential observed. For colonic fermentation, the release of phenolics was enhanced via microbiota present in the gut, commonly after 8 h of fermentation. Furthermore, the SCFA production was dominated by acetic acids, which exhibited significant similar changes among hydrolysates and UHT emulsions. Enzymatic hydrolysis of faba bean protein for 30 min resulted in higher antioxidant capacities and SCFAs, which could be more favorable for gut health. Nonetheless, the biotransformation of phenolic compounds after 48 h of colonic fermentation still needs deeper investigation. Overall, this study showed that enzymic hydrolysis of faba proteins not only improved the colloidal emulsion stability, but also released antioxidant capacity during in vitro digestibility and colonic fermentation. Gut microbiome functionality was also affected by hydrolysis for both proteins and emulsions as indicated by the impact of degree of hydrolysis on short chain fatty acid production. More studies are needed to further elucidate and differentiate the role of phenolics during faba protein processing and digestion stages in comparison to contributions of peptides, amino acids and microelements to digestion rates, antioxidant capacities and colonial short chain fatty acid production.

**Supplementary Materials:** The following supporting information can be downloaded at: https://www.mdpi.com/article/10.3390/nu15010089/s1, Table S1: The significant difference analysis result of Total SCFAs.

**Author Contributions:** Conceptualization, methodology, formal analysis, validation, and investigation, J.G., M.L., M.A.N., R.S., R.B. and H.A.R.S.; writing—original draft preparation, J.G., M.L. and H.A.R.S.; writing—review and editing, J.G., M.L., M.A.N., R.S., R.B. and H.A.R.S.; supervision, M.A.N., R.S., R.B. and H.A.R.S. All authors have read and agreed to the published version of the manuscript.

**Funding:** This research was funded by CSIRO Precision Health Future Science Platform, the University of Melbourne under the McKenzie Fellowship Scheme (grant no. UoM-18/21), the Future Food Hallmark Research Initiative Funds (grant no. UoM-21/23), and Collaborative Research Development Grant (grant no. UoM-21/23) funded by the Faculty of Veterinary and Agricultural Sciences, the University of Melbourne, Australia.

**Institutional Review Board Statement:** Not applicable.

**Informed Consent Statement:** Not applicable.

**Acknowledgments:** We would like to thank CSIRO Precision Health Future Science Platform for financially supporting this study. Additionally, we would like to thank the Honours/Master/PhD and postdoctoral researchers of the Hafiz Suleria group from the School of Agriculture and Food, Faculty of Veterinary and Agricultural Sciences, the University of Melbourne for their incredible support.

**Conflicts of Interest:** The authors declare no conflict of interest.

## References

1. Maalouf, F.; Hu, J.; O'Sullivan, D.M.; Zong, X.; Hamwieh, A.; Kumar, S.; Baum, M. Breeding and genomics status in faba bean (*Vicia faba*). *Plant Breed.* **2019**, *138*, 465–473. [CrossRef]
2. Ali, M. Functional properties of faba bean protein and effect of enzymatic hydrolysis on its antioxidant activity. *Zagazig J. Agric. Res.* **2019**, *46*, 2019. [CrossRef]
3. Karkanis, A.; Ntatsi, G.; Lepse, L.; Fernández, J.A.; Vågen, I.M.; Rewald, B.; Alsiņa, I.; Kronberga, A.; Balliu, A.; Olle, M.; et al. Faba Bean Cultivation—Revealing Novel Managing Practices for More Sustainable and Competitive European Cropping Systems. *Front. Plant Sci.* **2018**, *9*, 1115. [CrossRef] [PubMed]
4. Siah, S.; Konczak, I.; Wood, J.; Agboola, S.; Blanchard, C. Effects of Roasting on Phenolic Composition and In vitro Antioxidant Capacity of Australian Grown Faba Beans (*Vicia faba* L.). *Plant Foods Hum. Nutr.* **2014**, *69*, 85–91. [CrossRef] [PubMed]
5. Eckert, E.; Han, J.; Swallow, K.; Tian, Z.; Jarpa-Parra, M.; Chen, L. Effects of enzymatic hydrolysis and ultrafiltration on physicochemical and functional properties of faba bean protein. *Cereal Chem.* **2019**, *96*, 725–741. [CrossRef]
6. Khalil, A.; Mansour, E. The effect of cooking, autoclaving and germination on the nutritional quality of faba beans. *Food Chem.* **1995**, *54*, 177–182. [CrossRef]
7. Apaydin, H.; Ertan, S.; Ozekmekçi, S. Broad bean (*Vicia faba*)—A natural source of L-dopa–prolongs "on" periods in patients with Parkinson's disease who have "on-off" fluctuations. *Mov. Disord. Off. J. Mov. Disord. Soc.* **2000**, *15*, 164–166. [CrossRef]
8. Vucenik, I.; Shamsuddin, A.M. Protection Against Cancer by Dietary $IP_6$ and Inositol. *Nutr. Cancer* **2006**, *55*, 109–125. [CrossRef]
9. Laguna, L.; Picouet, P.; Guàrdia, M.D.; Renard, C.M.; Sarkar, A. In vitro gastrointestinal digestion of pea protein isolate as a function of pH, food matrices, autoclaving, high-pressure and re-heat treatments. *Lwt* **2017**, *84*, 511–519. [CrossRef]
10. Zhang, Q.; Cheng, Z.; Wang, Y.; Fu, L. Dietary protein-phenolic interactions: Characterization, biochemical-physiological consequences, and potential food applications. *Crit. Rev. Food Sci. Nutr.* **2020**, *61*, 3589–3615. [CrossRef]
11. Mihaylova, D.; Desseva, I.; Stoyanova, M.; Petkova, N.; Terzyiska, M.; Lante, A. Impact of In Vitro Gastrointestinal Digestion on the Bioaccessibility of Phytochemical Compounds from Eight Fruit Juices. *Molecules* **2021**, *26*, 1187. [CrossRef] [PubMed]
12. Hrčková, M.; Rusnáková, M.; Zemanovič, J. Enzymatic hydrolysis of defatted soy flour by three different proteases and their effect on the functional properties of resulting protein hydrolysates. *Czech J. Food Sci.* **2011**, *20*, 7–14. [CrossRef]
13. Ibrahim, E.; Ghani, M. The effect of enzymatic hydrolysis on the antioxidant activities and amino acid profiles of defatted chia (*Salvia hispanica* L.) flour. *Food Res.* **2020**, *4*, 38–50. [CrossRef] [PubMed]
14. Nawaz, M.A.; Tan, M.; Øiseth, S.; Buckow, R. An Emerging Segment of Functional Legume-Based Beverages: A Review. *Food Rev. Int.* **2020**, *38*, 1064–1102. [CrossRef]
15. Žilić, S.; Akıllıoğlu, G.; Serpen, A.; Barac, M.; Gökmen, V. Effects of isolation, enzymatic hydrolysis, heating, hydratation and Maillard reaction on the antioxidant capacity of cereal and legume proteins. *Food Res. Int.* **2012**, *49*, 1–6. [CrossRef]

16. Nawaz, M.A.; Buckow, R.; Jegasothy, H.; Stockmann, R. Enzymatic hydrolysis improves the stability of UHT treated faba bean protein emulsions. *Food Bioprod. Process.* **2022**, *132*, 200–210. [CrossRef]
17. Liu, C.; Bhattarai, M.; Mikkonen, K.; Heinonen, M. Effects of Enzymatic Hydrolysis of Fava Bean Protein Isolate by Alcalase on the Physical and Oxidative Stability of Oil-in-Water Emulsions. *J. Agric. Food Chem.* **2019**, *67*, 6625–6632. [CrossRef]
18. Ali, A.; Wu, H.; Ponnampalam, E.N.; Cottrell, J.J.; Dunshea, F.R.; Suleria, H.A.R. Comprehensive Profiling of Most Widely Used Spices for Their Phenolic Compounds through LC-ESI-QTOF-$MS^2$ and Their Antioxidant Potential. *Antioxidants* **2021**, *10*, 721. [CrossRef]
19. Ghribi, A.M.; Gafsi, I.M.; Sila, A.; Blecker, C.; Danthine, S.; Attia, H.; Bougatef, A.; Besbes, S. Effects of enzymatic hydrolysis on conformational and functional properties of chickpea protein isolate. *Food Chem.* **2015**, *187*, 322–330. [CrossRef]
20. Sánchez-Velázquez, O.A.; Cuevas-Rodríguez, E.O.; Mondor, M.; Ribéreau, S.; Arcand, Y.; Mackie, A.; Hernández-Álvarez, A.J. Impact of in vitro gastrointestinal digestion on peptide profile and bioactivity of cooked and non-cooked oat protein concentrates. *Curr. Res. Food Sci.* **2021**, *4*, 93–104. [CrossRef]
21. Gu, C.; Suleria, H.A.R.; Dunshea, F.R.; Howell, K. Dietary Lipids Influence Bioaccessibility of Polyphenols from Black Carrots and Affect Microbial Diversity under Simulated Gastrointestinal Digestion. *Antioxidants* **2020**, *9*, 762. [CrossRef] [PubMed]
22. Church, F.C.; Swaisgood, H.E.; Porter, D.H.; Catignani, G.L. Spectrophotometric assay using o-phthaldialdehyde for determination of proteolysis in milk and isolated milk proteins. *J. Dairy Sci.* **1983**, *66*, 1219–1227. [CrossRef]
23. Leong, T.S.; Walter, V.; Gamlath, C.J.; Yang, M.; Martin, G.J.; Ashokkumar, M. Functionalised dairy streams: Tailoring protein functionality using sonication and heating. *Ultrason. Sonochem.* **2018**, *48*, 499–508. [CrossRef]
24. Li, Q.; Wang, Z.; Dai, C.; Wang, Y.; Chen, W.; Ju, X.; Yuan, J.; He, R. Physical stability and microstructure of rapeseed protein isolate/gum Arabic stabilized emulsions at alkaline pH. *Food Hydrocoll.* **2019**, *88*, 50–57. [CrossRef]
25. Nwachukwu, I.D.; Aluko, R.E. Physicochemical and emulsification properties of flaxseed (*Linum usitatissimum*) albumin and globulin fractions. *Food Chem.* **2018**, *255*, 216–225. [CrossRef]
26. Rahmati, N.F.; Koocheki, A.; Varidi, M.; Kadkhodaee, R. Thermodynamic compatibility and interactions between Speckled Sugar bean protein and xanthan gum for production of multilayer O/W emulsion. *J. Food Sci. Technol.* **2018**, *55*, 1143–1153. [CrossRef]
27. Marsh, K.J.; Wallis, I.R.; Kulheim, C.; Clark, R.; Nicolle, D.; Foley, W.J.; Salminen, J. New approaches to tannin analysis of leaves can be used to explain in vitro biological activities associated with herbivore defence. *N. Phytol.* **2020**, *225*, 488–498. [CrossRef]
28. Nnamezie, A.A.; Famuwagun, A.A.; Gbadamosi, S.O. Characterization of okra seed flours, protein concentrate, protein isolate and enzymatic hydrolysates. *Food Prod. Process. Nutr.* **2021**, *3*, 14. [CrossRef]
29. Delfanian, M.; Kenari, R.E.; Sahari, M.A. Utilization of Jujube Fruit (*Ziziphus mauritiana* Lam.) Extracts as Natural Antioxidants in Stability of Frying Oil. *Int. J. Food Prop.* **2015**, *19*, 789–801. [CrossRef]
30. Puspita, M.; Déniel, M.; Widowati, I.; Radjasa, O.K.; Douzenel, P.; Marty, C.; Vandanjon, L.; Bedoux, G.; Bourgougnon, N. Total phenolic content and biological activities of enzymatic extracts from *Sargassum muticum* (Yendo) Fensholt. *J. Appl. Phycol.* **2017**, *29*, 2521–2537. [CrossRef]
31. Álvarez, R.; Araya, H.; Navarro-Lisboa, R.; Lopez de Dicastillo, C. Evaluation of polyphenol content and antioxidant capacity of fruits and vegetables using a modified enzymatic extraction. *Food Technol. Biotechnol.* **2016**, *54*, 462–467. [CrossRef] [PubMed]
32. Dangles, O.; Fenger, J.-A. The Chemical Reactivity of Anthocyanins and Its Consequences in Food Science and Nutrition. *Molecules* **2018**, *23*, 1970. [CrossRef] [PubMed]
33. Fossen, T.; Cabrita, L.; Andersen, O.M. Colour and stability of pure anthocyanins influenced by pH including the alkaline region. *Food Chem.* **1998**, *63*, 435–440. [CrossRef]
34. Sbroggio, M.F.; Montilha, M.S.; Figueiredo, V.R.G.D.; Georgetti, S.R.; Kurozawa, L.E. Influence of the degree of hydrolysis and type of enzyme on antioxidant activity of okara protein hydrolysates. *Food Sci. Technol.* **2016**, *36*, 375–381. [CrossRef]
35. Shahi, Z.; Sayyed-Alangi, S.Z.; Najafian, L. Effects of enzyme type and process time on hydrolysis degree, electrophoresis bands and antioxidant properties of hydrolyzed proteins derived from defatted *Bunium persicum* Bioss. press cake. *Heliyon* **2020**, *6*, e03365. [CrossRef] [PubMed]
36. Li, X.; Deng, J.; Shen, S.; Li, T.; Yuan, M.; Yang, R.; Ding, C. Antioxidant activities and functional properties of enzymatic protein hydrolysates from defatted *Camellia oleifera* seed cake. *J. Food Sci. Technol.* **2015**, *52*, 5681–5690. [CrossRef] [PubMed]
37. Zhu, L.; Chen, J.; Tang, X.; Xiong, Y.L. Reducing, radical scavenging, and chelation properties of in vitro digests of alcalase-treated zein hydrolysate. *J. Agric. Food Chem.* **2008**, *56*, 2714–2721. [CrossRef]
38. Chen, H.-M.; Muramoto, K.; Yamauchi, F.; Fujimoto, K.; Nokihara, K. Antioxidative Properties of Histidine-Containing Peptides Designed from Peptide Fragments Found in the Digests of a Soybean Protein. *J. Agric. Food Chem.* **1998**, *46*, 49–53. [CrossRef]
39. Samaei, S.P.; Ghorbani, M.; Tagliazucchi, D.; Martini, S.; Gotti, R.; Themelis, T.; Tesini, F.; Gianotti, A.; Toschi, T.G.; Babini, E. Functional, nutritional, antioxidant, sensory properties and comparative peptidomic profile of faba bean (*Vicia faba*, L.) seed protein hydrolysates and fortified apple juice. *Food Chem.* **2020**, *330*, 127120. [CrossRef]
40. Vasconcellos, F.C.S.; Woiciechowski, A.L.; Soccol, V.T.; Mantovani, D.; Soccol, C.R. Antimicrobial and antioxidant properties of-conglycinin and glycinin from soy protein isolate. *Int. J. Curr. Microbiol. Appl. Sci.* **2014**, *3*, 144–157.
41. Xu, B.; Chang, S.K.C. Isoflavones, Flavan-3-ols, Phenolic Acids, Total Phenolic Profiles, and Antioxidant Capacities of Soy Milk as Affected by Ultrahigh-Temperature and Traditional Processing Methods. *J. Agric. Food Chem.* **2009**, *57*, 4706–4717. [CrossRef] [PubMed]
42. Dai, H.; Leung, C.E.; Corradini, M.; Xiao, H.; Kinchla, A.J. Increasing the nutritional value of strawberry puree by adding xylo-oligosaccharides. *Heliyon* **2020**, *6*, e03769. [CrossRef] [PubMed]

43. Xu, B.; Chang, S.K.C. Total Phenolics, Phenolic Acids, Isoflavones, and Anthocyanins and Antioxidant Properties of Yellow and Black Soybeans as Affected by Thermal Processing. *J. Agric. Food Chem.* **2008**, *56*, 7165–7175. [CrossRef] [PubMed]
44. Cao, X.; Zhang, Y.; Zhang, F.; Wang, Y.; Yi, J.; Liao, X. Effects of high hydrostatic pressure on enzymes, phenolic compounds, anthocyanins, polymeric color and color of strawberry pulps. *J. Sci. Food Agric.* **2011**, *91*, 877–885. [CrossRef]
45. Nicoli, M.C.; Anese, M.; Parpinel, M.; Franceschi, S.; Lerici, C.R. Loss and/or formation of antioxidants during food processing and storage. *Cancer Lett.* **1997**, *114*, 71–74. [CrossRef]
46. Dias, F.F.G.; Augusto-Obara, T.R.; Hennebelle, M.; Chantieng, S.; Ozturk, G.; Taha, A.Y.; de Souza Vieira, T.M.F.; de Moura, J.M.L.N. Effects of industrial heat treatments on bovine milk oxylipins and conventional markers of lipid oxidation. *Prostaglandins Leukot. Essent. Fat. Acids* **2019**, *152*, 102040. [CrossRef]
47. Yoo, S.-H.; Chang, Y.H. Volatile Compound, Physicochemical, and Antioxidant Properties of Beany Flavor-Removed Soy Protein Isolate Hydrolyzates Obtained from Combined High Temperature Pre-Treatment and Enzymatic Hydrolysis. *Prev. Nutr. Food Sci.* **2016**, *21*, 338–347. [CrossRef]
48. Voss, G.B.; Osorio, H.; Valente, L.M.; Pintado, M.E. Impact of thermal treatment and hydrolysis by Alcalase and Cynara cardunculus enzymes on the functional and nutritional value of Okara. *Process Biochem.* **2019**, *83*, 137–147. [CrossRef]
49. Warsame, A.O.; Michael, N.; O'Sullivan, D.M.; Tosi, P. Identification and Quantification of Major Faba Bean Seed Proteins. *J. Agric. Food Chem.* **2020**, *68*, 8535–8544. [CrossRef]
50. Nisov, A.; Ercili-Cura, D.; Nordlund, E. Limited hydrolysis of rice endosperm protein for improved techno-functional properties. *Food Chem.* **2020**, *302*, 125274. [CrossRef]
51. Pan, Y.; Xie, Q.-T.; Zhu, J.; Li, X.-M.; Meng, R.; Zhang, B.; Chen, H.-Q.; Jin, Z.-Y. Study on the fabrication and in vitro digestion behavior of curcumin-loaded emulsions stabilized by succinylated whey protein hydrolysates. *Food Chem.* **2019**, *287*, 76–84. [CrossRef] [PubMed]
52. Akbari, N.; Milani, J.M.; Biparva, P. Functional and conformational properties of proteolytic enzyme-modified potato protein isolate. *J. Sci. Food Agric.* **2019**, *100*, 1320–1327. [CrossRef] [PubMed]
53. Zang, X.; Yue, C.; Wang, Y.; Shao, M.; Yu, G. Effect of limited enzymatic hydrolysis on the structure and emulsifying properties of rice bran protein. *J. Cereal Sci.* **2018**, *85*, 168–174. [CrossRef]
54. Nawaz, M.; Singh, T.; Stockmann, R.; Jegasothy, H.; Buckow, R. Quality Attributes of Ultra-High Temperature-Treated Model Beverages Prepared with Faba Bean Protein Concentrates. *Foods* **2021**, *10*, 1244. [CrossRef] [PubMed]
55. Ahmadian-Kouchaksaraei, Z.; Varidi, M.; Varidi, M.J.; Pourazarang, H. Influence of processing conditions on the physicochemical and sensory properties of sesame milk: A novel nutritional beverage. *LWT Food Sci. Technol.* **2014**, *57*, 299–305. [CrossRef]
56. Gülçin, İ.; Mshvildadze, V.; Gepdiremen, A.; Elias, R. The antioxidant activity of a triterpenoid glycoside isolated from the berries of Hedera colchica: 3-O-(β-d-glucopyranosyl)-hederagenin. *Phytother. Res. Int. J. Devoted Pharmacol. Toxicol. Eval. Nat. Prod. Deriv.* **2006**, *20*, 130–134.
57. Qamar, S.; Bhandari, B.; Prakash, S. Effect of different homogenisation methods and UHT processing on the stability of pea protein emulsion. *Food Res. Int.* **2019**, *116*, 1374–1385. [CrossRef]
58. Tan, M.; Nawaz, M.A.; Buckow, R. Functional and food application of plant proteins—A review. *Food Rev. Int.* **2021**, 1–29. [CrossRef]
59. Ribeiro, L.D.O.; Pinheiro, A.C.B.; Brígida, A.I.S.; Genisheva, Z.A.; Vicente, A.A.M.D.O.S.; Teixeira, J.A.C.; de Matta, V.M.; Freitas, S.P. In vitro gastrointestinal evaluation of a juçara-based smoothie: Effect of processing on phenolic compounds bioaccessibility. *J. Food Sci. Technol.* **2019**, *56*, 5017–5026. [CrossRef]
60. Ma, Y.; Yang, Y.; Gao, J.; Feng, J.; Shang, Y.; Wei, Z. Phenolics and antioxidant activity of bamboo leaves soup as affected by in vitro digestion. *Food Chem. Toxicol.* **2020**, *135*, 110941. [CrossRef]
61. Wojtunik-Kulesza, K.; Oniszczuk, A.; Oniszczuk, T.; Combrzyński, M.; Nowakowska, D.; Matwijczuk, A. Influence of in vitro digestion on composition, bioaccessibility and antioxidant activity of food polyphenols—A non-systematic review. *Nutrients* **2020**, *12*, 1401. [CrossRef] [PubMed]
62. Gonçalves, J.; Ramos, R.; Luís, A.n.; Rocha, S.; Rosado, T.; Gallardo, E.; Duarte, A.P. Assessment of the Bioaccessibility and Bioavailability of the Phenolic Compounds of *Prunus avium* L. by in Vitro Digestion and Cell Model. *ACS Omega* **2019**, *4*, 7605–7613. [CrossRef]
63. Zeng, Z.; Liu, C.; Luo, S.; Chen, J.; Gong, E. The Profile and Bioaccessibility of Phenolic Compounds in Cereals Influenced by Improved Extrusion Cooking Treatment. *PLoS ONE* **2016**, *11*, e0161086. [CrossRef]
64. Wang, S.; Amigo-Benavent, M.; Mateos, R.; Bravo, L.; Sarriá, B. Effects of in vitro digestion and storage on the phenolic content and antioxidant capacity of a red grape pomace. *Int. J. Food Sci. Nutr.* **2016**, *68*, 188–200. [CrossRef]
65. Maduwanthi, S.D.T.; Marapana, R. Total phenolics, flavonoids and antioxidant activity following simulated gastro-intestinal digestion and dialysis of banana (*Musa acuminata*, AAB) as affected by induced ripening agents. *Food Chem.* **2021**, *339*, 127909. [CrossRef] [PubMed]
66. Bouayed, J.; Hoffmann, L.; Bohn, T. Total phenolics, flavonoids, anthocyanins and antioxidant activity following simulated gastro-intestinal digestion and dialysis of apple varieties: Bioaccessibility and potential uptake. *Food Chem.* **2011**, *128*, 14–21. [CrossRef] [PubMed]
67. Sancho, R.A.S.; Pavan, V.; Pastore, G.M. Effect of in vitro digestion on bioactive compounds and antioxidant activity of common bean seed coats. *Food Res. Int.* **2015**, *76*, 74–78. [CrossRef]

68. Fernández, K.; Labra, J. Simulated digestion of proanthocyanidins in grape skin and seed extracts and the effects of digestion on the angiotensin I-converting enzyme (ACE) inhibitory activity. *Food Chem.* **2013**, *139*, 196–202. [CrossRef]
69. Seraglio, S.K.T.; Valese, A.C.; Daguer, H.; Bergamo, G.; Azevedo, M.S.; Nehring, P.; Gonzaga, L.V.; Fett, R.; Costa, A.C.O. Effect of in vitro gastrointestinal digestion on the bioaccessibility of phenolic compounds, minerals, and antioxidant capacity of *Mimosa scabrella* Bentham honeydew honeys. *Food Res. Int.* **2017**, *99*, 670–678. [CrossRef]
70. Bermúdez-Soto, M.J.; Tomás-Barberán, F.A.; García-Conesa, M.T. Stability of polyphenols in chokeberry (*Aronia melanocarpa*) subjected to in vitro gastric and pancreatic digestion. *Food Chem.* **2007**, *102*, 865–874. [CrossRef]
71. He, Z.; Yuan, B.; Zeng, M.; Chen, J. Effect of thermal processing and digestive protease on the antioxidant capacity of fruit juice-milk beverage model systems under simulated gastrointestinal digestion. *Int. J. Food Sci. Technol.* **2015**, *50*, 2306–2315. [CrossRef]
72. Chen, G.-L.; Chen, S.-G.; Zhao, Y.-Y.; Luo, C.-X.; Li, J.; Gao, Y.-Q. Total phenolic contents of 33 fruits and their antioxidant capacities before and after in vitro digestion. *Ind. Crops Prod.* **2014**, *57*, 150–157. [CrossRef]
73. Boutrou, R.; Henry, G.; Sánchez-Rivera, L. On the trail of milk bioactive peptides in human and animal intestinal tracts during digestion: A review. *Dairy Sci. Technol.* **2015**, *95*, 815–829. [CrossRef]
74. Chen, Y.; Lin, H.; Lin, M.; Zheng, Y.; Chen, J. Effect of roasting and in vitro digestion on phenolic profiles and antioxidant activity of water-soluble extracts from sesame. *Food Chem. Toxicol.* **2020**, *139*, 111239. [CrossRef] [PubMed]
75. Fawole, O.A.; Opara, U.L. Stability of total phenolic concentration and antioxidant capacity of extracts from pomegranate co-products subjected to in vitro digestion. *BMC Complement. Altern. Med.* **2016**, *16*, 358. [CrossRef]
76. Chotphruethipong, L.; Battino, M.; Benjakul, S. Effect of stabilizing agents on characteristics, antioxidant activities and stability of liposome loaded with hydrolyzed collagen from defatted Asian sea bass skin. *Food Chem.* **2020**, *328*, 127127. [CrossRef]
77. Wang, C.; Wu, H.; Liu, Z.; Barrow, C.; Dunshea, F.; Suleria, H.A.R. Bioaccessibility and movement of phenolic compounds from tomato (*Solanum lycopersicum*) during in vitro gastrointestinal digestion and colonic fermentation. *Food Funct.* **2022**, *13*, 4954–4966. [CrossRef]
78. Quatrin, A.; Rampelotto, C.; Pauletto, R.; Maurer, L.H.; Nichelle, S.M.; Klein, B.; Rodrigues, R.F.; Junior, M.R.M.; Fonseca, B.D.S.; de Menezes, C.R.; et al. Bioaccessibility and catabolism of phenolic compounds from jaboticaba (*Myrciaria trunciflora*) fruit peel during in vitro gastrointestinal digestion and colonic fermentation. *J. Funct. Foods* **2019**, *65*, 103714. [CrossRef]
79. Campos-Vega, R.; Vázquez-Sánchez, K.; López-Barrera, D.; Loarca-Piña, G.; Mendoza-Díaz, S.; Oomah, B. Simulated gastrointestinal digestion and in vitro colonic fermentation of spent coffee (*Coffea arabica* L.): Bioaccessibility and intestinal permeability. *Food Res. Int.* **2015**, *77*, 156–161. [CrossRef]
80. Gao, K.; Xu, A.; Krul, C.; Venema, K.; Liu, Y.; Niu, Y.; Lu, J.; Bensoussan, L.; Seeram, N.P.; Heber, D.; et al. Of the Major Phenolic Acids Formed during Human Microbial Fermentation of Tea, Citrus, and Soy Flavonoid Supplements, Only 3,4-Dihydroxyphenylacetic Acid Has Antiproliferative Activity. *J. Nutr.* **2006**, *136*, 52–57. [CrossRef]
81. Rowland, I.; Gibson, G.; Heinken, A.; Scott, K.; Swann, J.; Thiele, I.; Tuohy, K. Gut microbiota functions: Metabolism of nutrients and other food components. *Eur. J. Nutr.* **2018**, *57*, 34. [CrossRef] [PubMed]
82. Kawabata, K.; Yoshioka, Y.; Terao, J. Role of Intestinal Microbiota in the Bioavailability and Physiological Functions of Dietary Polyphenols. *Molecules* **2019**, *24*, 370. [CrossRef] [PubMed]
83. Watrelot, A.A.; Norton, E.L. Chemistry and Reactivity of Tannins in *Vitis* spp.: A Review. *Molecules* **2020**, *25*, 2110. [PubMed]
84. Cárdenas-Castro, A.P.; Zamora-Gasga, V.M.; Alvarez-Parrilla, E.; Ruíz-Valdiviezo, V.M.; Venema, K.; Sáyago-Ayerdi, S.G. In vitro gastrointestinal digestion and colonic fermentation of tomato (*Solanum lycopersicum* L.) and husk tomato (*Physalis ixocarpa* Brot.): Phenolic compounds released and bioconverted by gut microbiota. *Food Chem.* **2021**, *360*, 130051. [CrossRef] [PubMed]
85. Siah, S.D.; Konczak, I.; Agboola, S.; Wood, J.A.; Blanchard, C.L. In vitro investigations of the potential health benefits of Australian-grown faba beans (*Vicia faba* L.): Chemopreventative capacity and inhibitory effects on the angiotensin-converting enzyme, α-glucosidase and lipase. *Br. J. Nutr.* **2012**, *108*, S123–S134. [CrossRef] [PubMed]
86. Martineau-Côté, D.; Achouri, A.; Karboune, S.; L'Hocine, L. Faba Bean: An Untapped Source of Quality Plant Proteins and Bioactives. *Nutrients* **2022**, *14*, 1541. [CrossRef] [PubMed]
87. Zhu, L.; Li, W.; Deng, Z.; Li, H.; Zhang, B. The Composition and Antioxidant Activity of Bound Phenolics in Three Legumes, and Their Metabolism and Bioaccessibility of Gastrointestinal Tract. *Foods* **2020**, *9*, 1816. [CrossRef] [PubMed]
88. de Cosío-Barrón, A.C.G.; Hernández-Arriaga, A.M.; Campos-Vega, R. Spent coffee (*Coffea arabica* L.) grounds positively modulate indicators of colonic microbial activity. *Innov. Food Sci. Emerg. Technol.* **2020**, *60*, 102286. [CrossRef]
89. Wu, H.; Liu, Z.; Lu, P.; Barrow, C.; Dunshea, F.R.; Suleria, H.A. Bioaccessibility and bioactivities of phenolic compounds from roasted coffee beans during in vitro digestion and colonic fermentation. *Food Chem.* **2022**, *386*, 132794. [CrossRef]
90. Gullón, P.; Gullón, B.; Tavaria, F.; Vasconcelos, M.; Gomes, A.M. In vitro fermentation of lupin seeds (*Lupinus albus*) and broad beans (*Vicia faba*): Dynamic modulation of the intestinal microbiota and metabolomic output. *Food Funct.* **2015**, *6*, 3316–3322. [CrossRef]
91. Karataş, S.; Günay, D.; Sayar, S. In vitro evaluation of whole faba bean and its seed coat as a potential source of functional food components. *Food Chem.* **2017**, *230*, 182–188. [CrossRef] [PubMed]
92. Periago, M.J.; Martín-Pozuelo, G.; González-Barrio, R.; Santaella, M.; Gómez, V.; Vázquez, N.; Navarro-González, I.; García-Alonso, J. Effect of tomato juice consumption on the plasmatic lipid profile, hepatic HMGCR activity, and fecal short chain fatty acid content of rats. *Food Funct.* **2016**, *7*, 4460–4467. [CrossRef] [PubMed]

93. Gniechwitz, D.; Brueckel, B.; Reichardt, N.; Blaut, M.; Steinhart, H.; Bunzel, M. Coffee Dietary Fiber Contents and Structural Characteristics as Influenced by Coffee Type and Technological and Brewing Procedures. *J. Agric. Food Chem.* **2007**, *55*, 11027–11034. [CrossRef] [PubMed]
94. Pérez-Burillo, S.; Mehta, T.; Esteban-Muñoz, A.; Pastoriza, S.; Paliy, O.; Rufián-Henares, J. Effect of in vitro digestion-fermentation on green and roasted coffee bioactivity: The role of the gut microbiota. *Food Chem.* **2018**, *279*, 252–259. [CrossRef]

nutrients MDPI

*Article*

# Probing the Double-Layered Cotyledon Cell Structure of Navy Beans: Barrier Effect of the Protein Matrix on In Vitro Starch Digestion

**Duc Toan Do [1,*], Jaspreet Singh [1,2], Stuart Johnson [3] and Harjinder Singh [1]**

1 Riddet Institute, Massey University, Private Bag 11 222, Palmerston North 4442, New Zealand
2 School of Food and Advanced Technology, Massey University, Private Bag 11 222, Palmerston North 4442, New Zealand
3 School of Molecular and Life Sciences, Faculty of Science and Engineering, Curtin University, Bentley 6102, Australia
* Correspondence: t.do@massey.ac.nz

**Abstract:** The microstructure of legumes plays a crucial role in regulating starch digestion and postprandial glycemic responses. Starch granules are double encapsulated within the outer cell wall and the inner protein matrix of legume cotyledon cells. Despite progress in understanding the role of cell walls in delaying starch digestion, the role of the protein matrix has received little research attention. The aim of this study was to evaluate if the protein matrix and cell wall may present combined physical barriers retarding enzyme hydrolysis of intracellular starch. Intact cotyledon cells were isolated from navy beans and used to assess the barrier effect of the protein matrix on the digestion of starch under conditions simulating the upper gastrointestinal tract. The cells were pretreated with pepsin at 37 °C and pH 2.0 for 1, 4, or 24 h and without pepsin for 24 h (control) to facilitate removal of the intracellular protein matrix prior to cooking and simulated in vitro digestion. A longer pretreatment time resulted in a lower protein content of the cells and a higher initial rate and extent of starch hydrolysis. We suggest that in addition to the primary cell wall barrier, the protein matrix provides a secondary barrier restricting the accessibility of α-amylase to starch. This study provides a new fundamental understanding of the relationship between the structural organization of legume cotyledon cells and starch digestion that could inform the design of novel low glycemic index foods.

**Keywords:** cotyledon cells; in vitro starch digestion; legume structure; low glycaemic index; navy beans; protein matrix

**Citation:** Do, D.T.; Singh, J.; Johnson, S.; Singh, H. Probing the Double-Layered Cotyledon Cell Structure of Navy Beans: Barrier Effect of the Protein Matrix on In Vitro Starch Digestion. *Nutrients* **2023**, *15*, 105. https://doi.org/10.3390/nu15010105

Academic Editor: Alessandra Bordoni

Received: 6 December 2022
Revised: 19 December 2022
Accepted: 23 December 2022
Published: 26 December 2022



## 1. Introduction

Epidemiologic evidence supports the association of consumption of nutrient-dense, low glycaemic index legumes with a decreased risk of incident type 2 diabetes and improved glycaemic control in diabetic individuals [1–3]. Starchy legumes (e.g., beans, peas, chickpeas, lentils, etc.) contain a substantial proportion of slowly digestible starch and physically inaccessible resistant starch. These starch fractions are believed to slow the rate of carbohydrate release in vitro and, consequently, lower the postprandial blood glucose response in vivo [4–6].

An accumulating body of in vitro and in vivo evidence has firmly established a link between the microstructure of food legumes and the slow digestion properties of their starch [7–12]. The cotyledon tissue of starchy legumes comprises numerous cells within which starch granules are tightly embedded in protein matrices and all encapsulated by thick, strong cell walls [13]. In an earlier study, Berg et al. [7] showed that the entrapment of multiple starch granules within cotyledon cells reduced the in vitro hydrolysis rate and the extent of starch in cooked whole navy beans. This is because the resilient cell walls remained mostly intact throughout cooking and subsequent digestion and, thus, could act as physical barriers obstructing free access of α-amylase to the intracellular starch.

Recently, there has been a surge of interest in the use of intact, isolated cotyledon cells as a food model to gain a deeper mechanistic understanding of how microscale legume structure controls in vitro starch digestion [9–11,14]. Bhattarai, Dhital, Wu, Chen, and Gidley [15] studied the digestive behaviour of isolated legume cells using a dynamic in vitro rat stomach duodenum (DIVRSD) model. The authors proposed three possible mechanisms that may explain the limited hydrolysis of starch inside the cells. Firstly, cell wall intactness reduces the accessibility of digestive enzymes to starch during simulated digestion. Secondly, the swelling and gelatinisation of entrapped starch granules are restricted during cooking, resulting in the retention of the enzyme-resistant crystalline starch structure. Thirdly, the noncatalytic binding of α-amylase to cell wall components prevents its binding to the preferred starch substrate. New evidence further suggests that cell wall porosity controls the diffusion of digestive enzymes and the release of hydrolysed products, which can significantly affect the digestion of starch inside isolated cells from plant foods [16].

To date, most studies have identified the cell wall as a primary physical barrier to starch digestion [7,9,13,17]. Despite the fact that the cell wall coexists with the protein matrix within the cotyledon cell and both naturally encapsulate starch granules [13], the barrier effect exerted by the protein matrix on modulating starch digestion has received inadequate investigation. Rovalino-Córdova, Fogliano, and Capuano [18] and Rovalino-Córdova, Fogliano, and Capuano [19] revealed that the in vitro rate and/or extent of starch hydrolysis of intact red kidney bean cells was significantly reduced in the absence of one or more digestive proteases (i.e., pepsin, trypsin, and chymotrypsin) that were able to hydrolyse the protein matrix. This clearly indicates that the protein matrix represents an additional structural barrier to starch digestion [19]. It has been hypothesized that the compact protein network surrounding starch granules could impede α-amylase mobility within the cells and decrease the starch surface area available for amylase binding and catalysis [18].

The present study aims to extend our understanding of the role of the protein matrix in determining the kinetics of starch digestion using isolated navy bean cotyledon cells as a food model. Navy beans were selected as the legume variety for investigation due to previous evidence supporting the strong relationship between their cotyledon cell structures and starch digestion in vitro [7,11]. In this study, the cells were pretreated with pepsin at 37 °C, pH 2.0, and various incubation times to facilitate the removal of the protein barrier while retaining both the starch granules and the cell walls in an intact state. Accordingly, cell samples with different protein matrix microstructures were generated. Following the cooking of the pepsin-treated cells, the gastrointestinal (GI) fate of the encapsulated starch was assessed using a static in vitro digestion model. It was postulated that the enzymatic breakdown of the intracellular protein matrix could enhance starch hydrolysis efficiency.

## 2. Materials and Methods

### 2.1. Materials

One batch of raw navy beans (*Phaseolus vulgaris* L.) was procured from a local store in Palmerston North, New Zealand. Pepsin (from porcine gastric mucosa, ≥250 units/mg protein), pancreatin (from porcine pancreas, activity equivalent to 4× USP specifications), invertase (from baker's yeast (*S. cerevisiae*), Grade VII, ≥300 units/mg solid), porcine pancreatic α-amylase (from porcine pancreas, type VI-B), soluble potato starch (S-2630), and maltose (S-5885) were purchased from Sigma-Aldrich Ltd., St Louis, MO, USA. Amyloglucosidase (for total dietary fibre and starch assays, 3260 U/mL) was purchased from Megazyme International Ireland Ltd., Wicklow, Ireland. All other chemicals and reagents were of analytical grade. Reverse osmosis (RO) water was used for all experiments.

### 2.2. Preparation of Navy Bean Materials

#### 2.2.1. Isolation of Free Starch Granules

Free starch granules were isolated from navy bean seeds according to the method described by Berg et al. [7].

#### 2.2.2. Isolation of Cotyledon Cells

Raw, intact cotyledon cells were isolated without gelatinising starch by successive treatments of navy beans with acid and then alkali according to the method described by Do et al. [10]. Dried navy bean seeds were soaked in a 0.1 M hydrochloric acid (HCl) solution (pH~1.3) at room temperature (~20 °C) for 24 h. The hydrated beans were manually dehulled to remove the outer seed coats and hypocotyls, split into cotyledons, and rinsed with RO water to remove the remaining acids. The cotyledons were subsequently soaked in a 0.06 M sodium hydroxide (NaOH) solution (pH~12.5) in 1 L Schott bottles. These bottles were placed in a shaking incubator and shaken at 20 °C and 150 rpm for 24 h. The softened, alkali-treated cotyledons were gently mashed by a pestle and mortar to a consistent paste. The resultant paste was successively passed through 150 and 53 μm certified test sieves (Endecotts Ltd., London, England) by extensive washing with water. The cell extract was collected on the 53 μm sieve and ready for pretreatment with pepsin.

#### 2.2.3. Enzymatic Removal of Intracellular Protein Matrix

Enzymatic removal of the intracellular protein matrix was carried out by means of prehydrolysis of isolated navy bean cotyledon cells (INCs) with pepsin. The cell extract (~20 g) collected on the 53 μm sieve (as described in Section 2.2.2) was mixed with pepsin solution (3.2 g pepsin in 400 mL of 0.034 M sodium chloride (NaCl) buffer, pH 2.0) in 1 L Schott bottles. These bottles were shaken in a shaking incubator at 37 °C and 100 rpm for 1, 4, or 24 h. The pH of the cell slurries was occasionally checked and, if necessary, adjusted to 2.0 with 1.0 M HCl.

After the prehydrolysisn step, the cell slurries were transferred to 50 mL centrifuge tubes and centrifuged at 1500× $g$ for 10 min to recover the solids. The solid material was reslurried and washed with RO water before being recovered by centrifugation at 1500× $g$ for 10 min. The washing and centrifugation procedure was repeated five times to remove the remaining acids and protein digests. The pepsin-treated cells were dehydrated by rinsing in three changes of absolute ethanol (1 g of extract per 5 mL of ethanol) for 5 min each at room temperature and recovered by centrifugation at 1800× $g$ for 20 min. The cells were then spread onto clean glass plates and air-dried in a fume hood overnight at room temperature. The dried powder was bottled and stored at room temperature until further analysis.

INCs pretreated with pepsin for 1, 4, and 24 h were denoted as INC-1h, INC-4h, and INC-24h, respectively. A control sample (INC-Control) was prepared by treating the cell extract with pepsin-free NaCl buffer (0.034 M, pH 2.0) in the shaking incubator at 37 °C and 100 rpm for 24 h followed by ethanol dehydration and air drying. A native sample (INC-Native) was also prepared by subjecting the cell extract directly to the dehydration process without any chemical or enzymatic modification.

### 2.3. Determination of Physicochemical Properties

Moisture content was determined gravimetrically by drying the cell samples in an oven at 105 °C to a constant weight. Crude protein was analysed using the Dumas method (AOAC 968.06) [20] and a nitrogen-to-protein conversion factor of 6.25.

Total starch content was quantified using a total starch assay kit (K-TSTA). Total amylose content in starch was analysed using an amylose/amylopectin assay kit (K-AMYL). Both of the starch kits were obtained from Megazyme International Ireland Ltd., Wicklow, Ireland. The analysis was carried out according to the instructions given by the manufacturer. Chemical composition was expressed on a dry weight basis (dwb).

Swelling power and solubility of starch granules were determined after heating aqueous dispersions of INC, containing approximately 2% starch (*w*/*w*) at 90 °C for 30 min according to the method of Leach, McCowen, and Schoch [21].

### 2.4. Determination of α-Amylase Activity

Alpha-amylase activities of pancreatin (from porcine pancreas, 4× USP) and α-amylase (from porcine pancreas, type VI-B) were measured using the α-amylase enzymatic assay previously described by Bernfeld [22]. Enzyme solution was prepared by dissolving enzyme powder in RO water at a concentration of approximately 1 unit/mL of α-amylase. One-millilitre aliquots of the enzyme solution were added to 1 mL aliquots of soluble potato starch (1%, *w*/*v*) in 20 mM sodium phosphate buffer pH 6.9 in 15 mL Kimax screw-capped glass tubes. The capped tubes were mixed by swirling and placed in a water bath at 20 °C. After 3 min, the reaction was terminated by adding 1 mL of colour reagent solution (prepared by combining 8 mL of 5.3 M sodium potassium tartrate solution in 2 M NaOH with 20 mL of 96 mM 3,5-dinitrosalicylic acid and 12 mL of RO water) to each tube. The tubes were capped, mixed by swirling, and immediately incubated in a boiling water bath for 15 min. After cooling on ice for a few minutes, an additional 9 mL of RO water was added to each tube with inversion to mix the contents. The absorbance of the resulting coloured solution was measured spectrophotometrically at 540 nm. A blank assay was prepared by addition of 1 mL of the enzyme solution after adding the colour reagent and placing the tube in the boiling water bath. A standard calibration curve was prepared from a series of aqueous maltose solutions (0–2.0 mg/mL) and run parallel with the samples. The α-amylase activity was calculated and expressed in units of α-amylase per mg of enzyme powder. One unit (U) of amylase is defined as the amount of enzyme that liberates 1.0 mg of maltose from starch in 3 min at pH 6.9 and 20 °C.

### 2.5. In Vitro Starch Digestion

#### 2.5.1. Static In Vitro Starch Digestion Procedure

The static in vitro digestion protocol described by Dartois, Singh, Kaur, and Singh [23] was followed. This employed a two-stage method to simulate the gastric and small intestinal conditions for starch digestion. Simulated gastric and intestinal fluids (SGF and SIF) were prepared according to the US Pharmacopeia [24].

INCs were suspended in RO water in 400 mL glass beakers to obtain aqueous samples containing approximately 4% starch (*w*/*w*). The beakers were covered with aluminium foil and placed in a water bath at ~95 °C for 20 min to simulate cooking conditions. Cooked samples (~170 g) were subsequently cooled and introduced into 500 mL jacketed glass reactors. Temperature was maintained at 37 ± 1 °C by circulating water through the reactor jackets. pH was controlled using a pH meter and adjusted by manual addition of HCl (1.0 and 0.5 M) and/or NaOH (1.0 and 0.5 M). The content of each reactor was mechanically agitated using a magnetic stirrer bar at 300 rpm throughout digestion.

The reactor contents were first incubated with the SGF (17 mL) containing pepsin (enzyme/starch ratio, 1.765:100, *w*/*w*) at pH 1.2 for 30 min to simulate gastric digestion. After the gastric phase, the pH was adjusted to 6.8. The reactor contents were then incubated with the SIF containing pancreatin (enzyme/starch ratio, 1.3:100, *w*/*w*), amyloglucosidase (enzyme/starch ratio, 0.26:1, *v*/*w*), and invertase (enzyme/starch ratio, 1:1000, *w*/*w*) for 120 min to simulate small intestinal digestion.

Duplicate aliquots (0.5 mL) were withdrawn from the reactors after 1, 15, and 30 min of the gastric phase, and after 1, 5, 10, 15, 30, 60, 90, and 120 min of the small intestinal phase. The aliquots were immediately mixed with 2 mL of absolute ethanol in 15 mL centrifuge tubes to stop the enzymatic reaction. The resulting mixtures were vortex-mixed and centrifuged at 1800× *g* for 10 min. The ethanolic supernatants (0.1 mL) were incubated with 0.5 mL of amyloglucosidase and invertase in acetate buffer (0.1 mL amyloglucosidase and 3.75 mg invertase per 10 mL acetate buffer) at pH 5.2 and 37 °C for 10 min to completely convert soluble dextrins in the supernatants to glucose. The glucose released was quantified using a D-glucose assay kit (GOPOD-FORMAT, K-GLUC, Megazyme International Ireland Ltd., Wicklow, Ireland).

The percentage of starch hydrolysis at each sampling time point was used to construct hydrolysis curves and calculated using Equation (1):

$$\%SH = \frac{S_h}{S_i} \times 100 = 0.9 \times \frac{G}{S_i} \quad (1)$$

where $\%SH$ is the percentage of starch hydrolysis (%), $S_h$ is the amount of hydrolysed starch (g), $S_i$ is the initial amount of starch (g), and $G$ is the amount of glucose released (g). A factor of 0.9 is used to convert glucose to starch and is based on the molecular mass ratio of starch monomer to glucose (162/180 = 0.9).

The initial rate of starch hydrolysis for the first 10 min of reaction (R10), according to Ezeogu, Duodu, and Taylor [25], was calculated using Equation (2):

$$\boldsymbol{R10} = \frac{\boldsymbol{m}}{\boldsymbol{V} \times \boldsymbol{t}} \quad (2)$$

where $\boldsymbol{m}$ is the amount of starch hydrolysed (mg), $\boldsymbol{V}$ is the volume of reaction mixture at 10 min of the small intestinal digestion (mL), and $\boldsymbol{t}$ is the reaction time ($\boldsymbol{t}$ = 10 min).

### 2.5.2. Effect of Proteolytic Enzymes on In Vitro Starch Digestion

As illustrated in Figure 1, a set of four different experiments was designed to evaluate the indirect effect of proteolytic enzymes on in vitro digestion of starch in cooked samples of INC-Control. All experiments were performed in accordance with the in vitro digestion protocol suggested by Dartois et al. [23] (Section 2.5.1) but with different combinations of digestive proteases. All enzyme solutions were prepared freshly prior to analysis.

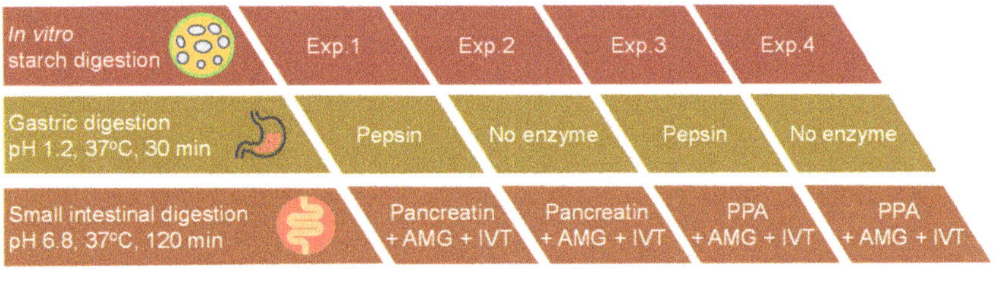


**Figure 1.** Schematic diagram of four different experiments for evaluating the indirect effect of proteolytic enzymes on in vitro starch digestion of cooked samples of control isolated navy bean cotyledon cells (INC-Control). Abbreviations: AMG—amyloglucosidase, IVT—invertase, PPA—porcine pancreatic α-amylase.

Four experiments (Exps.) were conducted as follows:

Exp.1: control with gastric pepsin (GP) in the SGF followed by pancreatin containing pancreatin proteases (PP) in the SIF.

Exp.2: only PP in the SIF.

Exp.3: only GP in the SGF.

Exp.4: without any protease.

For Exps. 3 and 4, porcine pancreatic α-amylase (PPA) was used in place of pancreatin (a commercial mixture of amylase, lipase, and protease from porcine pancreas) to achieve the unit of α-amylase activity per mL of the final digestion mixture (U/mL) similar to that of pancreatin. The α-amylase activities (U/mg solid) of enzymes were determined using the protocol of Benfield [22] as detailed in Section 2.4. The pancreatin and PPA used in this study were found to exhibit an α-amylase activity of 35.9 ± 0.8 and 17.2 ± 0.6 U/mg solid, respectively.

*2.6. Microscopy Analysis*

For light microscopy (LM), INCs were mounted onto glass microscope slides, suspended in water, sealed with coverslips, and then viewed under an Axiophot light microscope (Carl Zeiss, Jena, Germany) operating in brightfield mode using the objective of 20× magnification. Representative light micrographs of cells were captured using a Leica DFC320 camera equipped with the Leica software application suite LAS V3.8 (Leica Microsystems, Wetzlar, Germany). Digesta samples taken after 0 and 120 min of the small intestinal digestion were stained with 2% (*w*/*v*) Lugol's iodine solution and visualised under LM for detecting the presence of starch.

For scanning electron microscopy (SEM), INCs were directly mounted on double-sided adhesive tapes on aluminium stubs, sputter coated with gold (SCD 050, Balzers, Liechtenstein), and viewed under a scanning electron microscope (FEI Quanta 200 FEI Electron Optics, Eindhoven, the Netherlands). Representative electron micrographs of cell samples were captured with accelerating voltage of 25 kV and using the xT microscope software version 3.0.7 (FEI Quanta, Eindhoven, the Netherlands).

*2.7. Statistical Analysis*

Data were reported as means ± standard deviations for triplicate determinations unless otherwise specified. Tukey's test and analysis of variance (ANOVA) were used to assess the significance of differences ($p \leq 0.05$) between means using Minitab 18 software (Minitab Inc., State College, PA, USA).

## 3. Results and Discussion

*3.1. Microstructural Characteristics of Isolated Navy Bean Cotyledon Cells*

Isolated navy bean cotyledon cells (INC) were obtained using the sequential acid-alkali method (Do et al., 2019). This method enabled the easy separation of raw, intact cells without disrupting cell wall integrity and gelatinising starch. Representative light micrographs of INC (Figure 2A–E) confirmed good preservation of cell wall intactness and native starch granule structure after the cell isolation procedure and subsequent preincubation with pepsin. However, the cell samples contained some minor impurities, such as broken/damaged cells, free starch granules, and cell wall/protein fragments. Such impurities could have resulted from the breakage of the cell walls and consequent discharge of the cellular contents during the manual pestle-crushing process to separate the cells. In addition, the isolated navy bean starch (INS) sample was observed to be largely free of impurities (Figure 2F).

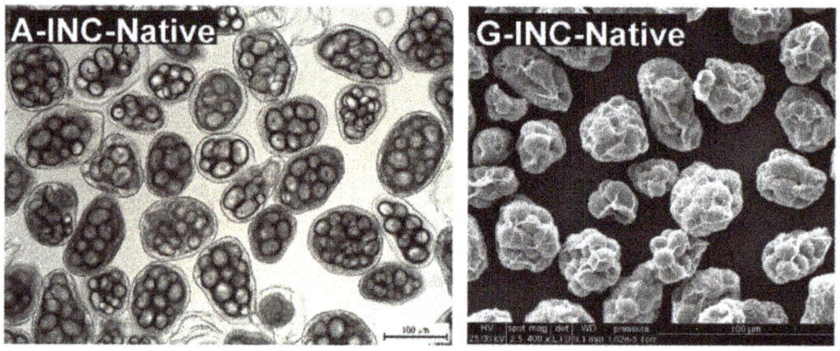


**Figure 2.** *Cont.*

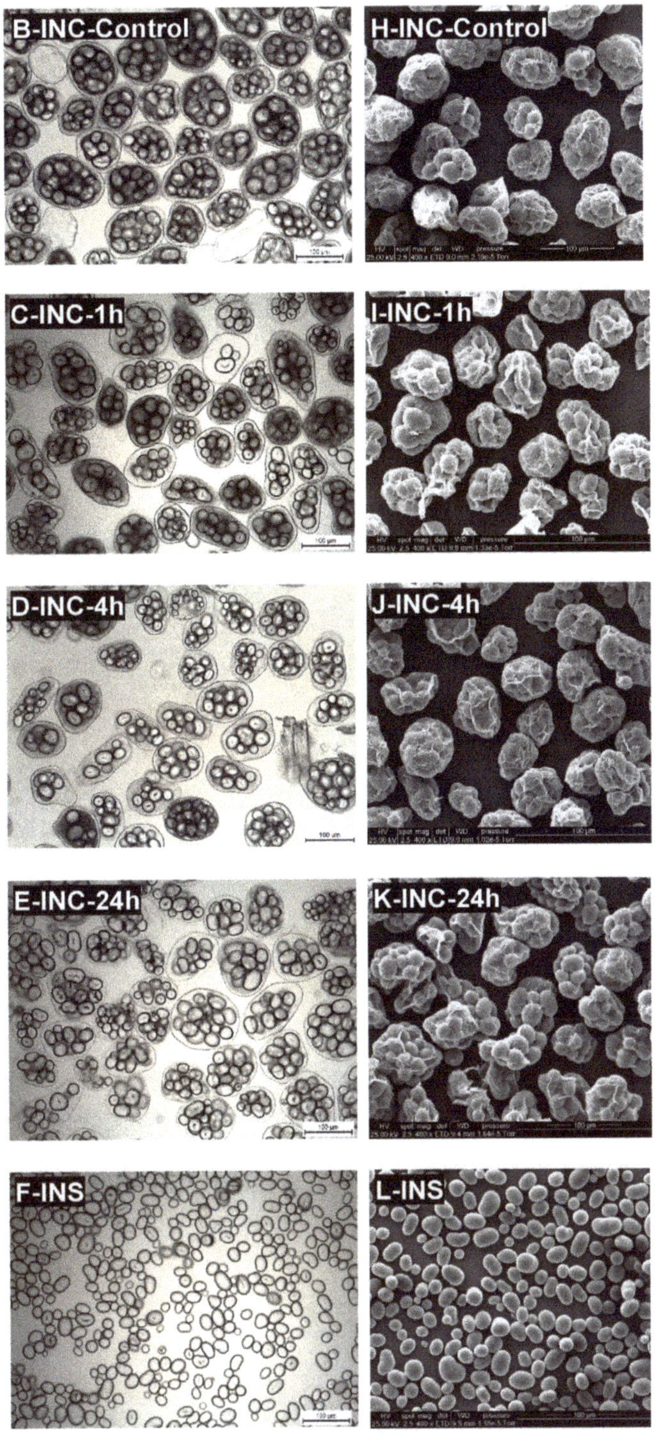


**Figure 2.** Representative brightfield light micrographs (**A–F**) and scanning electron micrographs (**G–L**) of isolated navy bean starch (INS), native INCs, control INCs, and INCs preincubated with pepsin for 1, 4, or 24 h at 37 °C and pH 2.0. Scale bar = 100 μm.

The microstructure of each native INC (Figure 2A) consists of multiple starch granules that are physically entrapped in a thick protein matrix and encased by an intact cell wall. The cotyledon cells and starch granules generally exhibit oval or round shapes. Similar microstructural observations of INCs have been reported previously [7,11]. No apparent morphological changes were detected in the control INCs (Figure 2B), indicating that the pepsin-free pretreatment of the native INCs did not significantly alter their cellular structures as expected.

Noticeably, the protein matrix appeared dark in contrast to the lighter background when viewed under LM (Figure 2A–D). The preincubation of the native INCs with pepsin at pH 2.0 and 37 °C resulted in the visual disappearance of the dark-coloured protein substance surrounding the starch granules, indicating hydrolysis of the protein matrix that had taken place inside the cells. As the preincubation time increased from 1 to 24 h, the protein matrix appeared progressively more transparent (lighter) (Figure 2C–E) and was no longer visually perceptible after 24 h (Figure 2E).

As is clearly shown by SEM micrographs in Figure 2, the INS (Figure 2L) displayed a smooth surface whereas the INCs (Figure 2G–K) possessed highly wrinkled surfaces, possibly due to extensive shrinkage and folding of the cell walls during the ethanol dehydration and air-drying process. The native starch granules were packaged into the INCs and wrapped around by the sheet-like cell walls while the protein matrix was not visually perceptible under SEM. These SEM observations agree well with those of previous reports [13,26]. Moreover, in line with the earlier LM observations, the cells seemed to have become progressively more transparent with longer exposure to pepsin, leading to greater visual clarity of the intracellular starch granules. It is possible that the protein matrix was partially removed from the cell cytoplasm when prehydrolysed with pepsin, leaving transparent cellular structures with distinct starch granules enclosed in loose cell walls.

### 3.2. Effect of the Intracellular Protein Matrix on Physicochemical Properties of Starch

The physicochemical properties of the INCs (control and pepsin-treated) and INS are presented in 1. The control INCs contained 64.3% starch and 17.6% protein (*w*/*w*, dwb). These values are within the range of previously published data [13]. Specifically, starch and protein are the two major components of legume cotyledon cells and constitute 57.2–69.0% and 17.3–20.3% of total dry cell mass, respectively [13].

Furthermore, it is evident from Table 1 that the preincubation with pepsin decreased the protein content while simultaneously increasing the total starch content of the INCs. As expected, a longer incubation time (1, 4, and 24 h) corresponded to a lower protein content (11.4, 9.2, and 2.7%) and a higher total starch content (70.5, 73.4, and 80.2%). The changes in the protein and starch contents were found to be statistically significant ($p \leq 0.05$). These quantitative results support the earlier LM observations of the visual disappearance of the protein matrix in pepsin-treated cells. Additionally, INS contained 91.9% and 0.3% of starch and residual protein, respectively. The amylose content was similar (~28%) across all samples ($p > 0.05$), indicating that neither the pretreatment duration nor pepsin had any effect on the amylose content.

Moreover, Table 1 shows that the swelling power (SP) and starch solubility (SS) of INC-Control did not differ significantly from those of INC-1h or INC-4h ($p > 0.05$). Only prolonged incubation with pepsin up to 24 h could result in significantly higher levels of SP and SS ($p \leq 0.05$). It has been suggested that the suppressed swelling and gelatinisation of starch granules entrapped in isolated cotyledon cells of legumes is attributed to limited water availability and spatial constraints inside cells [27,28]. This is perhaps a consequence of the physical confinement of starch granules in the cell wall/protein matrix as well as the competition for water and space with starch by nonstarch constituents such as proteins [13]. Therefore, as the prehydrolysis of INCs by pepsin progressed, more intracellular space that had been occupied by proteins was freed up. This might have allowed more water and free space to be made available inside cells for the swelling and dissolution of starch granules upon heating, resulting in greater leaching of soluble materials (mainly amylose) from the granules into the solution.

**Table 1.** Physicochemical properties of INS, control INCs, and INCs pretreated with pepsin for 1, 4, or 24 h.

| Navy Bean Material | AM (%) | TS (% dwb) | P (% dwb) | SP (g/g) | SS (%) |
|---|---|---|---|---|---|
| INC-Control | 28.8 ± 0.0 [a] | 64.3 ± 0.5 [e] | 17.6 ± 0.7 [a] | 5.2 ± 0.1 [d] | 5.5 ± 0.0 [c] |
| INC-1h | 28.6 ± 0.2 [a] | 70.5 ± 0.4 [d] | 11.4 ± 0.3 [b] | 5.4 ± 0.1 [cd] | 5.5 ± 0.0 [c] |
| INC-4h | 27.8 ± 0.3 [a] | 73.4 ± 0.2 [c] | 9.2 ± 0.3 [c] | 5.6 ± 0.0 [c] | 6.1 ± 0.1 [c] |
| INC-24h | 29.0 ± 0.0 [a] | 80.2 ± 0.5 [b] | 2.7 ± 0.2 [d] | 7.4 ± 0.2 [b] | 8.3 ± 0.3 [b] |
| INS | 28.0 ± 0.7 [a] | 91.9 ± 0.9 [a] | 0.3 ± 0.2 [e] | 12.6 ± 0.1 [a] | 16.9 ± 0.2 [a] |

[a,b,c,d,e] Values are means ± standard deviations of three determinations. Values with the same subscripts in a column do not differ significantly ($p > 0.05$). Abbreviations: AM—amylose content, TS—total starch content, P—protein content, SP—swelling power, SS—starch solubility.

It is also worth noting that, a vast proportion of the protein matrix (~85%) was effectively removed from the cells by pepsin after 24 h. Despite this, INC-24h exhibited considerably lower SP and SS compared to INS from which the cell structural barriers had been completely removed. This implies that the remaining proteins and the intact cell walls of INC-24h prevented complete starch gelatinisation together. From these findings, it appears likely that the limited SP and SS of INCs are linked to the barrier effects exerted by the cell wall and the protein matrix. These physical barriers combine to inhibit the swelling and gelatinisation of the intracellular starch granules during hydrothermal processing.

### 3.3. *Effect of the Intracellular Protein Matrix on In Vitro Starch Digestion*

Cooked samples of the control and pepsin-treated INCs, differing in their protein content, were subjected to in vitro gastric and small intestinal digestion. A cooked INS sample was included as a reference. Starch hydrolysis curves are shown in Figure 3. As clearly seen in this figure, virtually no hydrolysis occurred during 30 min of gastric digestion due to the absence of starch-hydrolysing enzymes. Starch was hydrolysed by pancreatic α-amylase during 120 min of small intestinal digestion. During this phase, a general trend was observed in all samples with an increased percentage of hydrolysis during the early stages before reaching a plateau towards the end of digestion.

As noted in Table 2, statistically significant differences in the digestion kinetic parameters were found among the samples ($p \leq 0.05$). The percentage of starch hydrolysed at 120 min of small intestinal digestion (H120, %) decreased in the following order: INS (85.1) > INC-24h (80.6) > INC-4h (77.6) > INC-1h (73.2) > INC-Control (70.6). In a somewhat similar pattern to that seen in H120, the initial rate of starch hydrolysis (R10, mg/mL/min) calculated for the first 10 min of reaction when it was highest decreased in the following order: INS (2.76) > INC-24h (2.11) ≈ INC-4h (2.07) > INC-1h (1.50) ≈ INC-Control (1.40). Evidently, both R10 and H120 progressively increased as the duration of the pepsin pretreatment increased from 1 to 24 h. These results demonstrate that the enzymatic removal of the protein matrix improves the rate of amylolysis at the initial stages of in vitro digestion and ultimately enhances the starch digestibility of INCs. This clearly proves that, aside from the intact cell wall, the protein matrix acts as an additional barrier to α-amylolysis of the enclosed starch. Similar findings have been previously reported for kidney bean cotyledon cells by Rovalino-Córdova et al. [19], although the conditions under which the cells were isolated by heating and then pretreated with proteases were different to those used in the present study.

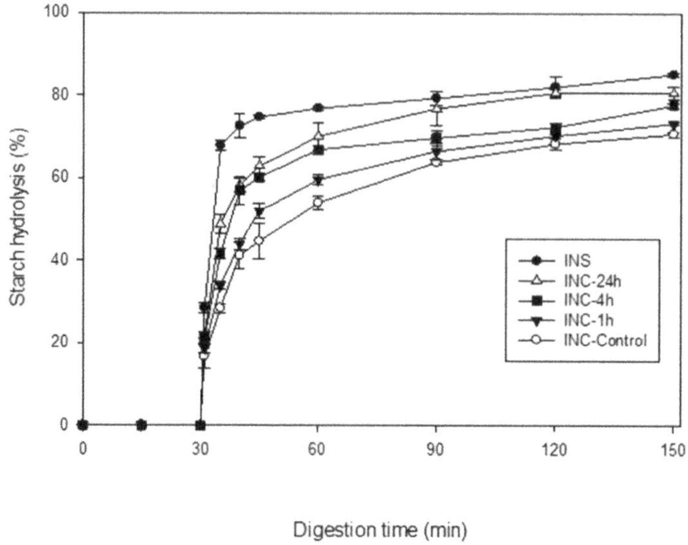


**Figure 3.** In vitro starch hydrolysis curves of cooked samples of control INCs (○) and INCs pretreated with pepsin for 1 h (▼), 4 h (■), and 24 h (△). INS (•) was included as a reference. Errors bars represent the standard deviations.

**Table 2.** Kinetic parameters of in vitro starch digestion of cooked samples of INS, control INCs, and pepsin-treated INCs.

| Navy Bean Materials | R10 (mg/mL/min) [f] | H120 (%) [g] |
|---|---|---|
| INC-Control | $1.40 \pm 0.11^{c}$ | $70.6 \pm 0.6^{e}$ |
| INC-1h | $1.50 \pm 0.03^{c}$ | $73.2 \pm 0.3^{d}$ |
| INC-4h | $2.07 \pm 0.13^{b}$ | $77.6 \pm 0.6^{c}$ |
| INC-24h | $2.11 \pm 0.13^{b}$ | $80.6 \pm 1.6^{b}$ |
| INS | $2.76 \pm 0.13^{a}$ | $85.1 \pm 0.4^{a}$ |

[a,b,c,d,e] Values are means ± standard deviations of three determinations. Values with the same subscripts in a column do not differ significantly ($p > 0.05$). [f] R10: initial rate of starch hydrolysis calculated for the first 10 min of reaction. [g] H120: values for percentage of starch hydrolysis determined experimentally at 120 min of small intestinal digestion.

Figure 4 shows representative light micrographs of INC samples before and after 120 min of the simulated small intestinal digestion. As can be seen from this figure, most cells maintained their structural integrity throughout in vitro digestion whereas only a minor portion of cells had broken and released their contents. In addition, a general trend was observed in all samples. Specifically, most cells became "emptier" after 120 min with the formation of empty "gap spaces" between the cellular contents and the peripheral cell walls. A possible explanation for this, which we have previously proposed [11], is that the starch hydrolysis by α-amylase progresses from the periphery towards the centre of the cells and, thus, causes "shrinking"/emptying of the cellular contents.

Staining of INCs with iodine enabled the visualisation of changes in the starch content. The blue–black colour intensity of amylose-iodine complexes is thought to be proportional to the amount of starch present in the cells, thus providing qualitative information on the extent of starch hydrolysis at specific digestion times. Emptying of the cellular contents and a decrease in iodine staining intensity were observed after 120 min in all samples. This effect occurred to a much greater extent in those with a longer duration of pepsin pretreatment. Evidently, in comparison with INCs preincubated with pepsin for only 1 h, those preincubated for 4 h or longer exhibited emptier cellular contents and greater reductions of iodine staining intensity after their exposure to small intestinal conditions, supporting their higher extents of amylolysis (H120) determined quantitatively. These findings are in line with previous reports investigating in vitro starch digestion kinetics of legume cotyledon cells [17,18,29].

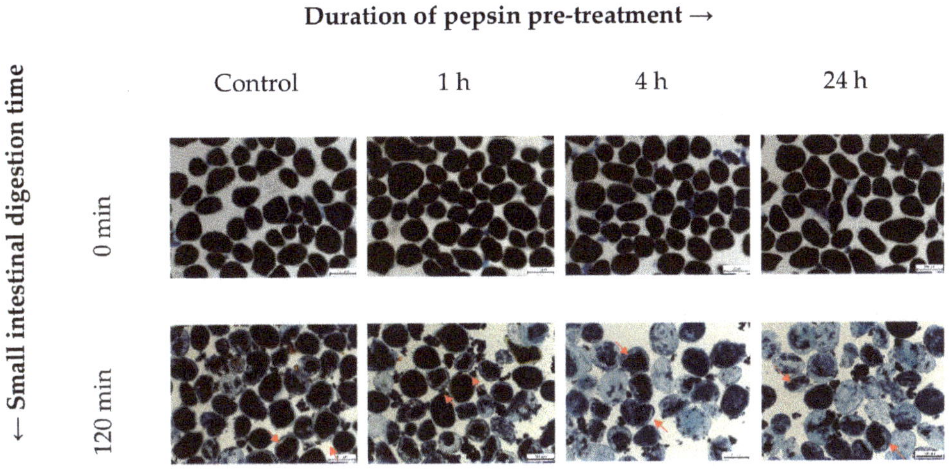


**Figure 4.** Representative brightfield light micrographs taken at 0 and 120 min of small intestinal digestion of control INCs and INCs pretreated with pepsin for 1, 4, and 24 h. The cells were stained with Lugol's iodine reagent for detecting the presence of starch. Red arrows indicate empty "gap spaces". Scale bar = 100 µm.

It is also noteworthy that the pretreatment of native INC with pepsin resulted in only partial protein hydrolysis. After 24 h, approximately 85% of protein digestion was achieved as calculated from Table 2. Despite the effective enzymatic removal of the bulk of the protein matrix, INC-24h had significantly lower values for R10 and H120 than INS (deprived entirely of cell structural barriers). This is mostly due to the retention of the intact cell walls in the INC-24h, providing a diffusion barrier to the passage of α-amylase. This evidence implies that the barrier effects of the protein matrix and the cell wall combine to inhibit in vitro starch digestion.

In an attempt to separate the net contribution towards the barrier effect of the protein matrix from that of the intact cell wall, Rovalino-Córdova et al. [19] incubated kidney bean cotyledon cells with proteases for 20 h in order to completely hydrolyse proteins prior to in vitro starch digestion with α-amylase. Contrary to expectations, it was noted that only 50% protein hydrolysis was achieved despite the prolonged protease pretreatment. Nevertheless, protease-treated cells showed a higher rate of starch hydrolysis at early digestion times compared to control cells (without protease pretreatment). These results are consistent with our findings showing the protein matrix as an additional barrier to in vitro starch digestion, but its individual contribution could not be quantitatively determined due to the incomplete protein prehydrolysis.

The starch–protein matrix, consisting of a compact protein matrix entrapping starch granules, is a distinctive microstructural feature of some natural and processed foods. Starch–protein interactions and physiochemical characteristics of the protein matrix (e.g., gluten network in pasta, disulphide-bonded kafirin network in sorghum, etc.) may play a pivotal role in retarding α-amylase digestion of starch, resulting in foods with slow starch digestion properties and potential to modulate glycaemic response [30–32]. However, due to the low content of sulphur amino acids in bean proteins, the slow starch digestion of INCs may be attributed primarily to the microstructural organisation of starch granules and proteins in the cell cytoplasm rather than the formation of disulphide bonds in proteins [19]. In fact, INCs can be represented as a double-layered encapsulation system. The protein matrix acts as an inner layer for coating starch granules while the cell wall acts as an outer layer for coating the starch granule–protein matrix. These dual encapsulation layers provide double protection against amylolytic degradation, perhaps through limiting the swelling/gelatinisation of starch granules (as evidenced by SP and SS data presented in Table 1, Section 3.2) as well as hindering α-amylase access/binding to starch. Therefore, the breakdown of either the cell wall or protein barriers can lead to a substantial increase in the rate and extent of starch hydrolysis as has been demonstrated in this and other studies [7,17–19,33].

### 3.4. Effect of Proteolytic Enzymes on In Vitro Starch Digestion

As demonstrated above, the protein matrix plays a crucial part in limiting in vitro digestion of starch in INCs. Since dietary proteins are susceptible to proteolysis catalysed by GP in the SGF and PP (mainly trypsin and chymotrypsin) in the SIF, it is necessary to further explore the indirect effect of these proteolytic enzymes on starch digestion.

Figure 5 shows in vitro starch digestion curves of cooked samples of INC-Control from four experiments differing in the combination of proteases in the digestive fluids: (Exp.1) GP followed by PP, (Exp.2) GP alone, (Exp.3) PP alone, and (Exp.4) no protease. As evident in this figure, all four combinations displayed a similar starch hydrolysis pattern. Specifically, virtually no starch was broken down during 30 min of gastric digestion. This was followed by a gradual rise in the percentage of hydrolysis during 120 min of small intestinal digestion. As shown in Table 3, the initial rate of amylolysis (R10, mg/mL/min) decreased in the following order: Exp.1 (1.40) $\approx$ Exp.3 (1.41) > Exp.2 (0.97) $\approx$ Exp.4 (1.04). Meanwhile, the extent of amylolysis (H120, %) decreased in the following order: Exp.1 (70.6) $\approx$ Exp.2 (70.8) $\approx$ Exp.3 (69.9) > Exp.4 (65.8).

The quantitative data clearly show that both R10 and H120 values were significantly reduced when cooked control INCs were digested in the absence of all proteases. This is consistent with previous work conducted by Rovalino-Córdova et al. [18,19]. These authors found a significant decrease in the rate and extent of α-amylase digestion of starch when kidney bean cotyledon cells were digested with the exclusion of all proteolytic enzymes (i.e., pepsin in the SGF, trypsin and chymotrypsin in the SIF). Wang, Li, Zhang, Wang, and Copeland [34] also found that the greatest starch digestibility of cooked rice was achieved through a combination of amylolytic enzymes (PPA and AMG) and proteolytic enzymes (pepsin, trypsin, and chymotrypsin). Conversely, the total exclusion of all proteases resulted in the lowest starch digestibility. These findings have demonstrated that efficient enzymatic hydrolysis of the protein matrix is necessary for improving starch digestibility, which has previously been described as a cooperative process [18]. In addition, digestion with only PP resulted in a higher R10 value but a similar H120 value compared to digestion with only GP, which may be suggestive of the effectiveness of different proteolytic enzymes in degrading the protein matrix. This is supported by the previous suggestion that PP are thought to be more efficient in hydrolysing dietary proteins than GP [19].

Overall, the results of this study have provided strong evidence that the natural presence of proteolytic enzymes in both the stomach and the small intestine of the human GI tract is necessary for efficient in vitro digestion of starch inside INCs. Initially, GP is likely to play a role in loosening the compact starch granule–protein matrix. Subsequently, extensive protein degradation by PP in the small intestine may facilitate enzyme mobility within the cell cytoplasm and enhance starch–amylase interactions. Considering the facilitating effect of proteolytic enzymes on starch digestion, the encapsulation of starch granules in double-layered INC with the protein matrix (inner layer) being surrounded by the cell wall (outer layer) seems to be advantageous in the sense that the protein matrix is shielded from proteases, making it more resistant to proteolysis while providing extra protection for starch.

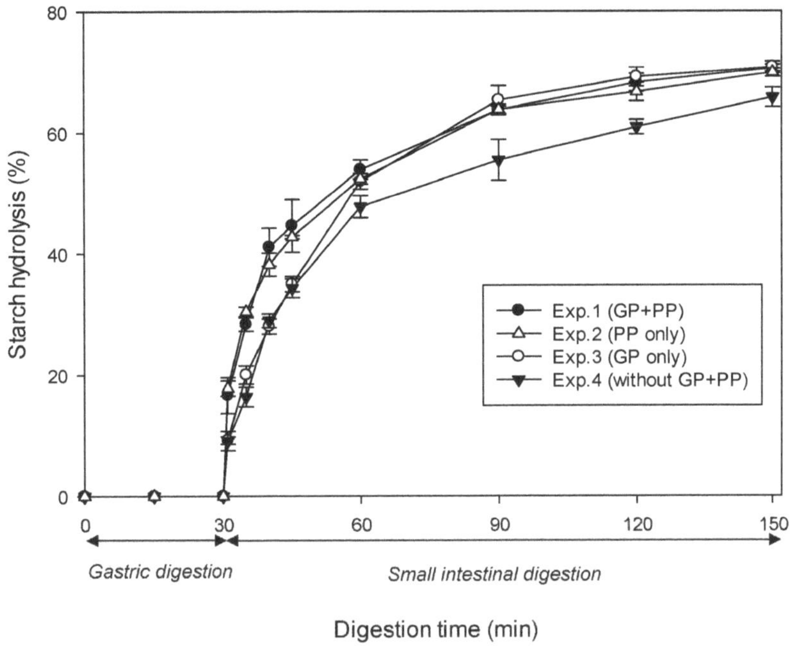


**Figure 5.** In vitro starch hydrolysis curves of cooked samples of control INCs from four experiments differing in protease combinations. Symbols (•, △, ○, ▼) represent the four experiments (see Experimental Section 2.5.2). Errors bars represent the standard deviations. Abbreviations: GP—gastric pepsin, PP—pancreatin proteases.

**Table 3.** Kinetic parameters of in vitro starch digestion of cooked samples of control INCs from four experiments differing in protease combinations.

| Experiments | R10 (mg/mL/min) [c] | H120 (%) [d] |
|---|---|---|
| Exp.1 (GP + PP) | 1.40 ± 0.11 [a] | 70.6 ± 0.6 [a] |
| Exp.2 (PP only) | 1.41 ± 0.07 [a] | 69.9 ± 0.7 [a] |
| Exp.3 (GP only) | 0.97 ± 0.04 [b] | 70.8 ± 0.9 [a] |
| Exp.4 (without GP + PP) | 1.04 ± 0.04 [b] | 65.8 ± 1.6 [b] |

[a,b] Values are means ± standard deviations of three determinations. Values with the same subscripts in a column do not differ significantly ($p > 0.05$). [c] R10: initial rate of starch hydrolysis calculated for the first 10 min of reaction. [d] H120: values for percentage of starch hydrolysis determined experimentally at 120 min of small intestinal digestion.

## 4. Conclusions

The present study has unveiled intriguing new insights into the role of the protein matrix in modulating the in vitro digestion of starch in navy bean cotyledon cells. We have shown that the entrapment of starch granules in the protein matrix may restrict their swelling and gelatinisation. In addition to the cell wall, the protein matrix presents a secondary physical barrier blocking the access/binding of α-amylase to starch. Proteolytic degradation of the protein matrix removes these restrictions, rendering starch more susceptible to α-amylolysis. Hence, efficient protein hydrolysis before or during simulated GI digestion is a necessary prerequisite for improving starch digestibility. Finally, as proposed by Do, Singh, Oey, and Singh [35], the unique cotyledon cell structure serves as a great source of inspiration for designing biomimetic materials. Encapsulation of starch granules within a protein core and a polysaccharide shell to form a double-layered structure is a novel strategy to fabricate novel food-grade particles for reduced glycaemic impact.

**Author Contributions:** Conceptualization, D.T.D., J.S. and H.S.; methodology, D.T.D., J.S. and H.S.; data curation, D.T.D.; writing—original draft preparation, D.T.D.; writing—review and editing, D.T.D., J.S., S.J. and H.S. All authors have read and agreed to the published version of the manuscript.

**Funding:** This research was funded by the Riddet Institute, Massey University, New Zealand under the CoRE Research Programme.

**Institutional Review Board Statement:** Not applicable.

**Informed Consent Statement:** Not applicable.

**Data Availability Statement:** Not applicable.

**Conflicts of Interest:** The authors declare no conflict of interest.

## References

1. Becerra-Tomás, N.; Díaz-López, A.; Rosique-Esteban, N.; Ros, E.; Buil-Cosiales, P.; Corella, D.; Estruch, R.; Fitó, M.; Serra-Majem, L.; Aros, F.; et al. Legume consumption is inversely associated with type 2 diabetes incidence in adults: A prospective assessment from the PREDIMED study. *Clin. Nutr.* **2018**, *37*, 906–913. [CrossRef] [PubMed]
2. Sievenpiper, J.; Kendall, C.; Esfahani, A.; Wong, J.; Carleton, A.; Jiang, H.Y.; Jenkins, D.J.A. Effect of non-oil-seed pulses on glycaemic control: A systematic review and meta-analysis of randomised controlled experimental trials in people with and without diabetes. *Diabetologia* **2009**, *52*, 1479–1495. [CrossRef] [PubMed]
3. Villegas, R.; Gao, Y.-T.; Yang, G.; Li, H.-L.; Elasy, T.A. Legume and soy food intake and the incidence of type 2 diabetes in the Shanghai Women's Health Study. *Am. J. Clin. Nutr.* **2008**, *87*, 162–167. [CrossRef] [PubMed]
4. Jenkins, D.; Ghafari, H.; Wolever, T.; Taylor, R.; Jenkins, A.L.; Barker, H.M.; Bowling, A.C. Relationship between rate of digestion of foods and post-prandial glycaemia. *Diabetologia* **1982**, *22*, 450–455. [CrossRef]
5. Tharanathan, R.; Mahadevamma, S. Grain legumes—A boon to human nutrition. *Trends Food Sci. Technol.* **2003**, *14*, 507–518. [CrossRef]
6. Tovar, J. Bioavailability of carbohydrates in legumes: Digestible and indigestible fractions. *Arch. Latinoam. Nutr.* **1996**, *44*, 36S–40S.
7. Berg, T.; Singh, J.; Hardacre, A.; Boland, M.J. The role of cotyledon cell structure during in vitro digestion of starch in navy beans. *Carbohyd. Poly.* **2012**, *87*, 1678–1688. [CrossRef]
8. Brummer, Y.; Kaviani, M.; Tosh, S.M. Structural and functional characteristics of dietary fibre in beans, lentils, peas and chickpeas. *Food Res. Internat.* **2015**, *67*, 117–125. [CrossRef]
9. Dhital, S.; Bhattarai, R.R.; Gorham, J.; Gidley, M.J. Intactness of cell wall structure controls the in vitro digestion of starch in legumes. *Food Funct.* **2016**, *7*, 1367–1379. [CrossRef]
10. Do, D.T.; Singh, J.; Oey, I.; Singh, H. Modulating effect of cotyledon cell microstructure on in vitro digestion of starch in legumes. *Food Hydrocol.* **2019**, *96*, 112–122. [CrossRef]
11. Do, D.T.; Singh, J.; Oey, I.; Singh, H.; Yada, R.Y.; Frostad, J.M. A novel apparatus for time-lapse optical microscopy of gelatinisation and digestion of starch inside plant cells. *Food Hydrocol.* **2020**, *104*, 105551. [CrossRef]
12. Noah, L.; Guillon, F.; Bouchet, B.; Buleon, A.; Molis, C.; Gratas, M.; Champ, M. Digestion of carbohydrate from white beans (*Phaseolus vulgaris* L.) in healthy humans. *J. Nutr.* **1998**, *128*, 977–985. [CrossRef] [PubMed]
13. Do, D.T.; Singh, J. Legume microstructure. In *Encyclopedia of Food Chemistry*; Varelis, P., Melton, L.D., Shahidi, F., Eds.; Elsevier: Amsterdam, The Netherlands, 2019; Volume 3, pp. 15–21. [CrossRef]
14. Pallaresares, A.P.; Gwala, S.; Pälchen, K.; Duijsens, D.; Hendrickx, M.; Grauwet, T. Pulse seeds as promising and sustainable source of ingredients with naturally bioencapsulated nutrients: Literature review and outlook. *Comp. Rev. Food Sci. Food Saf.* **2021**, *20*, 1524–1553. [CrossRef] [PubMed]
15. Bhattarai, R.R.; Dhital, S.; Wu, P.; Chen, X.D.; Gidley, M.J. Digestion of isolated legume cells in a stomach-duodenum model: Three mechanisms limit starch and protein hydrolysis. *Food Func.* **2017**, *8*, 2573–2582. [CrossRef] [PubMed]
16. Li, H.; Gidley, M.J.; Dhital, S. Wall porosity in isolated cells from food plants: Implications for nutritional functionality. *Food Chem.* **2019**, *279*, 416–425. [CrossRef] [PubMed]
17. Pallares, A.P.; Miranda, B.A.; Truong, N.Q.A.; Kyomugasho, C.; Chigwedere, C.M.; Hendrickx, M.; Grauwet, T. Process-induced cell wall permeability modulates the in vitro starch digestion kinetics of common bean cotyledon cells. *Food Func.* **2018**, *9*, 6544–6554. [CrossRef]
18. Rovalino-Córdova, A.M.; Fogliano, V.; Capuano, E. A closer look to cell structural barriers affecting starch digestibility in beans. *Carbohy. Poly.* **2018**, *181*, 994–1002. [CrossRef]
19. Rovalino-Córdova, A.M.; Fogliano, V.; Capuano, E. The effect of cell wall encapsulation on macronutrients digestion: A case study in kidney beans. *Food Chem.* **2019**, *286*, 557–566. [CrossRef]
20. AOAC. Method 968.06. Protein (crude) in animal feed. In *Official Methods of Analysis of the Association of Official Analytical Chemists (AOAC)*; Latimer, G.W., Horwitz, W., Eds.; AOAC International: Gaithersburg, MD, USA, 2005.
21. Leach, H.; McCowen, D.; Schoch, T. Swelling and solubility patterns of various starches, structure of starch granule. *Cereal Chem.* **1959**, *36*, 534–544.

22. Bernfeld, P. Amylases, α and β. *Methods Enzymol.* **1955**, *1*, 149–158.
23. Dartois, A.; Singh, J.; Kaur, L.; Singh, H. Influence of guar gum on the in vitro starch digestibility—Rheological and microstructural characteristics. *Food Biophys.* **2010**, *5*, 149–160. [CrossRef]
24. Pharmacopeia, U.S. Pharmacopeia, U.S. Pharmacopeia, simulated gastric fluid, TS, simulated intestinal fluid, TS, United States Pharmacopeial Convention. In *The National Formulary 9*; U.S. Pharmacopeia Board of Trustees: Rockville, MD, USA, 2000; Volume 24, p. 2235.
25. Ezeogu, L.; Duodu, K.; Taylor, J. Effects of endosperm texture and cooking conditions on the in vitro starch digestibility of sorghum and maize flours. *J. Cereal Sci.* **2005**, *42*, 33–44. [CrossRef]
26. Xiong, W.; Zhang, B.; Huang, Q.; Li, C.; Pletsch, E.A.; Fu, X. Variation in the rate and extent of starch digestion is not determined by the starch structural features of cooked whole pulses. *Food Hydrocol.* **2018**, *83*, 340–347. [CrossRef]
27. Fujimura, T.; Kugimiya, M. Gelatinization of starches inside cotyledon cells of kidney beans. *Starch-Stärke* **1994**, *46*, 374–378. [CrossRef]
28. Fujimuru, T.; Kugimiya, M. Estimation of intracellular water content during gelatinization of starches inside cotyledon cells of legumes. *J. App. Glycosci.* **1995**, *42*, 7–13.
29. Gwala, S.; Pallares, A.P.; Pälchen, K.; Hendrickx, M.; Grauwet, T. In vitro starch and protein digestion kinetics of cooked Bambara groundnuts depend on processing intensity and hardness sorting. *Food Res. Internat.* **2020**, *137*, 109512. [CrossRef]
30. Ezeogu, L.I.; Duodu, K.G.; Emmambux, M.N.; Taylor, J.R. Influence of cooking conditions on the protein matrix of sorghum and maize endosperm flours. *Cereal Chem.* **2008**, *85*, 397–402. [CrossRef]
31. Kim, E.H.J.; Petrie, J.R.; Motoi, L.; Morgenstern, M.P.; Sutton, K.H.; Mishra, S.; Simmons, L.D. Effect of structural and physico-chemical characteristics of the protein matrix in pasta on in vitro starch digestibility. *Food Biophys.* **2008**, *3*, 229–234. [CrossRef]
32. Li, H.T.; Li, Z.; Fox, G.P.; Gidley, M.J.; Dhital, S. Protein-starch matrix plays a key role in enzymic digestion of high-amylose wheat noodle. *Food Chem.* **2021**, *336*, 127719. [CrossRef]
33. Pallares, A.P.; Rousseau, S.; Chigwedere, C.M.; Kyomugasho, C.; Hendrickx, M.; Grauwet, T. Temperature-pressure-time combinations for the generation of common bean microstructures with different starch susceptibilities to hydrolysis. *Food Res. Internat.* **2018**, *106*, 105–115. [CrossRef]
34. Wang, S.; Li, P.; Zhang, T.; Wang, S.; Copeland, L. Trypsin and chymotrypsin are necessary for in vitro enzymatic digestion of rice starch. *RSC Adv.* **2017**, *7*, 3660–3666. [CrossRef]
35. Do, D.T.; Singh, J.; Oey, I.; Singh, H. Biomimetic plant foods: Structural design and functionality. *Trends Food Sci. Technol.* **2018**, *82*, 46–59. [CrossRef]

nutrients

*Article*

# Legume Intake, Body Weight, and Abdominal Adiposity: 10-Year Weight Change and Cross-Sectional Results in 15,185 U.S. Adults

**Larry A. Tucker**

College of Life Sciences, Brigham Young University, Provo, UT 84602, USA; tucker@byu.edu; Tel.: +1-801-422-4927

**Abstract:** There were three objectives: (1) evaluate the relationship between legume intake and weight change across the previous 10 years, (2) examine the cross-sectional associations between legume consumption, BMI, and abdominal adiposity, and (3) determine if the relationship between legume intake and the outcomes were influenced by multiple covariates, particularly fiber intake. The sample included 15,185 randomly selected adults representative of the U.S. population. Percent change in weight was used as the outcome measure for the 10-year analysis. BMI, and waist circumference, corrected for height, were employed as the outcomes for the cross-sectional analyses. Legume, fiber, and energy intakes were measured using the average of two 24-h dietary recalls. Legume intake was divided into three categories. Five demographic and five lifestyle covariates were controlled statistically. There was an inverse dose-response relationship between legume intake and percent weight change over the previous 10 years after adjusting for 9 of the covariates (F = 6.5, $p$ = 0.0028). However, after controlling for fiber with the other covariates, there were no differences across the three legume intake groups (F = 1.9, $p$ = 0.1626). The cross-sectional findings showed similar inverse dose-response results until fiber intake was controlled. Then the associations became non-significant. In conclusion, legume intake is a good predictor of percent weight change over the previous 10 years, and it is also a significant predictor of BMI and abdominal adiposity cross-sectionally. These relationships are strongly influenced by fiber consumption. Evidently, legumes have dietary advantages, especially high fiber levels, that seem to be valuable in the battle against weight gain and obesity.

**Keywords:** beans; fiber; pulses; carbohydrate; fat; protein; NHANES; obesity; waist

**Citation:** Tucker, L.A. Legume Intake, Body Weight, and Abdominal Adiposity: 10-Year Weight Change and Cross-Sectional Results in 15,185 U.S. Adults. *Nutrients* **2023**, *15*, 460. https://doi.org/10.3390/nu15020460

Academic Editors: Stuart Johnson and Rewati Bhattarai

Received: 23 November 2022
Revised: 25 December 2022
Accepted: 11 January 2023
Published: 16 January 2023



## 1. Introduction

Legumes have numerous nutritional characteristics that have the potential to decrease the risk of disease [1–3]. Meta-analyses show that legume intake is inversely associated with incident cardiovascular disease, including coronary heart disease and cardiometabolic risk factors [1,2,4–6]. Despite the many disease-prevention and health-enhancing qualities associated with legume intake, consumption levels tend to be modest in the U.S. and appear to be decreasing [7,8]. Based on purchasing data from grocery stores, chain supermarkets, club stores, and other retail outlets, Semba et al. concluded in 2021, "Although legumes are inexpensive, healthy, and a sustainable source of protein, per capita legume intake remains low in the U.S. and below dietary guidelines" [9].

One of the most common chronic diseases in the United States is obesity. It affects about 43% of the adult population, with another 31% classified as overweight [10]. Research indicates that regular legume consumption may help prevent obesity [11]. Legumes, particularly beans, include only small amounts of dietary fat and large amounts of dietary fiber. Moreover, the glycemic index of beans is low. They also contain high levels of plant protein. They are satiating, and they can enhance the gut microbiome. Consequently, beans and other legumes have many qualities that may help in the fight against weight gain and its consequences [11,12].

Although legumes have many nutritional characteristics that should make them a desirable food when the prevention of obesity is the goal, individual weight loss studies focusing on legumes have rarely resulted in favorable outcomes [13,14]. However, when combined using meta-analysis, results suggest that legume consumption may be a good choice for weight loss [2,14]. Clearly, individual investigations and meta-analyses differ in their findings and in their conclusions. Consequently, Williams et al. determined, "There is insufficient evidence to make clear conclusions about the protective effect of legumes on weight" [15].

More than 20 years ago, Papanikolaou et al. studied the relationship between bean consumption and a variety of outcomes using a cross-sectional design [16]. The present study and the investigation by Papanikolaou et al. have some similarities and many differences. Papanikolaou et al. focused on bean consumption, primarily baked beans, not legume intake. The earlier study compared bean consumers to non-consumers. Intake was not quantified, and weight change over time was not addressed. Only one dietary recall assessment was used, yet the Papanikolaou investigation found that bean eaters had lower body weights and lower waist circumferences than non-consumers [16].

According to the literature, few studies have examined the influence of legume intake on 10-year weight change in a large sample of adults representing the U.S. population. Moreover, investigations focusing on the relationship between legume intake and abdominal adiposity are scarce. Finally, few, if any, studies have examined the mediating influence of dietary fiber on the association between legume intake and measures of body mass and abdominal obesity. Hence, the present investigation was conducted to reduce some of the potential issues identified in other studies. In particular, a large representative sample of the U.S. population was utilized (n = 15,185). Moreover, changes in body mass over 10 years were assessed, and cross-sectional relationships were also examined.

Given previous research, there were three key objectives of this study: (1) evaluate the relationship between legume intake and weight change across the previous 10 years, (2) examine the cross-sectional associations between legume consumption, body mass, and abdominal adiposity, and (3) determine the extent to which the relationship between legume intake and the study outcomes were a result of differences in several covariates, particularly dietary fiber intake.

## 2. Materials and Methods

### *2.1. Study Design and Sample*

Study participants were chosen randomly to participate in the U.S. National Health and Nutrition Examination Survey (NHANES). Data were collected by NHANES over 8 years, from 2011 to 2018. Participants were assigned specific sample weights by NHANES based on census outcomes. Consequently, results from the present study can be generalized to the non-institutionalized adult population of the U.S.

The Ethics Review Board for the U.S. National Center for Health Statistics (NCHS) approved the data gathering procedures and publishing of the NHANES data online. The files posted online by NHANES contain no confidential information and are free to the public. The protocol codes indicating approval for NHANES data collected from 2011 to 2018 were: Protocol #2011–17 and Protocol #2018-01. Each subject provided written consent to take part in the national survey.

The present study's sample size varied based on if the focus was 10-year weight change or cross-sectional outcomes. A total of 10,137 subjects were part of the 10-year analyses, whereas 15,185 individuals were included in the cross-sectional analyses. The difference was because the 10-year analyses included subjects 36–75 years of age, whereas the cross-sectional part of the study included subjects 18–75 years old. If the 10-year analyses had included subjects as young as 18, then their initial body mass would have been based on their weight at 8 years of age. Therefore, a minimum subject age limit was set at 36 years for the 10-year weight change variable, so only fully grown individuals (those ≥26 years old) were included.

Subjects who were underweight (BMI < 18.5) were not included in the sample because of the risk of an eating disorder or severe illness (n = 270). Pregnant women (n = 177) and adults who fasted or who reported extremely low 24-h energy intakes (3 standard deviations or more below the mean: <492 kcal) were also excluded from the analyses (n = 361).

*2.2. Instrumentation and Measurement Methods*

The association between legume intake and 10-year percent weight change and cross-sectional BMI and abdominal adiposity were evaluated. Age, sex, race/ethnicity, economic status, year of assessment, energy intake, total physical activity, cigarette smoking, alcohol use, and fiber intake were used as covariates, with a special focus on fiber intake.

Diet. Legume consumption, fiber intake, and energy consumption were each indexed using the average of two 24-h dietary recall evaluations. Both dietary recalls collected comprehensive data about all the foods and beverages consumed during the 24-h prior to the interview (midnight to midnight). The first of the two dietary interviews were conducted in person in a private room in the NHANES mobile examination center. The second interview was conducted by telephone 3–10 days after the in-person assessment. The diet interview was based on the "What We Eat in America" partnership between the U.S. Department of Agriculture (USDA) and the U.S. Department of Health and Human Services. A variety of measuring guides, such as glasses, mugs, spoons, bowls, bottles, cups, thickness sticks, a ruler, etc., were available to help subjects accurately report food quantities. After finishing the initial in-person diet assessment, participants were given sample bowls, glasses, etc., and a food model booklet to help them during the upcoming telephone dietary evaluation.

The personnel conducting the diet recalls were multi-lingual. Interviewers were directed by scripts, and the computer-based program provided a standard interview process. The diet evaluations utilized a multi-pass format called the Automated Multiple Pass Method (AMPM), available online [17]. This dietary assessment utilized food probes to maximize recall of all foods and beverages consumed. The United States Department of Agriculture (USDA) National Food and Nutrient Database was used to measure food and nutrient content [18]. It is evaluated and updated regularly and stands as the gold standard within the United States [19].

Results derived from the NHANES 24-h diet recall protocol have been studied comprehensively, and predictive validity has been confirmed by many investigations. Many studies have shown that dietary findings derived from the recalls are predictive of a host of chronic and acute conditions [20–25].

Harvard nutritional epidemiologist, Walter Willett, indicates that a single 24-h dietary recall assessment is likely sufficient if the sample size is large. Additionally, he states that to estimate within-person variability, "it is statistically most efficient to increase the number of individuals in the sample, rather than to increase the number of days beyond 2 days per individual" (page 55) [26]. Taking into account that the current study included a sample of over 15,000 adults, each completing two 24-h dietary recalls, the assessment strategy utilized was suitable to secure quality estimates of dietary consumption.

Legume Intake. Total legume intake was used as the exposure variable. Because legume consumption tends to increase as total energy intake increases, legume intake was indexed as average grams consumed per 1000 kilocalories (kcal) over the two independent days of dietary evaluation. Included were dried beans (nonspecific), canned beans, chickpeas (nonspecific, dried, or canned), beans from fast-food, bean dip, black-eyed peas, kidney beans, white beans, black beans, fava beans, lima beans, navy beans, pink beans, pinto beans, Peruvian beans, soybeans, split peas, wasabi peas, mung beans, baked beans, refried beans, pork and beans, lentil curry, and other legumes.

Legume intake was divided into 3 categories: None, Low, and Moderate/High. The "None" group included adults reporting no legume consumption during the two dietary recall assessments. The Low group comprised adults reporting some legume consumption, but less than 47 g per 1000 kcal, on average, over the two days of dietary evaluation. The Moderate/High category included men and women who ate 47 g or more of legumes per 1000 kcal, on average, during the two days of dietary assessment. The Low and Moderate/High intake groups included all participants who reported eating legumes, which was about 24% of the sample (n = 3614), and these two groups were split precisely in half, so the Low group included 12% of the sample and the Moderate/High category also included 12% of the sample.

Fiber Intake. Dietary fiber was employed as a covariate to determine the extent that the relationship between legume intake and weight-related outcomes was influenced by differences in dietary fiber. Grams of soluble fiber and insoluble fiber were included in the total. Total dietary fiber was expressed as grams consumed per 1000 kcal.

Energy Intake. Total energy (kcal) consumption was based on all foods and beverages consumed during the previous 24 h (midnight to midnight). Energy intake was employed as a reference variable for legume and fiber consumption, as both were reported in grams per 1000 kcal. Participants who significantly underreported their energy consumption were not included in the sample. Specifically, under-reporting was defined as intakes that were 3 or more standard deviations below the mean (<492 kcal).

10-Year Percent Weight Change. A digital scale was used to weigh subjects in kilograms (kg). The scale was checked each day using precise weights. A standard paper gown was worn during the assessment. It included disposable pants, a shirt, and slippers. Under the gown, underwear was worn. To assess 10-year weight change, subjects reported their body weight from 10 years previous. That weight was subtracted from their present weight. Percent weight change was then calculated by dividing the difference between the two weights by the initial weight. For example, if a subject's current weight was 90 kg and initial weight was 80 kg, then the percent weight change was $(90 - 80)/80 = 12.5\%$.

According to a number of high-quality studies, self-reported body weight is closely related to measured weight. In a study by Stunkard et al. [27] using more than 1300 men and women, the two measures were almost perfectly correlated (r = 0.99, $p < 0.01$). Stunkard concluded, "Self-reported weights were remarkably accurate across all these variables in the American sample, even among obese people, and may obviate the need for measured weights in epidemiological investigations" (p. 1593). Similarly, investigations by other researchers studying the relationship between self-reported and measured body weights have resulted in strong correlations (r = 0.98) [28].

Studies comparing current body weights to recalled weights of the past have also produced very similar values. For example, when subjects were asked to remember their draft registration weights at age 25 when they were 45–55 years old, on average, the difference was only 2.2% [29]. Likewise, in a large sample, women were required to remember their body weight at age 18 when they were 25–42 years old. The two weights were highly correlated (r = 0.87, $p < 0.01$) and only differed by 1.4 kg on average [30].

In more recent investigations, Kyulo et al. examined the validity of recalled body weight over a 26-year period in more than 2700 older adults (mean age: 70 years). The average difference was 0.67 kg, and the two weights were highly correlated (r = 0.88, $p < 0.0001$) [31]. The authors concluded that the "validity of 26-year recall of body weight during adulthood was very high in an older sample" [31]. Finally, Tamakoshi et al. compared recalled body weights at 25 years of age with measured weights at age 25 in 2453 men aged 34–61 [32]. The recalled weights were strongly correlated with the measured weights (r = 0.85), and the difference was 1.3 kg.

Willett showed that measured BMIs and recalled values roughly 55 years earlier were correlated 0.92 ($p < 0.01$) in women and 0.91 ($p < 0.01$) in men. He concluded that "recalled weight from many years earlier appears to be highly valid" (p. 216), and differences between recalled and measured values "have minimal effect on epidemiologic measures of association" (p. 216) [33].

Body Mass Index (BMI). As height increases, body mass tends to increase. BMI is the standard index of body mass adjusted for differences in height. BMI was determined using the equation: weight in kilograms divided by height in meters squared, kg/m$^2$. BMI allows differences in body mass or weight to be compared independently of stature or height. In the present study, height was measured using a wall-mounted stadiometer.

Waist to Height Ratio (WHtR). Similar to body mass, as height increases, waist circumference tends to increase. Therefore, the present study used waist circumference as a measure of abdominal adiposity corrected for differences in height. This is known as the waist-to-height ratio. It is calculated by dividing waist circumference (cm) by height (cm).

Waist circumference is an excellent predictor of differences in abdominal adipose tissue indexed using magnetic resonance imaging, accounting for 91% of the variance [34]. Numerous investigations indicate that the waist-to-height ratio is a better predictor of cardiometabolic health than BMI and typical measures of waist circumference (WC) [35–37]. According to a review and meta-analysis by Ashwell et al., which included over 300,000 individuals from diverse populations, the WHtR is better than BMI and WC at predicting diabetes, dyslipidemia, CVD, hypertension, and all these combined [36].

Waist circumference was assessed by trained technicians. The specialists were graded regularly to ensure the validity and reliability of their measurements. A specially designed room in the mobile examination center was utilized for the measurements. A trained recorder helped with the waist measurements. The superior border of the ilium was used as the reference point for the horizontal placement of the measuring tape. A wall mirror, along with the assistant, was employed to confirm the horizontal placement of the tape. The goal was to place the tape snugly around the waist without compressing the tissue. The waist assessment was taken after a normal exhalation [38].

Smoking. Cigarette smoking was evaluated by asking participants to report the number of cigarettes smoked per day during the past month. Subjects who responded that they did not smoke any cigarettes in the past month were assigned zeros, and smokers were given a maximum of 95 cigarettes per day [39]. Cigarette smoking values were controlled statistically in this investigation.

Physical Activity. Total physical activity was also employed as a covariate. An interview was utilized to assess physical activity. Moderate and vigorous activities were evaluated separately. Activities that caused modest increases in heart rate and breathing speed were defined as moderate. Casual biking and walking were examples of moderate physical activity. On the other hand, significant increases in heart and/or breathing rate were recorded as vigorous physical activity. Jogging, running, or walking up a hill were used as examples of vigorous physical activity.

Specific questions were utilized to ascertain the time spent in moderate and vigorous physical activity, each asked separately. The focus was on "days per week" and "time per day." Days and minutes were combined using multiplication, resulting in minutes of moderate and minutes of vigorous activity per week. Participants reporting more than 8 h per day of moderate or more than 5 h per day of vigorous activity were given these values, respectively. The moderate and vigorous minutes were added together, producing the total time (minutes) spent in moderate and vigorous physical activity (MVPA) per week.

The NHANES PA questionnaire was based on the World Health Organization (WHO) "Global Physical Activity Questionnaire (GPAQ) [40]". The U.S. version was adapted for those using the English language. The GPAQ is currently used by over 50 countries to monitor physical activity within their borders. Research shows that the questionnaire results that focus on MVPA are significantly correlated with pedometer counts, Actigraph accelerometer counts, percent body fat, waist circumference, and VO2 max results. Indexed using the intraclass correlation (ICC), the test-retest reliability of the questionnaire across 10-days was 0.96 for moderate recreational physical activity and 0.90 for vigorous recreational physical activity [41].

Alcohol Use. A total of 3 categories were used to differentiate among alcohol users: Abstainers, Moderate drinkers, and Heavy drinkers. To be classified as an Abstainer, participants had to respond that they consumed no alcohol in the past year. For men, Moderate drinkers reported drinking more than 0 and less than 3 alcoholic beverages per day during the past 12 months. For women, Moderate drinkers reported drinking more than 0 but less than 2 drinks per day over the past year. For men, Heavy drinkers reported drinking 3 or more alcoholic drinks per day over the past year, whereas, for women, Heavy drinkers reported drinking 2 or more alcoholic beverages per day over the past year.

Economic Status. Another covariate was economic status. It was assessed indirectly using a question about housing, which placed subjects into one of 3 categories. Specifically, participants were asked if they were renting or buying their dwelling or other.

Race/Ethnicity. To control for differences in race/ethnicity, six categories established by NHANES were utilized: Mexican American, Non-Hispanic Black, Non-Hispanic White, Non-Hispanic Asian, Other Hispanic, or Other Race/Multi-Racial.

### 2.3. *Data Analysis*

Findings can be generalized to the U.S. adult population because NHANES employed a multi-level, random sampling strategy to secure participants. To achieve this, each statistical model incorporated clusters, strata, and individual sample weights.

Given the very large sample associated with the current investigation, statistical power would be projected to be exceptionally high. Nevertheless, because of the unique multi-stage random sampling approach used by NHANES, the degrees of freedom (df) were established by subtracting the number of strata (59) from the number of clusters (121), resulting in 62 df instead of more than 10,000 df in the denominator.

There were 3 outcome measures (10-year percent weight change, BMI, and WHtR) and 1 exposure variable (grams of legume intake per 1000 kcal). Adjustments were made for differences in 5 demographic covariates (age, sex, race/ethnicity, year of assessment, and economic status) and 5 lifestyle covariates (physical activity, cigarette smoking, alcohol use, energy intake, and fiber intake per 1000 kcal) to minimize their effect on the associations between legume intake and the outcome variables. Special attention was focused on the effect of controlling for differences in fiber intake on the relationships of interest.

The associations between legume intake expressed as a categorical variable with 3 levels (None, Low, and Moderate/High) and the 3 outcome measures, each a continuous variable, were evaluated using analysis of variance (ANOVA) and multiple regression. The SAS SurveyReg procedure was used to evaluate mean differences across the legume categories, taking into account the NHANES sampling weights. The covariates were controlled statistically using partial correlation, and adjusted means were compared across the 3 legume intake categories. The Variance Inflation Factor (VIF) was evaluated because of the possible issue of multicollinearity with fiber consumption and legume intake. VIF was 1.9 or lower, so multicollinearity was not a threat. SAS version 9.4 (SAS Institute, Inc., Cary, NC, USA) was used to analyze the data. Two-sided statistical tests were employed, and $p$-values were set at $<0.05$ to operationalize statistical significance.

## 3. Results

A total of 10,137 participants were included in the 10-year weight change calculations, and 15,185 were integrated into the cross-sectional analyses. The median age of the U.S. sample was approximately 45 years, and about 50% of the representative sample reported that they engaged in moderate or vigorous physical activity for about 1 h and 13 min or more of per week. Percentiles for each of the principal continuous variables are shown below in Table 1.

Table 1 indicates that U.S. adults gain significant amounts of weight over time. About 50% of U.S. adults gained more than 5% in body weight during the previous decade, whereas one in four adults gained almost 15% during the previous 10 years, and 10% gained more than 25% in weight. Additionally, fiber intake was low. As displayed in Table 1, the median U.S. adult intake was 7.8 g of fiber per 1000 kcal, and the 90th percentile was only 13.5 g per 1000 kcal per day.

**Table 1.** Percentile distributions of the key continuous variables representing U.S. adults (n = 15,185).

| Variable | Percentile | | | | |
|---|---|---|---|---|---|
| | 10th | 25th | 50th | 75th | 90th |
| Age (years) | 22.5 | 30.6 | 44.7 | 57.3 | 66.0 |
| 10-year weight change (%) | −10.6 | −2.9 | 5.5 | 14.6 | 25.9 |
| Waist to Height Ratio | 0.47 | 0.52 | 0.58 | 0.65 | 0.73 |
| Body Mass Index (kg/m$^2$) | 21.8 | 24.3 | 28.1 | 33.0 | 38.4 |
| Energy intake (kcal) | 1244 | 1574 | 2009 | 2553 | 3154 |
| Fiber intake (g/1000 kcal) | 4.4 | 5.8 | 7.8 | 10.4 | 13.5 |
| Physical Activity (MVPA: min/week) | 0 | 0 | 73 | 239 | 479 |
| Cigarettes (per month) | 0 | 0 | 0 | 0 | 290 |

Note. Table values include person-level weighted adjustments based on the sampling methods of NHANES, so values represent those of the U.S. adult population.

Table 2 illustrates that most U.S. adults eat few if any, legumes. Over the two days of dietary evaluation conducted by NHANES, intake was 0 g per 1000 kcal for about 76% of the U.S. population. Table 2 also indicates that almost half of U.S. adults reported that they abstain from drinking alcohol, and 48% reported that they own or are buying their dwelling place. Furthermore, by design, about 25% of adults participating in the NHANES survey were selected from each two-year data collection cycle between 2011 and 2018, and the present sample included a wide range of racial/ethnic groups, with nearly equal representation from women and men.

**Table 2.** Characteristics of the sample based on weighted values for each column for the categorical variables (n = 15,185).

| Categorical Variable | N | % | SE |
|---|---|---|---|
| **Sex** | | | |
| Women | 7608 | 50.1 | 0.56 |
| Men | 7577 | 49.9 | 0.56 |
| **Race/Ethnicity** | | | |
| Mexican American | 1473 | 9.7 | 0.98 |
| Other Hispanic | 957 | 6.3 | 0.60 |
| Non-Hispanic White | 9657 | 63.6 | 1.75 |
| Non-Hispanic Black | 1716 | 11.3 | 0.98 |
| Non-Hispanic Asian | 835 | 5.5 | 0.48 |
| Other Race/Multiracial | 547 | 3.6 | 0.28 |
| **Year of Assessment** | | | |
| 2011–2012 | 3720 | 24.5 | 1.28 |
| 2013–2014 | 3766 | 24.8 | 1.34 |
| 2015–2016 | 3796 | 25.0 | 1.18 |
| 2017–2018 | 3903 | 25.7 | 1.03 |
| **Economic Status (housing)** | | | |
| Renting | 3584 | 23.6 | 0.99 |
| Buying | 7289 | 48.0 | 1.54 |
| Other | 4312 | 28.4 | 1.08 |
| **Alcohol Use** | | | |
| Abstainer | 7046 | 46.4 | 1.12 |
| Moderate Drinker | 3872 | 25.5 | 0.92 |
| Heavy Drinker | 4267 | 28.1 | 0.65 |
| **Legume Intake** | | | |
| None | 11,571 | 76.2 | 0.66 |
| Low | 1807 | 11.9 | 0.43 |
| Moderate/High | 1807 | 11.9 | 0.52 |

Note: SE refers to the standard error of the percentage. The N and % columns refer to the number of subjects and the percentage of participants after the NHANES sample weights were applied. Weighted values are more meaningful than unweighted values because they can be generalized to the U.S. adult population.

As shown in Table 3, legume intake was a strong predictor of 10-year weight change. There was a consistent inverse dose-response relationship. As legume intake (grams per 1000 kcal) increased, the 10-year weight gain decreased. Adults with moderate to high intakes gained substantially less weight over the past decade than those reporting no legume intake. They also gained significantly less than adults reporting low consumption of legumes. Specifically, adults reporting no legume intake gained 10.5% of their initial body weight over the previous 10 years, whereas moderate/high consumers gained 8.5%. Percent weight gain was 23.5% greater in the non-consumers compared to the moderate/high consumers.

Adjusting for differences in the demographic covariates, age, sex, race, year of assessment, and economic status, as well as the lifestyle covariates, physical activity, smoking, alcohol use, and energy intake, had little effect on the findings. However, when dietary fiber intake (grams per 1000 kcal) was controlled with the other covariates, the association between legume consumption and 10-year percent weight change was nullified.

The cross-sectional analyses revealed similar findings to the 10-year results. With BMI as the outcome variable, there was an inverse dose-response relationship. As legume intake increased, BMI levels decreased. Specifically, women and men reporting no legume consumption had significantly higher BMI levels than their counterparts. Controlling for the lifestyle covariates along with the demographic factors had little influence on the mean differences. However, when adjustments were made for fiber intake and other covariates, BMI differences across the groups were no longer statistically significant.

**Table 3.** Mean differences in 10-year percent weight change, BMI, and the waist-to-height ratio, given 62 degrees of freedom, across categories of legume intake, after adjusting for the covariates.

| | Legume Intake (grams/1000 kcal) | | | | | | | |
|---|---|---|---|---|---|---|---|---|
| | None | | Low | | Moderate/High | | | |
| **Outcome Variable** | **Mean** | **95% CI** | **Mean** | **95% CI** | **Mean** | **95% CI** | **F** | ***p*** |
| **10-yr Percent Weight Change** | n = 7673 | | n = 1231 | | n = 1233 | | | |
| Model 1 | 10.5 [a] | 9.6–11.4 | 9.9 [a] | 8.3–11.5 | 8.5 [b] | 7.5–9.6 | 6.7 | 0.0023 |
| Model 2 | 10.5 [a] | 9.6–11.3 | 9.7 [a] | 8.2–11.3 | 8.5 [b] | 7.4–9.5 | 6.5 | 0.0028 |
| Model 3 | 10.3 | 9.4–11.1 | 9.9 | 8.4–11.5 | 9.3 | 8.3–10.3 | 1.9 | 0.1626 |
| **Body Mass Index** | n = 11,571 | | n = 1807 | | n = 1807 | | | |
| Model 1 | 29.7 [a] | 29.4–29.9 | 29.1 [b] | 28.6–29.7 | 28.9 [b] | 28.5–29.4 | 7.2 | 0.0015 |
| Model 2 | 29.5 [a] | 29.3–29.8 | 29.0 [b] | 28.4–29.5 | 28.8 [b] | 28.4–29.3 | 8.0 | 0.0008 |
| Model 3 | 29.5 | 29.2–29.7 | 29.1 | 28.6–29.7 | 29.3 | 28.9–29.8 | 0.8 | 0.4399 |
| **Waist to Height Ratio** | n = 11,571 | | n = 1807 | | n = 1807 | | | |
| Model 1 | 0.601 [a] | 0.596–0.605 | 0.590 [b] | 0.582–0.598 | 0.589 [b] | 0.582–0.595 | 9.7 | 0.0002 |
| Model 2 | 0.598 [a] | 0.584–0.612 | 0.589 [b] | 0.573–0.604 | 0.587 [b] | 0.572–0.602 | 9.4 | 0.0003 |
| Model 3 | 0.599 | 0.592–0.606 | 0.593 | 0.584–0.603 | 0.598 | 0.588–0.607 | 1.2 | 0.3221 |

Means on the same row with the same superscript letter are not significantly different. Means have been adjusted based on the covariates in the model. Model 1 included age, sex, race, year of assessment, and economic status as covariates. Model 2 included the covariates of Model 1 and also minutes of moderate to vigorous physical activity, number of cigarettes smoked per month, total energy intake, and alcohol use. Model 3 included the same covariates as Model 2 and also fiber intake per 1000 kcal. For the 10-yr weight change variable (Model 3) the mean difference between Low and Moderate/High intake was borderline significant ($p > 0.05$, $p < 0.10$).

With the waist-to-height ratio employed as the outcome variable, again, the association was inverse and dose-response. U.S. men and women who consumed legumes had less abdominal adiposity compared to those reporting no legume intake. Moreover, as legume consumption increased, the waist-to-height ratio tended to decrease. Controlling for the covariates had almost no influence on the mean differences. However, like the other associations, adjusting for differences in dietary fiber intake eliminated the relationship between legume consumption and abdominal adiposity.

Not shown in the table, legume intake was significantly related to fiber consumption. Specifically, with all the covariates controlled, mean fiber intake differed across the three legume categories (F = 174.1, $p < 0.0001$). Subjects reporting no legume intake averaged (±SE) 8.1 ± 0.1 g per 1000 kcal, whereas those with Low intakes averaged 9.6 ± 0.2 g per 1000 kcal, and U.S. adults with Moderate/High legume intakes averaged 11.5 ± 0.2 g of fiber per 1000 kcal. The dietary fiber difference between the 1st and 3rd categories of legume intake (11.5 vs. 8.1) was approximately 42%.

## 4. Discussion

The fundamental objective of the present study was to measure the relationship between legume intake and 10-year percent weight change in a large sample of randomly selected U.S. women and men. Additionally, identifying the cross-sectional associations between legume consumption and BMI and also abdominal adiposity were important aims. Finally, another key purpose was to determine the extent to which various covariates, particularly fiber intake, influenced these associations.

This study had four findings worthy of highlighting. (1) Ten-year percent weight change differed significantly and in a dose-response pattern across the three legume intake categories. (2) Cross-sectionally, mean levels of BMI and abdominal adiposity also differed significantly and in a dose-response manner across the categories of legume consumption. (3) Adjusting for nine demographic and lifestyle covariates had little effect on the legume relationships. (4) Controlling for differences in dietary fiber intake with the other covariates negated the associations between legume intake and the weight-related outcomes for the 10-year results and the cross-sectional findings.

The consistency of the key findings was notable. Whether the focus was on 10-year changes in weight or cross-sectional findings associated with BMI and abdominal adiposity, the results were similar: legume intake was a good predictor of the outcomes. Additionally, in each case, controlling for differences in fiber consumption nullified the relationships.

The average 10-year weight gain in this U.S. representative sample was high, but it was significantly higher in adults who did not consume legumes compared to their counterparts. For example, 10-year percent weight gain for adults reporting no legume intake was 10.5% (see Table 3) or about 9.45 kg (20.8 lbs) for a 90 kg (198 lb) person. On the other hand, participants reporting Moderate/High legume consumption gained 8.5% or 7.65 kg (16.8 lb) for a 90 kg (198 lb) person. Over 10 years, the difference was meaningful. Across several decades, the weight gain difference could be substantial.

Literature reviews indicate that the vast majority of randomized controlled trials have not found that regularly eating legumes results in weight loss [2,14]. Part of the problem could be that many legume intervention studies have employed small sample sizes. For example, researchers have conducted experiments focusing on legume intake and weight loss using 18 men [42], 20 men [43], 19 adults [44], and 21 men [45]. Unless within-group variation is minimal, more participants are usually required to generate enough statistical power to detect the mean differences between groups when small samples are employed.

Another issue that might partly explain the lack of significant treatment effects could be the duration of legume interventions. Weight loss is not easy for most adults. It takes time and practice. When legume-based interventions last only 3 weeks [43,46], or 4 weeks [47,48], or 5 weeks [49], weight loss differences between treatment and control groups are difficult to generate, especially when energy intake is designed to be iso-caloric, which many legume investigations have used [14].

Although randomized controlled trials are considered the gold standard by many, they have a few significant drawbacks. Treatments are often limited in duration and magnitude, so compliance does not suffer. Participants are usually unwilling to follow a special diet for long periods. Moreover, some subjects may not appreciate or tolerate eating large quantities of legumes. Many legume-focused randomized controlled trials have required participants to eat 250–275 g or more of legumes daily [43,45,50,51], so dietary compliance could be an issue. On the other hand, the present investigation did not require subjects to eat a particular diet. Instead, the usual diets of participants were assessed. Therefore, conformity to dietary treatment requests was not an issue, and reported intakes in the present investigation might better reflect long-term dietary patterns than those achieved with intervention studies. Other observational investigations have also found this to be the case [16].

Why would legume intake be a good predictor of 10-year weight change? What factors might account for the differences in BMI and waist-to-height ratio? There are multiple possible mechanisms. For example, legumes contain a significant amount of plant protein. As a result, legumes have a dual classification. They are officially a vegetable and also a protein food according to the U.S. Dietary Guidelines for Americans, 2020–2025 [52]. No other food group has this characteristic. Several investigations have determined that plant protein consumption reduces the risk of obesity [53,54].

Moreover, legumes tend to be low in fat. Dietary fat is energy dense, providing more than twice the energy per gram than carbohydrate and protein. In other words, gram for gram, legumes tend to be low in calories. Numerous investigations indicate that regular intake of low-fat foods decreases the risk of weight gain [55–58].

Additionally, legumes have a favorable impact on the gut microbiome [59,60]. A healthy digestive environment has been shown to diminish the development of several diseases, including obesity [61]. Regular intake of legumes tends to improve the gut microbiome, which could assist with maintaining a healthy weight.

Overall, legumes are a low glycemic index food [62]. That means that blood glucose concentrations are not raised by eating legumes compared to most carbohydrates. Legumes are among the foods that are encouraged to help manage glucose concentrations [13]. A number of investigations have shown that dietary patterns that focus on low-glycemic foods, such as legumes, reduce the risk of obesity and abdominal adiposity [63,64].

Further, some foods satiate better than others. Consumption of legumes tends to appease hunger and encourage cessation of eating. Multiple studies have found that eating legumes leads to consuming consumes less energy than other foods [65]. A meta-analysis of nine experiments showed that legume intake improved short-term satiety by over 30% when matched with other dietary choices [66].

Last, and perhaps of most importance, legumes are an excellent source of dietary fiber [67]. The present investigation targeted dietary fiber intake to determine its effect on the predictive utility of legumes. According to the USDA, legumes provide roughly 12–16 g of fiber per cup [12]. In contrast, a cup of cooked white rice contains less than 1 g of dietary fiber [12]. The high fiber content of legumes could be the driving factor accounting for the strong relationships found in this study. Dietary fiber is essentially calorie-free, but it helps to satiate. Moreover, soluble fiber can bind with fats and sugars and reduce their absorption and use in the body.

Adjusting for differences in dietary fiber intake not only caused the significant weight-related differences between the legume intake groups to be nullified, but it also had other meaningful statistical effects. First is the F-ratio, which reflects the magnitude of the between-group differences (variance) over the within-group variance (error). For the models focusing on 10-yr percent weight change, Model 2 (which included all the covariates except fiber intake) showed an F-ratio of 6.5. When fiber intake was added to Model 2 (thus becoming Model 3), the F-ratio decreased to 1.9, a decrease of more than 3-fold. For the Body Mass Index, the decrease in the F-ratio was 10-fold, another large and meaningful decrease. For the waist-to-height ratio, the decrease was almost 8-fold. Overall, controlling for differences in fiber intake led to substantial changes in the magnitude of the ratio of the between-group variance over the within-group variance.

For the 10-yr percent weight change variable, after adjusting for all the covariates (except fiber intake), the mean difference between the two extreme legume intake groups (None vs. Moderate/High) was 2 percentage points of body weight. That is a meaningful difference in weight gain, especially when considering the size of the samples and that the weight gain difference was associated with just one variable: legume consumption.

Also meaningful is the fact that the weight gain difference between the two legume intake groups (None vs. Moderate/High) was decreased by 50% when fiber intake was added to the variables controlled in the model. Similarly, when focusing on the waist-to-height ratio variable, the weight gain difference between the two legume intake groups was decreased by over 90% when fiber intake was added to the other covariates. It appears that many of the differences in weight change, BMI, and abdominal adiposity between the legume intake groups were a function of differences in fiber intake.

Studies consistently indicate that a high-fiber diet decreases the risk of overweight and obesity and promotes a leaner body [68–70]. In the current investigation, subjects in the Moderate/High legume intake category consumed 11.2 g of fiber per 1000 kcal per day. Those in the middle group averaged 9.6 g, and adults reporting no legume intake ate 8.1 g of fiber per 1000 kcal per day ($F = 174.1$, $p < 0.0001$). Clearly, legume intake goes hand-in-hand with fiber intake.

Findings indicated that fiber consumption in the U.S. adult population is far below the levels recommended in the Dietary Guidelines for Americans (2020–2025) [52]. Specifically, results (Table 1) showed that the median U.S. adult intake is about 7.8 g of fiber per 1000 kcal. The latest Guidelines recommend that adults eat 14 g or more per 1000 kcal. Therefore, most U.S. adults are consuming at least 79% less dietary fiber than they should be. This could be one reason weight gain and obesity are serious issues within the U.S.

There were several limitations connected with this investigation. It was an observational study. There was no treatment or intervention. Hence, causation cannot be inferred. Regular consumption of legumes could result in less weight gain over time and lower levels of BMI and abdominal fat. However, other factors besides legume consumption could account for the differences discovered in this investigation. Similarly, eating legumes regularly might be an indicator of other behaviors or conditions that influence weight gain. Ten covariates were controlled in this study. There are always others that could account for some of the relationships identified in this study. Additionally, legume consumption was assessed using two 24-h dietary recalls. More recall assessments would have been better and would have reduced the misclassification of subjects within the legume intake categories. Moreover, weight change in the present investigation was partly based on weight recall. Although many investigations have found weight recall to be accurate and valid, the recall includes more errors than the direct measurement of body weight. Overall, reduced measurement error from the two dietary assessments and the weight recall evaluations would have likely increased the magnitude of the weight-related differences between the legume intake groups because error variance does not predict well.

The current investigation also had many merits. There were more than 15,000 participants. Individuals were chosen randomly, so the findings apply to the non-institutionalized U.S. adult population. Six racial/ethnic groups were sampled, and women and men 18–75 years old were included, so participants represented a broad spectrum of the United States. Legume intake was measured using the average of two, not just one, 24-h dietary recall assessments. The technicians and specialists who worked directly with the participants were well-trained. They were evaluated regularly and recertified systematically to ensure that high-quality measurement methods were practiced.

## 5. Conclusions

The scientific literature indicates that legumes are a special food. They have an array of characteristics that make them nourishing and healthful. Legumes are full of nutrients, but they are not calorically dense. Despite their many qualities, they are not eaten regularly by Americans. The present study found that U.S. adults who include legumes in their diets have significantly less 10-year weight gain than their counterparts and lower BMIs and leaner waists. Overall, it appears that legumes have dietary attributes associated with successful weight control in U.S. adults.

**Funding:** This research received no external funding.

**Institutional Review Board Statement:** The study was conducted according to the guidelines of the Declaration of Helsinki, and The Ethics Review Board (ERB) of the National Center for Health Statistics, approved the NHANES data collection protocol. The ethical approval codes for NHANES data collection from 2011–2018 were: Protocol #2011-17 and Protocol #2018-01. Each subject was required to give written consent to take part in the national survey.

**Informed Consent Statement:** Written informed consent was obtained from all subjects involved in the study.

**Data Availability Statement:** All data sets associated with the results of the present study can be found online as part of the National Health and Nutrition Examination Survey (NHANES). There is no charge for the data sets. They can be downloaded at the following website: https://wwwn.cdc.gov/nchs/nhanes/Default.aspx (date accessed: 5 January 2023).

**Conflicts of Interest:** The author declares no conflict of interest.

## References


1. Afshin, A.; Micha, R.; Khatibzadeh, S.; Mozaffarian, D. Consumption of nuts and legumes and risk of incident ischemic heart disease, stroke, and diabetes: A systematic review and meta-analysis. *Am. J. Clin. Nutr.* **2014**, *100*, 278–288. [CrossRef] [PubMed]
2. Viguiliouk, E.; Mejia, S.B.; Kendall, C.W.; Sievenpiper, J.L. Can pulses play a role in improving cardiometabolic health? Evidence from systematic reviews and meta-analyses. *Ann. N. Y. Acad. Sci.* **2017**, *1392*, 43–57. [CrossRef] [PubMed]
3. Nchanji, E.B.; Ageyo, O.C. Do Common Beans (*Phaseolus vulgaris* L.) Promote Good Health in Humans? A Systematic Review and Meta-Analysis of Clinical and Randomized Controlled Trials. *Nutrients* **2021**, *13*, 3701. [CrossRef]
4. Papandreou, C.; Becerra-Tomás, N.; Bulló, M.; Martínez-González, M.; Corella, D.; Estruch, R.; Ros, E.; Arós, F.; Schroder, H.; Fitó, M.; et al. Legume consumption and risk of all-cause, cardiovascular, and cancer mortality in the PREDIMED study. *Clin. Nutr.* **2018**, *38*, 348–356. [CrossRef] [PubMed]
5. Becerra-Tomás, N.; Papandreou, C.; Salas-Salvadó, J. Legume Consumption and Cardiometabolic Health. *Adv. Nutr. Int. Rev. J.* **2019**, *10*, S437–S450. [CrossRef] [PubMed]
6. Bouchenak, M.; Lamri-Senhadji, M. Nutritional Quality of Legumes, and Their Role in Cardiometabolic Risk Prevention: A Review. *J. Med. Food* **2013**, *16*, 185–198. [CrossRef] [PubMed]
7. Mitchell, D.C.; Lawrence, F.R.; Hartman, T.J.; Curran, J.M. Consumption of Dry Beans, Peas, and Lentils Could Improve Diet Quality in the US Population. *J. Am. Diet. Assoc.* **2009**, *109*, 909–913. [CrossRef]
8. Perera, T.; Russo, C.; Takata, Y.; Bobe, G. Legume Consumption Patterns in US Adults: National Health and Nutrition Examination Survey (NHANES) 2011–2014 and Beans, Lentils, Peas (BLP) 2017 Survey. *Nutrients* **2020**, *12*, 1237. [CrossRef]
9. Semba, R.D.; Rahman, N.; Du, S.; Ramsing, R.; Sullivan, V.; Nussbaumer, E.; Love, D.; Bloem, M.W. Patterns of Legume Purchases and Consumption in the United States. *Front. Nutr.* **2021**, *8*, 732237. [CrossRef]
10. Fryar, C.D.; Carroll, M.D.; Afful, J. *Prevalence of Overweight, Obesity, and Severe Obesity among Adults Aged 20 and over: United States, 1960–1962 through 2017–2018*; National Center for Health Statistics: Hyattsville, MD, USA, 2020.
11. Rebello, C.J.; Greenway, F.L.; Finley, J.W. A review of the nutritional value of legumes and their effects on obesity and its related co-morbidities. *Obes. Rev.* **2014**, *15*, 392–407. [CrossRef]
12. U.S. Department of Agriculture, FoodData Central. 2019. Available online: https://fdc.nal.usda.gov (accessed on 5 January 2023).
13. Jenkins, D.J.; Kendall, C.W.; Augustin, L.S.; Mitchell, S.; Sahye-Pudaruth, S.; Blanco Mejia, S.; Chiavaroli, L.; Mirrahimi, A.; Ireland, C.; Bashyam, B.; et al. Effect of legumes as part of a low glycemic index diet on glycemic control and cardiovascular risk factors in type 2 diabetes mellitus: A randomized controlled trial. *Arch. Intern. Med.* **2012**, *172*, 1653–1660. [CrossRef] [PubMed]
14. Kim, S.J.; de Souza, R.J.; Choo, V.L.; Ha, V.; Cozma, A.I.; Chiavaroli, L.; Mirrahimi, A.; Blanco Mejia, S.; Di Buono, M.; Bernstein, A.M.; et al. Effects of dietary pulse consumption on body weight: A systematic review and meta-analysis of randomized controlled trials. *Am. J. Clin. Nutr.* **2016**, *103*, 1213–1223. [CrossRef]
15. Williams, P.G.; Grafenauer, S.; O'Shea, J.E. Cereal grains, legumes, and weight management: A comprehensive review of the scientific evidence. *Nutr. Rev.* **2008**, *66*, 171–182. [CrossRef] [PubMed]
16. Papanikolaou, Y.; Fulgoni, V.L., 3rd. Bean Consumption Is Associated with Greater Nutrient Intake, Reduced Systolic Blood Pressure, Lower Body Weight, and a Smaller Waist Circumference in Adults: Results from the National Health and Nutrition Examination Survey 1999-2002. *J. Am. Coll. Nutr.* **2008**, *27*, 569–576. [CrossRef] [PubMed]
17. U.S. Department of Agriculture, Food Surveys Research Group: AMPM, USDA Automated Multiple-Pass Method. Available online: https://www.ars.usda.gov/northeast-area/beltsville-md-bhnrc/beltsville-human-nutrition-research-center/food-surveys-research-group/docs/ampm-usda-automated-multiple-pass-method/ (accessed on 5 January 2023).
18. U.S. Department of Agriculture, Food and Nutrient Database for Dietary Studies (FNDDS). Available online: https://data.nal.usda.gov/dataset/food-and-nutrient-database-dietary-studies-fndds (accessed on 5 January 2023).
19. Ahuja, J.K.; Moshfegh, A.J.; Holden, J.M.; Harris, E. USDA Food and Nutrient Databases Provide the Infrastructure for Food and Nutrition Research, Policy, and Practice. *J. Nutr.* **2012**, *143*, 241S–249S. [CrossRef]

20. Tucker, L.A. Dietary Fiber and Telomere Length in 5674 U.S. Adults: An NHANES Study of Biological Aging. *Nutrients* **2018**, *10*, 400. [CrossRef]
21. Tucker, L. Fruit and Vegetable Intake and Telomere Length in a Random Sample of 5448 U.S. Adults. *Nutrients* **2021**, *13*, 1415. [CrossRef]
22. Wilkinson, K.; Tucker, L.; Davidson, L.; Bailey, B. Milk-Fat Intake and Differences in Abdominal Adiposity and BMI: Evidence Based on 13,544 Randomly-Selected Adults. *Nutrients* **2021**, *13*, 1832. [CrossRef]
23. Wolffenbuttel, B.H.R.; Heiner-Fokkema, M.R.; Green, R.; Gans, R.O.B. Relationship between serum B12 concentrations and mortality: Experience in NHANES. *BMC Med.* **2020**, *18*, 307. [CrossRef]
24. Deng, F.E.; Shivappa, N.; Tang, Y.; Mann, J.R.; Hebert, J.R. Association between diet-related inflammation, all-cause, all-cancer, and cardiovascular disease mortality, with special focus on prediabetics: Findings from NHANES III. *Eur. J. Nutr.* **2017**, *56*, 1085–1093. [CrossRef]
25. Christensen, K.; Lawler, T.; Mares, J. Dietary Carotenoids and Non-Alcoholic Fatty Liver Disease among US Adults, NHANES 2003–2014. *Nutrients* **2019**, *11*, 1101. [CrossRef] [PubMed]
26. Willett, W. *Nutritional Epidemiology*, 2nd ed.; Oxford University Press: New York, NY, USA, 1998; 514p.
27. Stunkard, A.J.; Albaum, J.M. The accuracy of self-reported weights. *Am. J. Clin. Nutr.* **1981**, *34*, 1593–1599. [CrossRef] [PubMed]
28. Perry, L.; Leonard, B. Obesity and mental health. *JAMA* **1963**, *183*, 808.
29. Rhoads, G.; Kagan, A. The relation of coronary disease, stroke, and mortality to weight in youth and in middle age. *Lancet* **1983**, *321*, 492–495. [CrossRef]
30. Troy, L.M.; Hunter, D.J.; Manson, J.E.; Colditz, G.; Stampfer, M.J.; Willett, W.C. The validity of recalled weight among younger women. *Int. J. Obes. Relat. Metab. Disord. J. Int. Assoc. Study Obes.* **1995**, *19*, 570–572.
31. Kyulo, N.L.; Knutsen, S.F.; Tonstad, S.; Fraser, G.E.; Singh, P.N. Validation of recall of body weight over a 26-year period in cohort members of the Adventist Health Study 2. *Ann. Epidemiol.* **2012**, *22*, 744–746. [CrossRef]
32. Tamakoshi, K.; Yatsuya, H.; Kondo, T.; Hirano, T.; Hori, Y.; Yoshida, T.; Toyoshima, H. The accuracy of long-term recall of past body weight in Japanese adult men. *Int. J. Obes.* **2003**, *27*, 247–252. [CrossRef]
33. Willett, W. *Nutritional Epidemiology*, 3rd ed.; Oxford University Press: Oxford, UK, 2013; 529p.
34. Ross, R.; Leger, L.; Morris, D.; de Guise, J.; Guardo, R. Quantification of adipose tissue by MRI: Relationship with anthropometric variables. *J. Appl. Physiol.* **1992**, *72*, 787–795. [CrossRef]
35. Ashwell, M. Plea for simplicity: Use of waist-to-height ratio as a primary screening tool to assess cardiometabolic risk. *Clin. Obes.* **2012**, *2*, 3–5. [CrossRef]
36. Ashwell, M.; Gunn, P.; Gibson, S. Waist-to-height ratio is a better screening tool than waist circumference and BMI for adult cardiometabolic risk factors: Systematic review and meta-analysis. *Obes. Rev.* **2011**, *13*, 275–286. [CrossRef]
37. Kahn, H.S.; Bullard, K.M. Beyond Body Mass Index: Advantages of Abdominal Measurements for Recognizing Cardiometabolic Disorders. *Am. J. Med.* **2015**, *129*, 74–81.e2. [CrossRef] [PubMed]
38. NHANES. Anthropometry Procedures Manual. Available online: https://www.cdc.gov/nchs/data/nhanes/nhanes_11_12/Anthropometry_Procedures_Manual.pdf (accessed on 5 January 2023).
39. NHANES. Data Documentation, Codebook, and Frequencies. Available online: https://wwwn.cdc.gov/Nchs/Nhanes/2013-2014/PAQ_H.htm (accessed on 5 January 2023).
40. Armstrong, T.; Bull, F. Development of the World Health Organization Global Physical Activity Questionnaire (GPAQ). *J. Public Health* **2006**, *14*, 66–70. [CrossRef]
41. Cleland, C.L.; Hunter, R.F.; Kee, F.; Cupples, M.E.; Sallis, J.F.; Tully, M.A. Validity of the Global Physical Activity Questionnaire (GPAQ) in assessing levels and change in moderate-vigorous physical activity and sedentary behaviour. *BMC Public Health* **2014**, *14*, 1255. [CrossRef]
42. Abete, I.; Parra, D.; Martinez, J.A. Legume-, Fish-, or High-Protein-Based Hypocaloric Diets: Effects on Weight Loss and Mitochondrial Oxidation in Obese Men. *J. Med. Food* **2009**, *12*, 100–108. [CrossRef]
43. Anderson, J.W.; Story, L.; Sieling, B.; Chen, W.J.L.; Petro, M.S.; Story, J. Hypocholesterolemic effects of oat-bran or bean intake for hypercholesterolemic men. *Am. J. Clin. Nutr.* **1984**, *40*, 1146–1155. [CrossRef]
44. Nestel, P.; Cehun, M.; Chronopoulos, A. Effects of long-term consumption and single meals of chickpeas on plasma glucose, insulin, and triacylglycerol concentrations. *Am. J. Clin. Nutr.* **2004**, *79*, 390–395. [CrossRef]
45. Veenstra, J.; Duncan, A.; Cryne, C.; Deschambault, B.; Boye, J.; Benali, M.; Marcotte, M.; Tosh, S.; Farnworth, E.; Wright, A. Effect of pulse consumption on perceived flatulence and gastrointestinal function in healthy males. *Food Res. Int.* **2010**, *43*, 553–559. [CrossRef]
46. Jimenez-Cruz, A.; Bacardi-Gascon, M.; Turnbull, W.H.; Rosales-Garay, P.; Severino-Lugo, I. A Flexible, Low-Glycemic Index Mexican-Style Diet in Overweight and Obese Subjects With Type 2 Diabetes Improves Metabolic Parameters During a 6-Week Treatment Period. *Diabetes Care* **2003**, *26*, 1967–1970. [CrossRef]
47. Anderson, J.W.; Gustafson, N.J.; Spencer, D.B.; Tietyen, J.; Bryant, C.A. Serum lipid response of hypercholesterolemic men to single and divided doses of canned beans. *Am. J. Clin. Nutr.* **1990**, *51*, 1013–1019. [CrossRef]
48. Marinangeli, C.P.F.; Jones, P.J.H. Whole and fractionated yellow pea flours reduce fasting insulin and insulin resistance in hypercholesterolaemic and overweight human subjects. *Br. J. Nutr.* **2010**, *105*, 110–117. [CrossRef]

49. Pittaway, J.K.; Ahuja, K.D.K.; Robertson, I.K.; Ball, M.J. Effects of a Controlled Diet Supplemented with Chickpeas on Serum Lipids, Glucose Tolerance, Satiety and Bowel Function. *J. Am. Coll. Nutr.* **2007**, *26*, 334–340. [CrossRef] [PubMed]
50. Abeysekara, S.; Chilibeck, P.D.; Vatanparast, H.; Zello, G.A. A pulse-based diet is effective for reducing total and LDL-cholesterol in older adults. *Br. J. Nutr.* **2012**, *108* (Suppl. 1), S103–S110. [CrossRef] [PubMed]
51. Tonstad, S.; Malik, N.; Haddad, E. A high-fibre bean-rich diet versus a low-carbohydrate diet for obesity. *J. Hum. Nutr. Diet.* **2013**, *27* (Suppl. 2), 109–116. [CrossRef]
52. U.S. Department of Agriculture, U.S. Department of Health and Human Services, Dietary Guidelines for Americans, 2020–2025. Available online: https://www.dietaryguidelines.gov/ (accessed on 5 January 2023).
53. Park, K.-B.; Park, H.A.; Kang, J.-H.; Kim, K.; Cho, Y.G.; Jang, J. Animal and Plant Protein Intake and Body Mass Index and Waist Circumference in a Korean Elderly Population. *Nutrients* **2018**, *10*, 577. [CrossRef] [PubMed]
54. Lin, Y.; Mouratidou, T.; Vereecken, C.; Kersting, M.; Bolca, S.; de Moraes, A.C.F.; Cuenca-García, M.; Moreno, L.A.; González-Gross, M.; Valtueña, J.; et al. Dietary animal and plant protein intakes and their associations with obesity and cardio-metabolic indicators in European adolescents: The HELENA cross-sectional study. *Nutr. J.* **2015**, *14*, 10. [CrossRef] [PubMed]
55. Astrup, A.; Grunwald, G.; Melanson, E.; Saris, W.; Hill, J. The role of low-fat diets in body weight control: A meta-analysis of ad libitum dietary intervention studies. *Int. J. Obes.* **2000**, *24*, 1545–1552. [CrossRef]
56. Bray, G.A.; Popkin, B.M. Dietary fat intake does affect obesity! *Am. J. Clin. Nutr.* **1998**, *68*, 1157–1173. [CrossRef]
57. Nelson, L.H.; Tucker, L.A. Diet Composition Related to Body Fat in a Multivariate Study of 203 Men. *J. Am. Diet. Assoc.* **1996**, *96*, 771–777. [CrossRef]
58. Tucker, L.A.; Kano, M.J. Dietary fat and body fat: A multivariate study of 205 adult females. *Am. J. Clin. Nutr.* **1992**, *56*, 616–622. [CrossRef]
59. Singh, B.; Singh, J.P.; Shevkani, K.; Singh, N.; Kaur, A. Bioactive constituents in pulses and their health benefits. *J. Food Sci. Technol.* **2016**, *54*, 858–870. [CrossRef]
60. Gullón, P.; Gullón, B.; Tavaria, F.; Vasconcelos, M.; Gomes, A.M. In vitro fermentation of lupin seeds (Lupinus albus) and broad beans (Vicia faba): Dynamic modulation of the intestinal microbiota and metabolomic output. *Food Funct.* **2015**, *6*, 3316–3322. [CrossRef] [PubMed]
61. Tseng, C.H.; Wu, C.Y. The gut microbiome in obesity. *J. Formos. Med. Assoc.* **2019**, *118* (Suppl. 1), S3–S9. [CrossRef] [PubMed]
62. Rizkalla, S.W.; Bellisle, F.; Slama, G. Health benefits of low glycaemic index foods, such as pulses, in diabetic patients and healthy individuals. *Br. J. Nutr.* **2002**, *88* (Suppl. 3), S255–S262. [CrossRef] [PubMed]
63. Joslowski, G.; Halim, J.; Goletzke, J.; Gow, M.; Ho, M.; Louie, J.C.-Y.; Buyken, A.E.; Cowell, C.T.; Garnett, S.P. Dietary glycemic load, insulin load, and weight loss in obese, insulin resistant adolescents: RESIST study. *Clin. Nutr.* **2015**, *34*, 89–94. [CrossRef] [PubMed]
64. Larsen, T.M.; Dalskov, S.-M.; van Baak, M.; Jebb, S.A.; Papadaki, A.; Pfeiffer, A.F.; Martinez, J.A.; Handjieva-Darlenska, T.; Kunešová, M.; Pihlsgård, M.; et al. Diets with High or Low Protein Content and Glycemic Index for Weight-Loss Maintenance. *N. Engl. J. Med.* **2010**, *363*, 2102–2113. [CrossRef]
65. Mollard, R.; Zykus, A.; Luhovyy, B.L.; Nunez, M.F.; Wong, C.L.; Anderson, G.H. The acute effects of a pulse-containing meal on glycaemic responses and measures of satiety and satiation within and at a later meal. *Br. J. Nutr.* **2011**, *108*, 509–517. [CrossRef]
66. Li, S.S.; Kendall, C.W.; de Souza, R.J.; Jayalath, V.H.; Cozma, A.I.; Ha, V.; Mirrahimi, A.; Chiavaroli, L.; Augustin, L.S.; Blanco Mejia, S.; et al. Dietary pulses, satiety and food intake: A systematic review and meta-analysis of acute feeding trials. *Obesity* **2014**, *22*, 1773–1780. [CrossRef]
67. Sathe, S.K.; Deshpande, S.S.; Salunkhe, D.K.; Rackis, J.J. Dry beans of *phaseolus*. A review. Part 2. Chemical composition: Carbohydrates, fiber, minerals, vitamins, and lipids. *Crit. Rev. Food Sci. Nutr.* **1984**, *21*, 41–93. [CrossRef]
68. Tucker, L.A.; Thomas, K.S. Increasing Total Fiber Intake Reduces Risk of Weight and Fat Gains in Women. *J. Nutr.* **2009**, *139*, 576–581. [CrossRef]
69. Pereira, M.A.; Ludwig, D.S. Dietary Fiber and Body-Weight Regulation. Observations and mechanisms. *Pediatr. Clin. N. Am.* **2001**, *48*, 969–980. [CrossRef]
70. Anderson, J.W.; Baird, P.; Davis, R.H., Jr.; Ferreri, S.; Knudtson, M.; Koraym, A.; Waters, V.; Williams, C.L. Health benefits of dietary fiber. *Nutr. Rev.* **2009**, *67*, 188–205. [CrossRef] [PubMed]

MDPI
St. Alban-Anlage 66
4052 Basel
Switzerland
www.mdpi.com

*Nutrients* Editorial Office
E-mail: nutrients@mdpi.com
www.mdpi.com/journal/nutrients